MAPPING IN THE CLOUD

MAPPING in the CLOUD

Michael P. Peterson

THE GUILFORD PRESS
New York   London

Library of Congress Cataloging-in-Publication Data

Peterson, Michael P.
  Mapping in the cloud / Michael P. Peterson.
      pages cm
    Includes bibliographical references and index.
    ISBN 978-1-4625-1041-2 (pbk.)
    ISBN 978-1-4625-1403-8 (hardcover)
    1. Cartography—Computer network resources. 2. Cloud
computing. I. Title.
  GA102.4.E4P49 2014
  526.0285′46782—dc23
                                                    2014003174

Preface

It is difficult to overstate the importance of maps as a form of communication about the world. They help us understand both our surroundings and the space beyond our direct perception. Maps influence how we think about the world and how we act within it. They connect us to our environment.

Each of us is a mapmaker, or cartographer, in the sense that we all make mental maps. Sometimes, we even draw these maps for others to help explain how to find a particular location. The making of maps and the analysis of the underlying spatial information have evolved into a science and are valuable skills for many different types of work. Cloud-based mapping tools allow for the creation of very sophisticated maps that can be easily made available to others.

This book seeks to balance the conceptual and practical aspects of maps and mapping. It is as important to know the meaning of maps, their background, and their development as it is to know how to make maps with the tools currently available. This book alternates between chapters that deal with the concepts of cartography and cloud-based methods of mapmaking. There's a package of resources including examples, exercises, and links to other sources of information for each even-numbered chapter available at www.guilford.com/peterson-materials. This site will be updated regularly with new resources.

Making maps in the cloud requires using a server. Although almost any computer can be transformed into a server, it is easier and more secure to use a web-hosting service. These services generally charge under $10 a month. Many web-hosting services offer a free plan with up to 1500 mb of disk space, which is more than sufficient for the exercises in this book. Two such web-hosting services are 000webhost.com and podserver.info. All that is required to request a free account is an email address.

Most web-hosting services are accessed through a browser and use a standard graphical user interface called the cPanel, which uses an icon interface to the many utility programs on the server. All exercises in this book make use of just three of these: FileManager, MySQL, and phpMyAdmin. Most work involves uploading and editing files with FileManager.

The exercises for the individual chapters are provided as zip files. Each zip file consists of a single folder that contains all the files necessary to complete the exercises for a particular chapter. FileManager can be used to upload the zip file into a public_html directory and will automatically unzip the file to create a folder. The folders are called code02, code04, code06, code08, code10, and so forth. Each folder contains an index.htm file that has links to all of the exercises for that chapter. Creating an index.htm file in the base public_html directory with links to each code folder provides access to the exercises for all the chapters. A central file can access the index.htm file for all students in a class. Being able to see what other students have done is a major advantage of using a server-based approach in the classroom.

Cloud-hosting systems such as Amazon Web Services (AWS) can also be used to host web pages. These services provide scalable and reliable hosting based on clustered load-balanced servers. At this point, cloud hosting is generally more cumbersome to access and manage than a web-hosting service. It is usually necessary to configure the server and install all of the software needed to serve web pages. The charging mechanism is also more complicated, and while free "instances" such as AWS T1.micro are available, it is likely that the user will eventually be charged for using a cloud-hosting service, and the charges can easily range in the hundreds of dollars.

It is important to work through these exercises in order to understand how the code works and how it can be modified. In working through the exercises, remember to make incremental changes to the working code. If your altered code does not work, return to a working version.

A book of this kind would not have been possible without the help of many individuals. During the course of its writing, I taught and presented multiple workshops. While the exercises were mostly developed for classes at the University of Nebraska at Omaha, I also taught courses at the Vienna University of Technology in Austria and the University of Applied Sciences in Karlsruhe, Germany, the National Institute of Design in Ahmadabad, India, and the University of Canterbury in Christchurch, New Zealand. Workshops were given at the Universiti Pendidikan Sultan Idris in Tanjung Malim, Malaysia, the Getty Conservation Institute in Los Angeles, the Brazilian Cartographic Congress in Rio de Janeiro, Brazil, and the District University of Bogotá, Colombia. A number of students in these courses and workshops suggested improvements to both the text and the exercises.

Prof. Georg Gartner at the Vienna University of Technology in Austria and Prof. Bill Cartwright from the Melbourne Institute of Technology in Australia were instrumental in early discussions of cartography and the Internet. Both suggested topics that should be included. The many interactions with colleagues through the Maps and the Internet Commission of the International Cartographic Association were also important in developing an outline for the book. A number of valuable discussions with Prof. Rex Cammack helped with the outline of the book. His help in focusing the organization of the book is much appreciated.

Students in Omaha who suggested a number of improvements to the text include Ed Zuelke and Nicholas Petersen. Prof. Leslie Rawlings provided a multipoint map example of Wayne, Nebraska. A number of students in advanced seminars helped with specific exercises, including Patrick Butler, Andrew Clouston, Michael DeBoer, Konal Dobson, Kevin Fandry, Charles Fortier, Paul Hunt, Gregory Jameson, Kelly Koepsell, Bruce Muller, Gabriel Pereda, Rob Shepard, Bill Shrader, and Spencer Trowbridge. Manuela Schmidt designed some of the illustrations. Interns that came to Omaha from universities in Karlsruhe, Mainz, and Munich, Germany, assisted in various aspects of the book. Manuela Schmidt designed some of the illustrations. Matthias Uhler worked on the PHP and MySQL exercises and developed a server-based solution for their implementation. Matthias Krisam and Alexander Stobbe developed a method to input points into a database. Other interns and students from Germany that helped with various aspects include Andreas Hiebsch, Stefan Stark, Christoph Weiss, and Paul Weiser. Finally, Dr. Ben Appleton at Google in Australia needs to be thanked for reviewing a draft of the book.

Prof. Shunfu Hu from Southern Illinois University provided valuable comments on an earlier draft of the book, as did Prof. Scott Freundschuh from the University of New Mexico and Prof. Henry Bulley at Central Connecticut State University.

Finally, I could not have completed this book without the help of my wife, Kathy. Not only did she suggest improvements to the book but she mostly understood when my mind was somewhere else. I would also like to thank my children, Sarah and Amelia, for their support.

The companion website *www.guilford.com/peterson-materials* links to the exercises and other instructional materials.

Contents

CHAPTER 15

CHAPTER 16

CHAPTER 17

CHAPTER 18

CHAPTER 19

Mobile Mapping   361

CHAPTER 20

Local Mapping   376

CHAPTER 21

Maps That Move   387

The companion website *www.guilford.com/peterson-materials*
links to the exercises and other instructional materials.

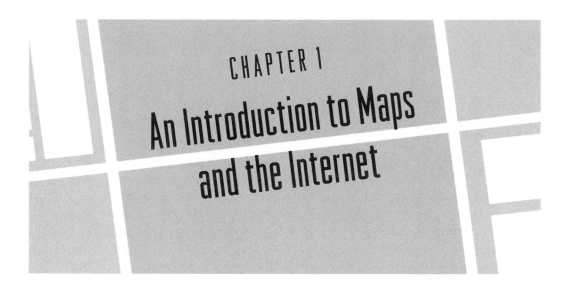

CHAPTER 1
An Introduction to Maps and the Internet

The Internet! Is that thing still around?

—Homer Simpson

1.1 Introduction

It is impossible to know when the first map was made. Ancient pictures of animals on cave walls from 40,000 years ago indicate that these early humans could represent objects from the real world in abstract form as a series of lines and shapes (Kelly and Thomas et al. 2010, p. 303). As representations of the environment from above, maps require a more complex form of abstraction and subsequent representation. Maps from Mesopotamia from about 4300 years ago are still in existence (Thrower 2007, p. 14). Drawn on clay tablets, these maps depict land boundaries and physical features, such as rivers and mountains. The necessity for humans to communicate information about the environment makes it likely that maps were drawn much before this time, perhaps with a stick in sand, to be destroyed long ago by the agents of weathering.

Although we don't know when the first map was made, the most recent was created just a fraction of a second ago. It was a map made by computer and transmitted through the Internet as electronic impulses. No longer restricted to clay tablets or even paper, maps are now sent almost instantly from place to place. The number of maps that are distributed through the Internet is phenomenal. Within a few years after the introduction of the World Wide Web in the early 1990s, individual websites were responding to millions of daily requests for maps. An important milestone was reached sometime during the 1990s when more maps were being transmitted through the Internet than were being printed on paper.

The Internet was revolutionary for maps, similar to the invention of printing.

Maps were not duplicated in mass until the mid-1400s. Before this time, they were reproduced by hand and very few existed. As a result of printing, more people could use maps and thus had a better understanding of the world. Like printing, the Internet increased the availability of maps, but the Internet went further and combined the printing and distribution in a single step. One of the major benefits of this new age of mapping is that maps can be made available to the user in a fraction of the time required to distribute maps on paper.

The Internet also changed the way maps are used. Maps became interactive, allowing the user to select the information to depict and how it was shown. They also became more timely. Traffic maps, for example, are updated continuously throughout the day. Most importantly, maps can be accessed through a panning and zooming interface that overcomes the problem of producing maps in sections. It was once said that a map user's area of interest was always at the intersection of four map sheets. Internet maps can be continuous with no sectional breaks. All of these changes resulted in a user experience that engages the map user on a different level than with maps on paper.

Cartography, the art and science of making and communicating with maps, has evolved through the years to incorporate new tools and methods of map presentation. The discipline can trace its origins to Greek scholars over 2000 years ago, with a history marked by a number of intellectual accomplishments that have furthered our understanding of the world and its representation. Beginning in the early 1960s, the computer started to be used to make maps. Initially, paper maps were "digitized"—literally converted to numbers. Eventually, these digitized maps would be distributed between computers through the Internet. Along the way, new and related areas of study developed including remote sensing—the use and analysis of imagery taken by aircraft and satellites—and geographic information systems (GIS)—the input, manipulation, and analysis of geographic information by computer. Today, cartography and these new areas of study are under the umbrella of "geospatial technology." A central activity in all areas of geospatial technology is the making of maps.

Research related to cartography has expanded dramatically since the 1950s (see Figure 1.1). The left side of the illustration shows developments related to air photos and satellite remote sensing. A large amount of work has been directed at finding ways to automate the entire process of acquiring imagery of the Earth and extracting the relevant information. Photogrammetry deals mainly with determining elevation using stereo imagery, while the goal of multispectral pattern recognition is the automated identification of features on the Earth's surface based on reflectance differences throughout the electromagnetic spectrum.

Thematic cartography, with origins extending back to at least the 1700s, is the basis for a multitude of different areas of research represented on the right side in Figure 1.1. An academic interest in cartographic communication can be identified by the mid-1950s, oriented mostly toward thematic maps, but with the overall goal of improving maps as a form of communication. From there, cartographic research developed in a number of different directions.

Geographic Information Systems (GIS), influenced in part by research in remote sensing and image processing, emerged in the 1960s and 1970s based on the comparison and analysis of thematic layers. Further integrating the computer,

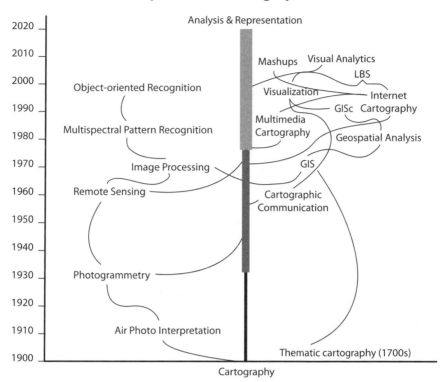

The Expansion of Cartographic Research

FIGURE 1.1. Development of research related to cartography since 1900. The expansion of research since the 1950s includes such areas as Geographic Information Systems (GIS), Geographic Information Science (GISc), and Location-Based Services (LBS).

areas such as geospatial analysis, multimedia cartography, cartographic visualization, location-based services, and visual analytics emerged during the 1990s. At the same time, developments related to the Internet gave rise to research in the distribution of maps through electronic networks. All of these interrelated areas of research, sometimes put under the umbrella of geospatial science, increase the body of knowledge related to maps and mapping.

These areas of research indicate that there are still many unanswered questions about how to best collect, process, analyze, and display geographic information. They also indicate the enormous amount of interest in using maps for a wide variety of applications. Maps are at the core of many different areas of research and development, and a common element of all of these areas of work is the intersection between maps and the Internet.

1.2   The Internet

The Internet has been described in many ways. In the simplest sense, it may be thought of as a system for transferring files between computers. These files, manipulated as numbers and ultimately stored and transferred in binary 0s and 1s, may consist of text, pictures, graphics, sound, animations, movies, or even computer programs. Defined in terms of hardware, the Internet may be regarded as a physical collection of computers, routers, and high-speed communication lines. In terms of software, it is a network of networks that are based on the Transmission Control Protocol/Internet Protocol (TCP/IP)—a particular method for moving packets of data. In terms of content, the Internet is a collection of shared resources. Finally, and most importantly, from a human standpoint, the Internet is a large and ever-expanding community of people who use its resources and contribute to its content.

The beginnings of the Internet can be found in ARPANet, a computer network created for the Advanced Research Projects Agency and funded by the United States Department of Defense. The purpose of the network was to create a redundantly linked system of computers that would continue to communicate even after a limited nuclear attack. The initial Network Control Protocol (NCP) was first implemented in 1969 between Stanford University, the University of California at Santa Barbara, and the University of Utah.

The ARPANet model specified that data communication would always occur between a source and a destination computer. Further, the connection between any two computers is assumed to be unreliable and could disappear at any moment. It is therefore necessary to automatically reroute messages around nonworking computers. If a certain connection between two computers is inoperative, the initial computer reroutes the message to another computer that attempts to deliver the message. This important attribute of the Internet has proven to be especially useful in modern times to circumvent repressive governmental controls on communication.

Sending data from computer to computer requires that it be put in an "envelope," called an Internet Protocol (IP) packet, with an appropriate "address." A single file is divided into pieces before it is sent. All of these pieces may take a different route to the destination—to be reassembled again when they arrive at their destination. Dividing the file into parts speeds data transmission and assures that the pieces can be rerouted around nonworking computers.

Increasing demand on the network at the beginning of the 1980s forced the U.S. government to commission the National Science Foundation (NSF), an agency of the U.S. government, to oversee the network. ARPANet switched from the NCP protocol to the currently used TCP/IP (Transmission Control Protocol/Internet Protocol) on January 1, 1983 and became NSFNET. Many view this date as the beginning of the Internet. The first graphical World Wide Web browser, NCSA Mosaic, was introduced almost exactly 10 years later (see Figure 1.2).

NSFNET was primarily designed to distribute the power of five supercomputers at major universities. These universities and other institutions were connected with a high-speed Internet "backbone." This significantly increased the speed of the Internet.

Internet service providers (ISPs) expanded the network toward the end of the 1980s to include telephone access from homes. High-speed, broadband access through cable television companies became widespread toward the end of the 1990s. Many people throughout the world now access the web through their mobile phones. For countries that lack telephone/cable infrastructure, the mobile network

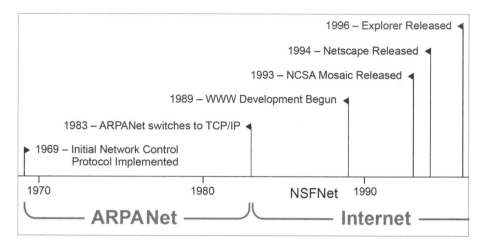

FIGURE 1.2. Time line of early Internet development. Initial work was begun in 1969. The introduction of NCSA Mosaic in 1993 spurred the development of other browsers and the widespread use of the World Wide Web.

represents the only way to access the resources of the Internet, and the speeds are often faster than the wired connections in the more developed parts of the world.

The Internet is a global computer network linking computers in different continents that are thousands of miles apart. It does not rely on a single computer, nor is it managed by any one entity. Rather, it is a system of computers and networks that are linked together in a cooperative, noncentralized collaboration.

1.2.1 World Wide Web

Conceived at the European Particle Physics Laboratory (CERN) located near Geneva, Switzerland, and introduced in 1991, the World Wide Web (WWW) was intended to assist researchers in high-energy physics by linking related documents. The developers wanted to create a seamless network in which textual information on high-energy physics from any source could be accessed in a simple and consistent way. Sir Tim Berners-Lee, director of the World Wide Web Consortium (W3C), played a major role in designing the system.

As originally conceived and implemented, the World Wide Web consisted only of text. Two university students in the United States, Mark Andreesen and Eric Bina, working at a supercomputer laboratory at the University of Illinois, recognized that the web would have limited acceptance as a text-only system. They added the display of graphics, sound and video with the Mosaic web browser introduced in March of 1993 (see Figure 1.3). Describing it as a "consistent and easy-to-use hypermedia-based interface into a wide variety of information sources" (Poole 2005, p. 2), the program was made freely available through the Internet and was widely used within a matter of months. Mosaic transformed the Internet by making it widely accessible.

In 1994, Tim Berners-Lee publicly chided Andreesen and Bina for developing Mosaic. He argued that it would result in a "flood of new users who would do

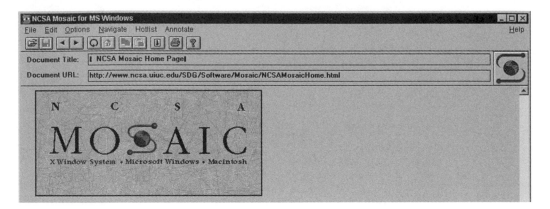

FIGURE 1.3. The Mosaic Internet browser was introduced in March 1993. Incorporating the display of graphics, it was the first widely used World Wide Web browser. The user interface has been imitated by all successive browsers. Courtesy of the National Center for Supercomputing Applications (NCSA) and the Board of Trustees of the University of Illinois.

things like post photos of nude women" (Maney 2003). Mosaic went on to become a "killer-app," a distinction given to programs that become quickly adopted by a large number of users. While Mosaic was soon replaced by other browsers including Netscape, Explorer, Firefox, and Chrome, the basic user interface that was implemented by the program has remained unchanged.

Interaction through the Internet is made possible through an interface that allows the user's client computer to interact with a program running on a server. Long-distance access to the resources of a remote computer is the basis of the Internet. Every time we use the Internet, we are interacting with a distant computer—increasingly located in massive data centers.

Figure 1.4 depicts the amount of Internet traffic that can be attributed to various protocols. Although absolute web traffic is still increasing, its proportion relative to video content is decreasing. Internet video streaming has become a major way of distributing television programming and movies and is now dominating Internet traffic. Video transmissions require a large amount of data in contrast to web, email, data, and file sharing protocols.

1.2.2   Maps through the World Wide Web

On-demand web maps began appearing soon after the introduction of the Mosaic browser in 1993. One of the first of these online mapping programs was developed by Steve Putz (1994) at the Xerox Palo Alto Research Center (PARC). His Map Viewer program allowed the user's client computer to create maps from a geographic database. Each interaction with Map Viewer would request a map from a server that was zoomed in on a specific point (see Figure 1.5). The server would respond with a new map that was embedded into a new web page.

A still-operational example of this type of interaction is the Earth and Moon Viewer available through the Fourmi Laboratory in Switzerland (see Figure 1.6). The site displays views of the Earth from the Sun, the Moon, or orbiting satellites, and includes the overlay of current cloud patterns of the earth derived from

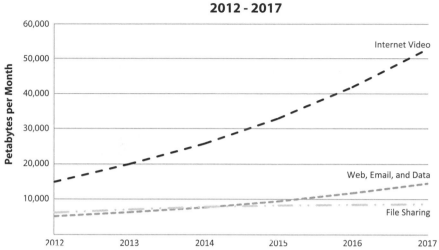

FIGURE 1.4. Internet traffic from various protocols. While World Wide Web traffic is still increasing, its proportion of Internet traffic is declining in comparison to video. Internet streaming and video downloads are dominating Internet traffic. Data from Cisco 2012.

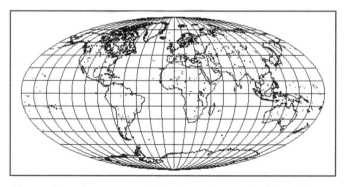

FIGURE 1.5. Xerox Parc Map Viewer was an early example of an interactive web map. The user interacted with a program on a server through the Common Gateway Interface (CGI). The user was able to generate a map of the world at different scales. The resultant map was converted into a graphic file and inserted into a web page.

2010 Jul 6 18:40 UTC

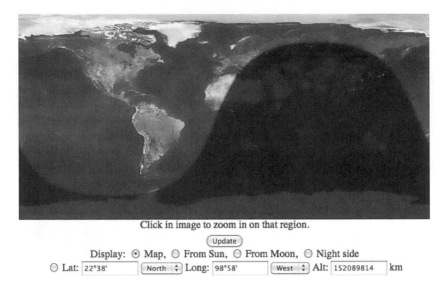

Click in image to zoom in on that region.

(Update)

Display: ⊙ Map, ○ From Sun, ○ From Moon, ○ Night side

○ Lat: 22°38'   [North ⬍] Long: 98°58'   [West ⬍] Alt: 152089814   km

FIGURE 1.6. The Earth and Moon interactive mapping site from Fourmi Laboratory in Switzerland. In continuous operation since the mid-1990s, the site uses a Common Gateway Interface (CGI) to allow the web user to interact with the Earth program on a remote computer. Search: Earth and Moon Viewer.

satellite imagery. With this interface, the user selects the type of map to view and the server inserts the desired map into a new web page.

Almost exactly three years after the introduction of Mosaic, a new era in online mapping was introduced with MapQuest's user-defined street maps in 1996. Using the same client-server model, the site responded to user requests by inserting the desired map into a new web page. MapQuest quickly became the largest publisher of maps on the Internet. By 1999, MapQuest was responding to 20 million daily map requests (Peterson 2001). Developed by a map publishing company called GeoSystems, MapQuest was a major business success and was soon purchased by the Internet giant AOL in 1999 for $1.1 billion (Rohde 2000, p. 64).

MapQuest held the distinction of having the greatest market share among mapping sites until 2009 when it was replaced by Google Maps. Competing against mapping services from Yahoo!, Microsoft (Bing), and eventually Google beginning in 2005, the phrase "I'll MapQuest it" was used to describe how to find a location, even if using a site other than MapQuest. All of these online mapping sites include the option of finding the shortest route between locations. A "satellite view," often taken from an airplane, is also included. Google extended the concept of the online map into a new type of search engine, thereby making it possible to search for features on the map.

1.2.3 Maps Outside the Browser

Not all maps on the Internet are presented through a browser. GeoBrowsers, such as Google Earth (see Figure 1.7) and NASA World Wind, are stand-alone Internet

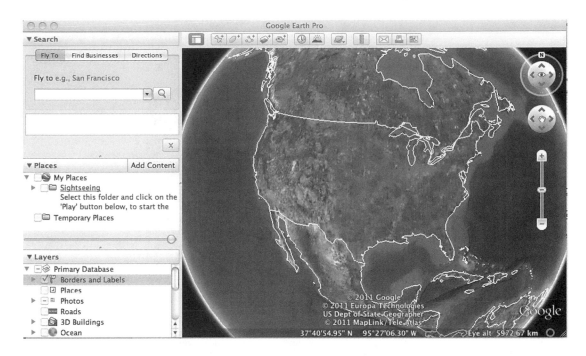

FIGURE 1.7. Google Earth is a digital globe application that operates outside the browser. Many of its functions have been integrated within the browser-based Google Maps. Copyright 2013 Google.

mapping programs. Incorporating the third dimension and implementing flythroughs that simulate flying over a terrain, GeoBrowsers are an increasingly important aspect of online mapping. They standardize the user experience by avoiding differences between browsers, and they streamline the interaction with air photographs and satellite imagery. Associated scripting languages allow users to customize the display and include user-defined information.

Mobile operating systems like Apple's iOS for the iPhone and iPad, and Google's Android have fostered a whole new software industry (see Figure 1.8). The programs written for these devices, called *apps*, generally accomplish a specific task, thus the iconic statement: "There's an app for that." More programs are being written for these operating systems than for any other computing platform, and they are either free or sell for less than US $5. Hundreds of thousands of apps have been written for both Apple's iOS and the Android operating systems.

Mobile phones and tablet computers represent another type of Internet medium for maps. Many have the added advantage of being able to display the current position of the user. Apps that provide spatial information use the positioning capability of the mobile device provided by (1) global positioning system (GPS) satellites; (2) Wi-Fi network IDs; or (3) a system that triangulates a location based on differing signal strengths from surrounding cell phone towers. Location-aware apps indicate the current position on a map and can provide information about the surrounding location such as stores and restaurants. Some apps also allow users to map the locations of family members and friends.

FIGURE 1.8. Apple's iOS operating system on the iPad. Copyright 2013 Apple.

1.2.4 Mapping in the Cloud

The term *cloud* is increasingly applied to the use of distributed data centers. Up to this point, data centers have been used primarily for the storage and retrieval of data. Cloud computing goes a step beyond this function by using the processing power of data centers. The National Institute of Standards and Technology (NIST) (Mell and Grance 2009) defines cloud computing as "a model for enabling convenient, on-demand network access to a shared pool of configurable computing resources (e.g., networks, servers, storage, applications and services) that can be rapidly provisioned and released with minimal management effort or service provider interaction." The U.S. government now mandates that all agencies either migrate to cloud computing or explain why it is more efficient to not use cloud computing. One of its major advantages is the added speed in processing and the ability to deal with large data sets (Yang 2011). It is also argued that the data is more secure and that cloud computing conserves electricity because it consolidates resources.

Cloud computing has major implications for online mapping. Most Internet maps are now stored at data centers. It is also where the programs reside that analyze and manipulate the underlying information. Finally, user data to augment and update the maps is centralized at these centers. The cloud is clearly the future for online mapping.

1.3   The Internet and Map Use

Although we have witnessed a revolution in the distribution of maps, many people still have difficulty using them. With paper maps, it was estimated that more than half of the population did not have basic map reading competency. This translated to a particular difficulty in performing tasks involving navigation. Map illiteracy may be related to a lack of specific education in using maps, differences in the way people conceive of space and spatial relationships, or the way maps are made and presented.

It may be impossible to determine exactly why some people have difficulty using maps. But, whatever the reason, the result is clear: Many people have poorly formed mental representations of the local environment and especially the space beyond their direct experience. Their conceptions of the world, both far and near, are restricted, and they are isolated by their limited conceptions of the world around them.

Before the Internet, the paper medium was the predominant form of map distribution. However, that medium does not facilitate interaction, an ingredient that many associate with learning. Although some are able to "make a connection" with the map on paper and mentally visualize what the map represents, others have difficulty doing so. Education may help overcome this barrier for some, but others might always have difficulty with this form of map presentation. Interaction may help make maps more useful for a majority of the population.

It has been argued that the incorporation of interaction increases the use of maps (Peterson 1995). No longer restricted to a single view, the user is encouraged to explore alternative methods of representation—different views that may better conform to the user's mode of learning. The views that are presented may be targeted to specific users. The most difficult task in navigating with maps may be determining the current position. The addition of a symbol to show users their current location removes this requirement.

The Internet has already improved the distribution of maps. If maps are presented properly, Internet maps also have the potential for improving the quality of maps as a form of communication, thereby changing both the mental representations that people have of the world and how people mentally process ideas about relationships in our environment. This will provide more people with a better and more complete conception and understanding of the planet on which we live.

1.4   The Cost of Maps

Increasing the availability of maps is the best way to increase their use. But making and storing maps is expensive, and their distribution adds significantly to their cost.

Governments and the private sector have invested large sums of money to produce all manner of maps. Decisions about what to map is driven by both governmental and commercial interests. Governments make maps based on whose interests will be served. In general, making maps of the country, the world, and even other celestial bodies is considered to be in the "national interest." Businesses make maps based on the amount of return on the investment.

There is also an important interplay between government and commercial mapping. Maps that are made by governments, often at enormous expense to the taxpayer, are given to the private sector to be repackaged and sold as commercial products. Most maps that are sold to the public are based on government data.

The change in how maps are distributed to people has been dramatic. As late as the early 1990s, nearly all maps were distributed on paper. The only maps that most people could view were on this medium. By the end of the decade, the Internet had become the predominant form of map distribution. The most remarkable aspect of this transformation is that maps are now distributed at no charge. The free distribution of maps by the government is justified because there are no printing and warehousing costs and taxpayers have already paid for the making of the maps.

Businesses have developed a new model for financing the production and distribution of maps. Initially, online maps were financed by small advertisements that appeared on the web page with the map. Companies like Google began to use maps to augment their search engine and provide a way for people to find the location of businesses. The map itself is provided for free because it simply serves as a background to display sponsored information.

While most maps are now distributed freely, the costs associated with their use have risen sharply. Computers and computer software are needed to make the digital maps usable. While it can be argued that the digital products are now more functional than the previous paper version, it cannot be overlooked that these maps are only usable by individuals with computers, computer software, a fast network connection, and the appropriate training. The access to maps has been increased for some, but they remain beyond the reach of many. In addition, the demand and necessity for paper maps has been reduced, making it even more difficult for people without a computer to find a map.

For better or worse, the Internet is firmly entrenched in our society and our way of life. How we communicate and access all types of information has drastically changed. Maps, and geographic information in general, are a big part of this change.

1.5   About This Book

This book and the associated online materials introduce maps and map making in the era of the Internet. The text introduces and explains the basic concepts, while the associated electronic resources illustrate and expand on these ideas. The materials are designed to augment each other.

Knowledge may be divided into theory and practice. According to Van de Ven and Johnson (2006), understanding "the relationship between theory and practice is a persistent and difficult problem" particularly for those scholars who develop "knowledge that can be translated into skills that advance the practice of the professions" (p. 802). They point to several studies that academic research has become less useful for solving practical problems and that the "gulf between theory and practice in the professions is widening."

Broadly defined, theory consists of concepts and ideas that have evolved through time along with the historical progression of these ideas. Practice, the type of information that can help you make maps, should be guided by theory. Underlying this notion is that practical knowledge (knowledge of how to do things) is based on research knowledge (scientific knowledge and scholarship more broadly). The gap between theory and practice is seen as a knowledge transfer problem in the sense that it is difficult to bring theoretical knowledge into the practical domain. Van de Ven and Johnson (2006, p. 802) point out that knowledge of theory and practice is sometimes viewed as "distinct kinds of knowledge. Each reflects a different ontology (truth claim) and epistemology (method) for addressing different questions."

More popular distinctions between the two forms of knowledge simply state that they are distinct based on their usefulness: Theory is useless, and practical knowledge is useful. To further the distinction, it is often implicitly claimed that if knowledge is useful, then it cannot be theory. Finally, the distinction is often seen as a way of separating individuals who work in both domains with those dealing in theory given a higher status.

Whatever the exact relationship between theory and practice, it is clear that both types of knowledge are interrelated. To facilitate the development of both areas, it is important to alternate between theory and practice (see Figure 1.9). This is accomplished in the book through the close juxtaposition of the two.

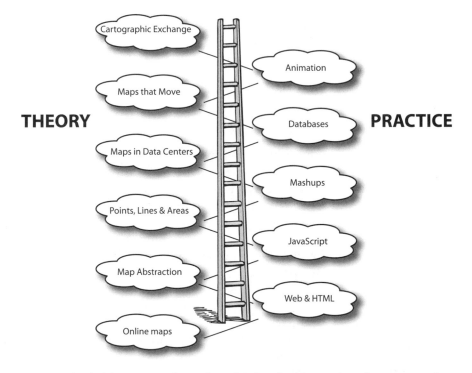

FIGURE 1.9. The ladder approach used in this book. Alternating chapters on theory and practice reinforce the major concepts.

The chapters that deal with theory examine how maps are processed in the brain, how mapping developed over time, the abstraction process in mapping, map input and the global positioning system, map layers and GIS, databases, and mobile and animated mapping. The "practice" chapters in this book involve making maps using online mapping systems. Code samples are included that implement the examples described in the book. Working through these examples will help you understand how easy it is to make maps using the available online tools, and how these maps can be made available to others.

If used in a classroom setting, the exercises are best done in a computer laboratory. Whether arranged by row or in circular "pod" fashion, a computer laboratory facilitates interaction with both computers and people. Sometimes the computers are provided, but students are increasingly bringing their own laptops to connect to the Internet and download class materials. The exercises can be challenging. As you try to follow them, you will at one point think you know how something is done and then, in the very next moment, feel completely lost. This is a normal experience in such a learning environment. While you simultaneously feel that you are lurching forward and falling behind, remember that the most important interaction for learning is not with the computer but with the other people in the classroom.

This book also addresses the quality of online maps. Unfortunately, many poor maps are available through the Internet. Some are simply illegible. Many are poorly designed, either from the standpoint of graphic quality or user interaction—or both. One way to evaluate the quality of online maps is to compare the available Internet maps to each other. A map gallery introduced in Chapter 3 outlines a format for displaying maps side by side to highlight differences in the maps. Interaction is a major component of Internet map design, and this book stresses ways of introducing interactivity to maps and discriminating good forms of interactivity from bad.

1.6   A Look Ahead

The next chapter takes a closer look at the Internet map landscape and how to create a website through a web-hosting service. Chapter 3 examines mental maps and how we process and store spatial information. Chapter 4 introduces the hypertext markup language (HTML) for the making of a map gallery page that provides links to maps on the web. These files are then uploaded to the Web hosting site. Maps as a mirror of civilization and the development of mapping through time is the topic of Chapter 5. The remaining chapters examine how maps are input, stored, and manipulated by computer, how they are integrated on web pages, and how they are distributed.

Beginning with Chapter 4, the remaining even-numbered chapters in this book include a file of code in the form of zip files (see Figure 1.10). These code files can be uploaded to the webserver as a zip file and unpacked. The files are called code04.zip, code06.zip, code08.zip, and so on. The zip file unpacks to a folder that contains an `index.htm` file that has links to all of the other files in that folder. It

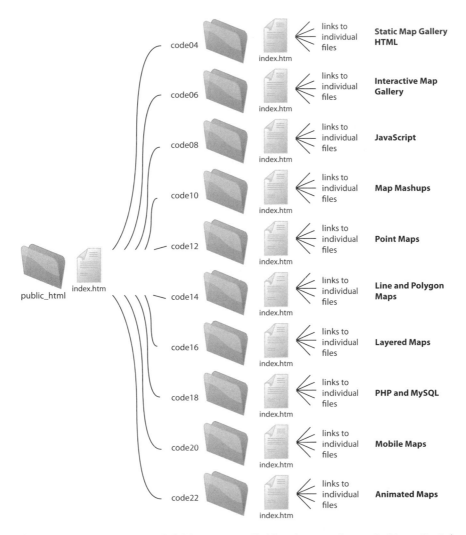

FIGURE 1.10. The public_html folder stores all files for a website. A file called `index.htm` is displayed when the site is accessed. It links to the folder of code for all of the even-numbered chapters in this book. Each of these folders also contains an `index.htm` file that has links to the examples.

is not necessary to create a link to the index.htm file because it is automatically displayed when a reference is made to the folder, as in this example: http://maps.unomaha.edu/Cloud/.

Some knowledge of programming is helpful to understand the exercises, but the book is designed for students with little or no programming experience. In most cases, all that is needed is the ability to work with a text editor and a willingness to experiment. Changes might involve such things as defining a new center for the map, a different zoom level, or the text for pop-up markers. Completing the exercises only requires understanding the code well enough to determine what needs to be changed to make the map you need.

1.7   Questions

1. What motivated the development of the World Wide Web?

2. How has Internet traffic changed since the World Wide Web was introduced? What now consumes most Internet bandwidth?

3. What was the importance of the Mosaic browser for maps? Why did Tim Berners-Lee, the inventor of the World Wide Web, disapprove of its introduction?

4. When were maps first distributed in large numbers through the Internet, and what major advance made this possible?

5. Describe some major milestones in the development of maps and the Internet.

6. How did the Internet change the way maps are used?

7. What is the fundamental difference between general reference and thematic maps?

8. Describe some areas of study that have their basis in thematic mapping.

9. How were initial interactive maps created for web pages?

10. How did MapQuest change our relationship with maps?

11. Describe the map-use problem with paper maps. Has the Internet made maps easier to use?

12. What makes mapping so expensive?

1.8   References

Anderson, Chris, and Michael Wolf (2010, Sept.) The Web Is Dead. Long Live the Internet. *Wired*. [http://www.cisco.com]

Cisco (2013) Cisco Visual Networking Index: Forecast and Methodology, 2012–2017. Cisco White Paper.

Kelly, Robert, and David Thomas (2010) *Archaeology*. Belmont, CA: Wadsworth.

Maney, Kevin (2003, Mar. 9) 10 Years Ago, Who Knew What His Code Would Do? *USA Today*.

Mell, P., and Grance, T. (2009) *The NIST Definition of Cloud Computing* Ver. 15. [online]. NIST.gov. Available from: http://csrc.nist.gov/groups/SNS/cloud-computing/ [Accessed 8 Jan. 2012].

Peterson, Michael P. (1995) *Interactive and Animated Cartography*. Upper Saddle River, NJ: Prentice Hall.

Peterson, Michael (2001) Maps and the Internet in R. B. Parry (ed.) *The Map Library in the New Millennium*. pp. 88–102. London: Library Association Publishing.

Poole, Hilary W. (2005) The Internet: A Historical Encyclopedia. Santa Barbara, CA: MTM Publishing.

Putz, Steve (1994) Interactive Information Services Using World-Wide Web Hypertext. *Computer Networks and ISDN Systems* 27(2): 273–280. Berlin: Elsevier Science.

Rohde, Laura (2000, Jan. 24) AOL Hammers Out Deals. *InfoWorld*. P. 64.

Thrower, Norman J. W. (2007) *Civilization: Cartography in Culture and Society.* Chicago: University of Chicago Press.

Van de Ven, Andrew H., and Paul E. Johnson (2006) Knowledge for Theory and Practice. *Academy of Management Review* 31(4): 802–821.

Yang, Chaowei, et al. (2011, July) Spatial Cloud Computing: How Can the Geospatial Sciences Use and Help Shape Cloud Computing? *International Journal of Digital Earth* 4(4): 305–329.

CHAPTER 2
The Internet Map Landscape

A good map is both a useful tool and a magic carpet to far away places.

—Anonymous

2.1 Introduction

Landscapes come in many forms. Physical landscapes consist of tropical valleys and windswept desert plains. Human landscapes are formed by different methods of farming or patterns of urban housing. The term *information landscape* is used to refer to the landscape of information that we traverse when we venture into a hypermedia environment, a landscape that also has its fertile plains and deserts.

Maps became a major component of the online information landscape in 1993 once graphics could be added to World Wide Web pages. All types of maps became easily available, from subway maps of cities to maps of the Moon. So many maps were available that map users could choose between alternative representations of the same environment to determine the one that best suited their needs. Some interactive sites made it possible to center the map on an area of interest and include features requested by the user. Given these possibilities, it is easy to understand why the distribution of maps through the Internet became so popular.

This chapter examines the Internet map landscape and serves as an introduction to Internet map types and graphic file formats. It also covers how to find maps through search engines and the new era of tile-based, cloud mapping. We begin by looking at static maps.

2.2 Maps on Paper

Many of the maps available through the Internet have been scanned from paper maps. Although the scanning of maps represents a quick way to transform a map into digital form, the scanned versions are often of poor quality. At times, so little care is taken in the scanning process that the text and graphics on the back of the paper map are visible (see Figure 2.1). More importantly, although scanned paper maps can become interactive by making features on them clickable, they normally are simply static objects with no underlying function.

The limitations of scanning aside, scanned maps contain a considerable amount of information. If scanned well, they can be printed for later use. Some sites incorporate a magnifying glass tool to zoom into parts of the maps, allowing them to be viewed in much greater detail than would be possible with the unaided eye. This technique has been used effectively with older maps that were originally drawn with pen and ink. Previously only available in museums, these historic documents can be viewed and studied in minute detail.

A distinct advantage of a printed map is resolution. A typical high-resolution printer can resolve between 1200 and 4800 dots per inch (dpi; 472 to 1890 dots per cm). In contrast, a computer monitor can only display about 65–165 dpi (25.6 to 65 dots per cm). Small, mobile devices such as mobile phones and some laptops have managed resolutions above 326 dpi (128 dots per cm), and it is likely that this resolution will eventually migrate to larger displays. Computer monitors are also limited in size, typically up to 24 in (60.1 cm) in diagonal measure. Printed maps and photographs can be much larger than this.

Although the spatial resolution and the size of computer monitors will increase, it is unlikely that they will ever have the detail or size that is currently available through printing. We accept the lower resolution and smaller size of computer

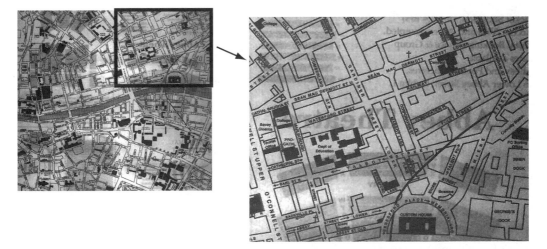

FIGURE 2.1. Text and graphics on the backside of this map are visible in the scanned version. Inserting a black piece of paper behind the map before scanning would have reduced the problem.

maps because the display is dynamic. Dynamic maps are either frequently updated, or they implement some type of interactivity. In some cases, the map can even be animated. A dynamic map is the trade-off for not having the resolution, size, or portability of a printed map.

A comparison between the online and paper map would not be complete without examining the commercial aspects. Printing on paper is expensive and printing in color even more so. Once the infrastructure is in place, it is simply less expensive to place graphics on the web, particularly those in color, than it is to print them on paper. When the additional costs of shipping and distribution are factored into the printed product, the cost advantages of distributing maps through the Internet are even more apparent.

While printing is more expensive, a major advantage of the printed map is that it is a physical product that cannot be easily copied or duplicated, especially the larger versions. The inability to easily copy the paper map gives it greater commercial value. By contrast, maps on the Internet can be readily copied and therefore have almost no commercial value by themselves. Their value is not as a physical product but in the information they convey.

The willingness to pay for maps varies. In many countries, maps of urban areas and roads are made available for free by state and local governments. More detailed maps that would be required for hiking or other activities still need to be purchased. In Europe, the purchase of maps is more common, and all types of maps are available—sometimes in specialized stores that sell nothing but maps and atlases. In other parts of the world, mostly in Asia, Africa, and South America, paper maps may be very difficult to find, or governments may not allow their sale.

GPS navigation devices have become increasingly popular, either implemented as a stand-alone device or through a mobile phone. The map display is limited in size, and navigation often relies on spoken directions. Part of the reason for the simplified map is that the driver cannot provide the amount of attention necessary to read a more detailed map. It is the location-finding capability of the product, not the quality of the map display, that makes these devices popular.

2.3   Computer Maps

Some maps available through the Internet are initially drawn by hand on paper and subsequently scanned. Most of the newer maps, however, were made with the help of computers. These computer maps are either individually drawn using general-purpose, graphic design software or made automatically with specialized mapping programs. Computer-generated maps have been designed with the limitations of the computer display in mind and are usually much more legible than scanned paper maps.

2.3.1   Graphic Processing

Graphic design programs, such as Adobe™ Illustrator™, effectively emulate the hand-drawing process while providing graphic tools that go far beyond pen and ink (see Figure 2.2). They offer a high level of graphic control over the resultant

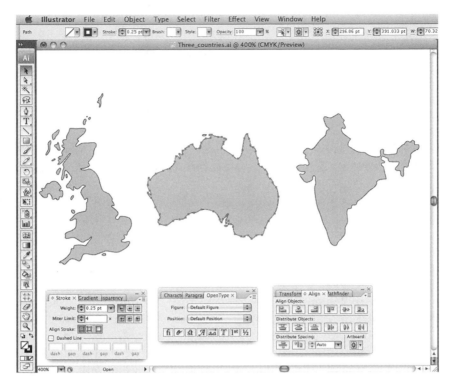

FIGURE 2.2. The graphical illustration program Adobe™ Illustrator™. Illustrator, along with Photoshop™, are examples of graphic processing programs. In contrast to word processing, very few people have a working knowledge of graphic processing. Copyright 2013 Adobe; Adobe product screenshot reprinted with permission from Adobe Systems Incorporated.

map and can produce a printed product that exceeds the quality of anything produced by hand. The cartographer is able to manually generalize features and precisely place labels and other map elements. Maps produced using this method are the closest we have to hand-produced maps. They are especially useful for illustrations that emphasize particular features or distributions. Many of the illustrations in this book have been produced with Adobe Illustrator, a commonly used program for these types of maps.

The ability to work with graphic design programs is not widespread and is mostly restricted to graphic artists. This is unfortunate because communicating through graphics of all kinds should be on the same level as communicating with words. While knowledge of word processing is very common with the use of programs such as Microsoft Word, the ability to "process" graphics with the graphics-oriented programs available is far less developed.

The use of graphics in communication varies by culture. In a study done on travel books between countries, it was found that those published in North America had the fewest maps and illustrations (Suzuki and Wakabayashi 2005). In place of maps, detailed but often incomprehensible textual instructions were provided to find particular locations. The study demonstrated that long passages of text in the English-language editions could have been communicated more effectively through maps. This overreliance on text occurs because most authors do not know how to create graphics or how to use graphical illustration programs. Moreover, book publishers find the cost of producing graphics to be prohibitive, and they have concerns about publishing maps and illustrations for which the copyright has not been determined.

2.3.2   Automated Mapping

Most maps are now made automatically from an underlying database. Specialized programs have been written to produce maps that look like those designed by humans. Although many cartographers believe that it is impossible to program all of the decisions that are part of the map design process, and that these maps will always be an inferior form of representation, most map users are not aware of these concerns. The automation of mapping is firmly entrenched, and it is unlikely that we will go back to a time when all maps are made by hand, whether by pen and ink or with the help of graphic design software.

The automation of mapping would seem to be a simple task. Maps are, after all, merely a collection of points, lines, and shapes drawn with the appropriate labeling. In actuality, the placement of all these features in proper juxtaposition, with minimal overlap and with the proper emphasis on desired features, is an enormously complicated task. Any scaled, and therefore generalized, depiction of the world involves a huge amount of interrelated decision making; these decisions are largely artistic and therefore difficult to automate. The cartographer makes map design decisions based on what "looks good" to a trained eye, but a computer cannot look at the map and make a similar judgment. We can describe the basic steps of the cartographic process, including such aspects as the selection of features, their classification, simplification, exaggeration, and symbolization, but it is a tremendous challenge to tell the computer how it should all be done and produce both a useful and visually pleasing product.

The process of label placement is perhaps the easiest way to illustrate the complexity of the cartographic decision-making process. Figure 2.3 shows how the city of Omaha is labeled in six different scaled depictions, from small scale to large scale. Notice how the label for Omaha is never in the same place relative to the city. Beginning with the smallest scale, the label placement starts on the right side of the city, over the state of Iowa. The label then moves above the city, to the left, above again, to the right, and to the left again. The positioning of the label in each map is based on calculations that show where the label can best fit, while allowing other features to be shown. It is a complex interplay of a decision-making process that seemingly has been successfully automated. A trained cartographer would not be satisfied with any of the automated solutions.

2.3.3   Raster Maps

Most graphic files distributed through the Internet are based on a format in which the image is represented as a grid of "picture elements" called pixels (see Figure 2.4). Each grid square is assigned a color that is represented in the computer as a

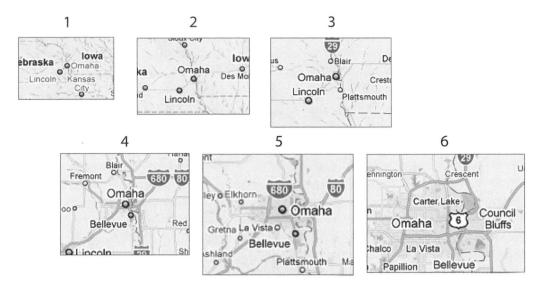

FIGURE 2.3. The placement of the label for Omaha is compared in six differently scaled depictions from Google Maps. Complex computer algorithms are used to place labels based on the inclusion and positioning of other map features. The end result of this automated label placement would not be acceptable to a trained cartographer. Copyright 2013 Google.

number. This grid of pixels is referred to as a raster. Raster data encodes an image with a matrix of pixels arranged in rows and columns. This format matches the display of the computer monitor, so no complicated conversion is necessary to bring the graphic to the screen. A variety of grid-based, graphic formats are available that vary in the number of colors they represent and in how they compress the image for faster transmission.

One of the first graphic file formats developed for the Internet was GIF (graphics interchange format). Limited to a maximum of 256 shades, or colors, GIF files became a standard way of distributing graphic files beginning in 1987—even before the introduction of the World Wide Web. Limiting the number of colors reduces the size of the files. While the 256 colors that are used can be from a palette of millions, a standardized subset of 206 "safe" colors developed that could transition between different computers and be relatively identical. The format is particularly well suited for graphics composed of lines that usually have few colors. Although the Lempel-Ziv-Welch (LZW) compression method used by GIF is a proprietary format (Godse 2008, p. 39), GIF is considered a public standard that is widely adopted and supported by all web browsers.

A problem with rasterized lines and lettering is a stair-step appearance on diagonal segments. The technical word for this effect is *aliasing*. A procedure called *anti-aliasing* is used to minimize the stair-step effect. Essentially, anti-aliasing involves introducing lighter-shaded pixels around the edges of the darker pixels to "soften" the appearance of the stairs. The introduction of these pixels is visible on the inset in Figure 2.4.

Figure 2.5 presents an enlargement of a color map. Each pixel is given a color that is represented in numerical form and depicted on the screen of the computer

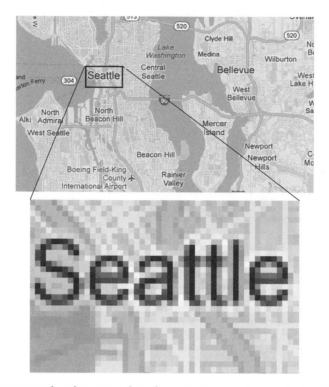

FIGURE 2.4. An example of a raster data format. A procedure called anti-aliasing inserts lighter pixels around lines to limit the stair-step appearance on diagonal lines, particularly visible with the "S" in Seattle on the enlargement. Copyright 2013 Google.

monitor as intensities of the three primary additive colors—red, green, and blue. If each primary color is represented with one byte, or 8 bits, there would be 256 possible shades of red, green, and blue (RGB). In combination, the possible permutations of the different primary intensities would produce 2^{24} different colors, or 16.7 million.

Both JPEG (JPG) and PNG use multiple bytes for each pixel to represent color. JPEG (Joint Photographic Experts Group) is the most common Internet graphic format. It is the best suited format for pictures because it can display a large number of colors, 16.7 million, and it incorporates image compression methods that significantly reduce file sizes. However, the compression algorithm results in a loss of detail that is particularly evident with lines and lettering (see Figure 2.6). Although this image compression is not very noticeable on pictures, the loss of sharpness is apparent on maps that are typically composed of lines. For this reason, the JPEG format should not be used for maps—particularly with higher levels of image compression. Nevertheless, the Internet is dominated by maps in the JPEG format.

Portable network graphics (PNG) offers a good compromise for maps because it produces small files without affecting the representation of lines. This is accomplished by using a lossless form of data compression. PNG file sizes are comparable to GIF, though generally not as small as JPEG. The major online mapping sites use the JPEG format for satellite images and the PNG format for maps.

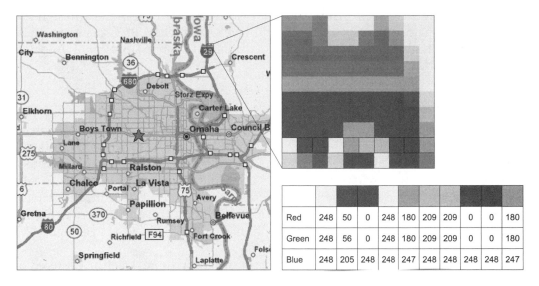

Red	248	50	0	248	180	209	209	0	0	180
Green	248	56	0	248	180	209	209	0	0	180
Blue	248	205	248	248	247	248	248	248	248	247

FIGURE 2.5. A map represented in the form of a grid with individual "pixels." This color image is coded with the three primary colors of red, green, and blue. Each primary is given intensities between 0 and 255. A small section of the map has been enlarged. The values for the three primary colors are given for 10 pixels. Copyright 2013 MapQuest—Portions copyright 2013 NAVTEQ, Intermap.

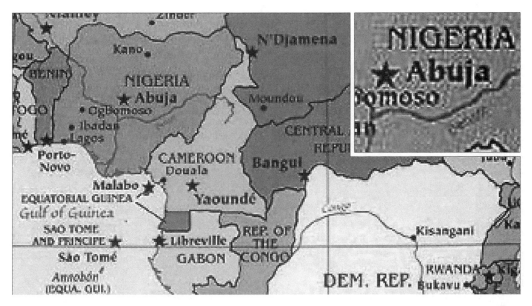

FIGURE 2.6. A portion of a map in the JPEG format. JPEG compression produces artifacts around lines including the lettering that becomes particularly apparent by zooming in on the map, or through printing. Courtesy of University of Texas Libraries.

The raster format is the primary method for distributing graphics through the Internet. The major browsers all support the GIF, JPEG, and PNG raster formats. Because its method of compression does not degrade the representation of lines, the PNG format is the best choice for maps.

2.3.4 MapQuest Era

MapQuest introduced its online mapping site in 1996 based on a large database that included most streets in North America. By 1997, it was producing 700,000 maps a day and was the leading online map provider until 2009 (Dougherty 2009). Using a standard client–server technology, the user's client computer would make a request for a specific map; then MapQuest servers would respond to the request by drawing the map from a database of points and lines, converting this to a grid-based raster format and delivering the resultant map within a web page. Figure 2.7 shows the initial growth in MapQuest usage.

Each request for an additional map, at a different zoom level or centered at another point, would result in another server request that would produce another map that would be embedded in another web page that would update the page on the user's computer. Although the process was fairly fast, there was always a wait for the server to respond. Any zoom or pan required waiting for the server to produce another map that was inserted into another web page. Response times would also be subject to Internet traffic so that a request for a map might take considerably longer when traffic was heavy. Maps would be produced more quickly during the

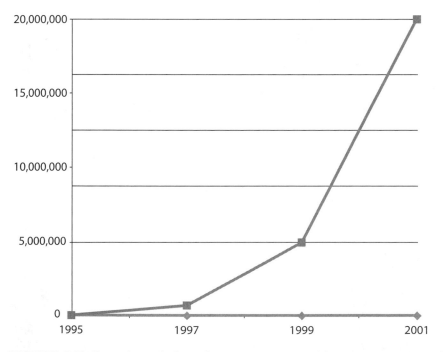

FIGURE 2.7. Growth in daily online maps generated by Mapquest. Introduced in 1996, MapQuest usage grew quickly, reaching 20 million maps a day by 2001.

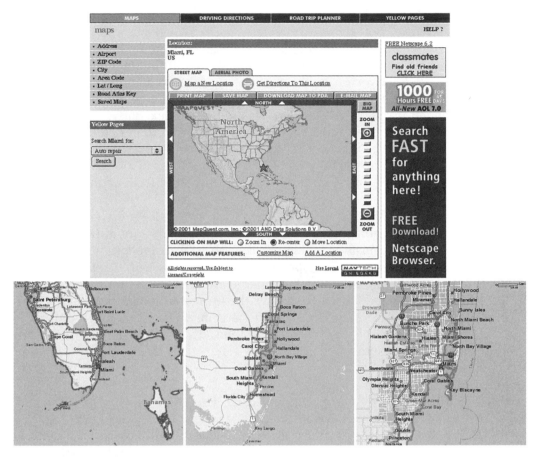

FIGURE 2.8. A 2001 version of the MapQuest web page. Dominated by ads, the map constitutes only a small part of the web page. The maps of southern Florida are at three different zoom levels. A total of 10 zoom levels were available. Copyright 2001 MapQuest—Portions copyright 2001 NAVTEQ, Intermap or copyright 2001 MapQuest.

overnight hours. Users found this variability in response times to be more annoying than the generally slow response. Figure 2.8 shows a 2001 version of MapQuest along with three maps of southern Florida at different scales.

2.3.5 Vector Maps

In contrast to the pixels of the raster format, the vector format defines a graphic with coordinates. This format is better suited for maps and can be easily zoomed with no loss of detail. Maps in this format also take advantage of the higher resolution of printers. Screen resolutions are often only 100 dpi. Even a medium resolution printer can resolve 600 dpi, and high-end typesetting devices, called image-setters, have resolutions up to 4800 dpi. The vector format takes advantage of these higher resolutions by scaling the coordinates to fit the size of the available raster dimensions.

With Adobe's Portable Document Format™ (PDF), based on a programming language called Postscript™, lines and lettering are defined as *x* and *y* coordinates and can be redrawn at each different zoom level (see Figure 2.8). Notice the quality of the lettering and rivers, both defined as vectors, in the zoomed version. A further advantage of most vector formats is that they can incorporate a raster image. As can be seen in the background shading in Figure 2.9, elevation is depicted with differently shaded pixels. The PDF format is used for the exchange of all types of documents, including government tax forms. But, PDF files that incorporate graphics are often large and take longer to download. In addition, the Adobe Acrobat plug-in must be loaded by the browser before the file can be displayed. Many users instinctively hit the back button when encountering a PDF file to abort the lengthy display process.

Adobe's Flash format is also based on vectors. It produces very compact files, particularly in comparison to raster formats. Flash files are only a fraction of the size of a corresponding PDF. The plug-in for Flash is also smaller and loads more quickly than the corresponding Adobe Acrobat plug-in used to display PDF files. The format is used primarily for interaction and animation, and for the delivery of video. Although designed for screen display, the format allows users to take advantage of a printer's higher resolution. Flash files are presented through the plug-in so it is not possible to save the files separately. Adobe Flex, a Flash-based development environment for producing web content, is designed to create a consistent user experience no matter what browser is used. It can also make applications that are independent of the browser.

PDF and Flash are both proprietary formats. The major open standard format for vector graphics is scalable vector graphics (SVG), under development by the World Wide Web (W3C) consortium since 1999. Based on the extensible markup language (XML), SVG files contain textual codes for defining graphics and can be written with a text editor. Once dependent on a plug-in, most browsers now have some degree of native support for SVG. Wikimedia Commons, the graphic library for Wikipedia, has adopted SVG as a standard for all vector-based diagrams and maps. File sizes are typically a third larger than PDF and more than twice the size of the Flash format.

FIGURE 2.9. Lines can be defined with the Portable Document Format (PDF), as depicted here with the rivers and lettering. Pixels are used in the background to show relief. PDF and Flash files can include both vector and raster graphics. Courtesy of University of Texas Libraries.

2.3.6 Tiled Web Maps

Google Maps, introduced in 2005, revolutionized the online mapping landscape. Known for its search engine, Google effectively added a map-based search through Google Maps. By not including ads around the map, as MapQuest did, they left more room for the map on the computer screen. More importantly, from a map user's perspective, Google Maps streamlined how we interact with maps introducing what is now referred to as a multi-scale, panable "MSP" map.

The delivery of a Google map is based on two major ideas: (1) image tiling; and (2) Asynchronous JavaScript and XML (Ajax). Image tiling had been used since the early days of the World Wide Web. In comparison to text, images require more storage and therefore take longer to download. A solution is to divide the image into smaller segments, or tiles, and send each tile individually through the Internet. These smaller files often travel faster because each can take a different route to the destination computer (Sample and Ioup 2010). On the receiving end, the tiles are reassembled in their proper location on the web page. With a moderately fast Internet connection, all of these steps occur so quickly that the user rarely notices that the image is actually composed of square pieces. With slower connections, the individual tiles are clearly evident.

The second major innovation introduced by Google Maps is the incorporation of Ajax, a new form of client–server interaction. This was the culmination of many years of effort to reshape how clients communicate with the server. Essentially, Ajax maintains a continuous connection with the server, exchanging small messages in the background even when the user has not made a specific request. This leads to faster server responses. Ajax might be thought of as an application that works in the background of a browser page to anticipate what the user might want, ready to communicate with the server to respond to a request. Operations in Google Maps that are particularly assisted by Ajax include zooming and panning, the most common form of interaction with maps.

In a dramatic transformation, all of the other major online map providers—MapQuest, Yahoo, and Microsoft (Bing)—converted from the standard client–server to the Ajax, tile-based method of map delivery within a short time after Google Maps was introduced in 2005.

2.3.7 Cost of Map Storage

Figure 2.10 depicts a series of Google maps at different levels of detail (LOD). All tiles are 256×256 pixels and require about 15 kb apiece to store in the PNG format. Table 2.1 shows the number of tiles that are used in a tile-based mapping system for 20 LODs, or zoom levels, and the associated storage requirements and storage costs. With 20 LODs, approximately 1 trillion tiles are needed for the world. At an average of 15 kb per tile, the total amount of memory required is 20 petabytes, or 20,480 terabytes. No single computer currently has this much storage capacity.

The total cost of map and satellite data storage has not been made public by any company that delivers information in this way. It can be estimated based on a cost of about US $100 per terabyte (see Table 2.1), an approximate price of a hard-drive and computer connection. To store the entire one trillion tiles on

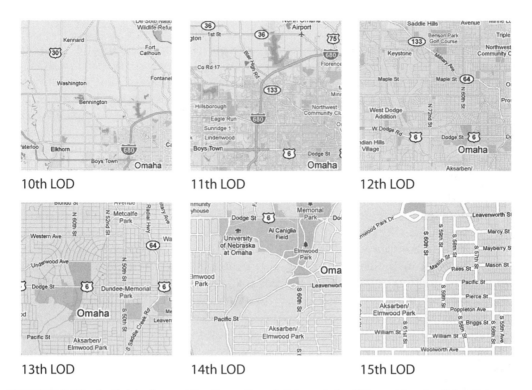

FIGURE 2.10. Individual map tiles from Google Maps at six different levels of detail (zoom levels). In 2005, Google introduced a tiling system to deliver online maps. Over a trillion tiles are used for Google's 20 zoom levels. In 2010, Google abandoned raster-based tiles and converted to a vector approach. Copyright 2013 Google.

disk drives would cost about $2 million (US $100 × 20,480 terabytes). In order to achieve faster response times, there is strong indication that data centers use faster random-access memory (RAM) to cache the map tiles. If the entire map of the world were stored in RAM, it would cost more than US $629 million. Data centers likely use a combination of disk drives and RAM to store the files, resulting in a map storage cost at each data center of approximately US $2 million to US $629 million. A still faster storage option would be to use a graphical processing unit (GPU). This device is specifically designed to store and manipulate images and transfer image data much faster than computer memory. Map storage on GPUs would be at least twice as expensive as on RAM, or nearly US $1.3 billion for the whole world.

These data storage requirements and costs are only for the map. The satellite view, with tiles in the JPEG format, requires approximately the same amount of storage space. Google also stores multiple versions of the satellite image from different dates. Each Google data center would likely have a copy of the map and satellite images, as well as any other map that is provided, such as the Terrain view. Combining all of these data storage costs indicates the importance placed on maps by Google and other companies.

2.3.8 Vector Tiled Maps

In 2010, Google switched from raster to vector tiles for some types of map delivery. One impetus for the change was the increased use of maps on mobile devices. Limited by a monthly data plan, mobile users noticed that any application involving maps quickly consumed this monthly allocation. A less data-intensive method was needed to distribute map information. Although the satellite and terrain views are still sent as raster tiles, Google adopted a vector approach for its map layer. Rather than sending tiles as pre-rendered PNG files, the map information is sent as a series of lines and shapes. While the primary benefit of this vector approach is a reduction in the amount of data to create a map, it also facilitated changing the map style by specifying different colors for features such as roads (see Figure 2.11).

Another major advantage of vector tiles is that it allows text to be separated from the underlying map. Previously, the cities in China and Japan were written with local characters, and it was not possible to change the text to a Western script because the text was "burned into" each raster tile. By separating the text, it is now possible to switch between "English" and local scripts (see Figure 2.12).

In making the transition to a new form of map delivery, Google needed to maintain support for the older, raster-based method on some devices. For example,

TABLE 2.1. The Number of Tiles, Storage Requirements, and Estimated Storage Costs Used by a Tile-Based Online Mapping System to Represent the World at 20 Different Levels of Detail (LOD) or Zoom Levels

Levels of detail (LOD)	Number of tiles	Ground distance per pixel in meters	Storage requirements at 15 kilobytes per tile	Disk storage costs at $100 per terabyte	RAM memory storage costs at $30 per gigabyte
1	4	78,272	60 kilobytes (kb)	$0.000006	$0.002
2	16	39,136	240 kb	$0.00002	$0.007
3	64	19,568	968 kb	$0.0001	$0.03
4	256	9,784	3.75 megabytes (mb)	$0.0004	$0.11
5	1,024	4,892	15 mb	$0.001	$0.44
6	4,096	2,446	60 mb	$0.006	$1.76
7	16,384	1,223	240 mb	$0.02	$7.03
8	65,536	611.50	960 mb	$0.09	$28.13
9	262,144	305.75	3.75 gigabytes (gb)	$0.37	$112.50
10	1,048,576	152.88	15 gb	$1.46	$450.00
11	4,194,304	76.44	60 gb	$5.86	$1,800.00
12	16,777,216	38.22	240 gb	$23.44	$7,200.00
13	67,108,864	19.11	968 gb	$93.75	$28,800.00
14	268,435,456	9.55	3.75 terabytes (tb)	$375	$115,200.00
15	1,073,741,824	4.78	15 tb	$1,500	$460,800.00
16	4,294,967,296	2.39	60 tb	$6,000	$1,843,200.00
17	17,179,869,184	1.19	240 tb	$24,000	$7,372,800.00
18	68,719,476,736	0.60	960 tb	$96,000	$29,491,200.00
19	274,877,906,944	0.30	3.75 petabytes (pb)	$384,000	$117,964,800.00
20	1,099,511,627,776	0.15	15 pb	$1,536,000	$471,859,200.00
Total	**1,466,015,503,700**		**20,480 terabytes or 20 petabytes**	**$2,048,000**	**$629,145,600**

| Unaltered Google Map | Google Map after Styling |

FIGURE 2.11. Google Maps before and after styling. The map-styling option allows individual features of the map to be drawn with a different color. It demonstrates that features in the map are drawn as vectors instead of being pre-rendered as a series of raster tiles. Search: google styled map wizard. Copyright 2013 Google.

the Maps application on Apple's iPad continued to display the map as raster tiles. The map from the iPad for a part of Tokyo in Figure 2.13 includes the Japanese script. It is not possible to substitute the script with another language because it is integrated within the raster map as pixels. Apple introduced its own mapping program in 2012 in response to the more advanced mapping application on Google's

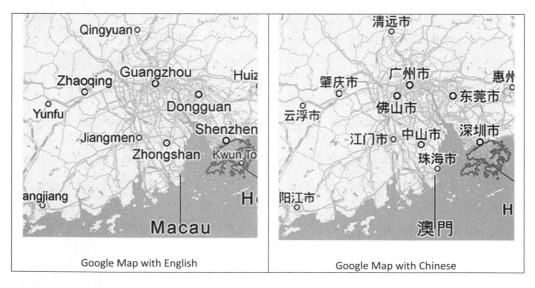

| Google Map with English | Google Map with Chinese |

FIGURE 2.12. Two Google maps of the Pearl River Delta of China near Hong Kong. The maps demonstrate that the text is separated from the underlying map. Copyright 2013 Google.

FIGURE 2.13. A Google map from the Maps application on Apple's iPad. The Japanese text is part of the map and cannot be changed to another language. Copyright 2013 Google.

Android operating system, and this program included the continuous scaling of text.

2.3.9 Online GIS Mapping

The growth of geographic information systems (GIS) has been phenomenal since the 1980s. These systems essentially combine a database with query and mapping capabilities. GIS is used at all levels of government, nonprofit organizations, and the private sector. By the late-1990s, web interfaces began to be developed for these systems (see Figure 2.14). Generally, online GIS systems only implemented a subset of the overall GIS functionality.

In online GIS, the user is presented with a web page that allows a query to be made of a database. The resulting map is then displayed within a web page, and further controls are provided to perform another query. Response times are typically slower than with online mapping systems because the map is usually drawn in the background and then converted to a raster image using the older client–server approach. Additionally, many online GIS systems make use of a single computer server—usually an older computer that often has a slower processor than newer client computers. Rather than operate their own servers, some are turning to servers in the cloud. The advantage is more flexibility in server environments and greater data security. Cloud-based solutions are also usually less expensive.

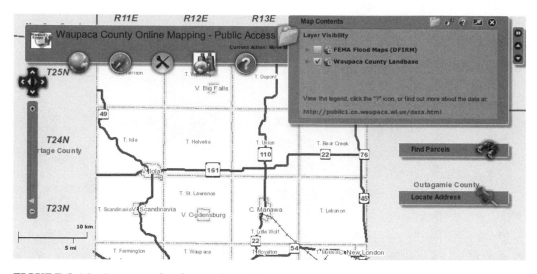

FIGURE 2.14. An example of an online GIS web page. Many metropolitan areas and counties have implemented online GIS systems to allow the user to find property tax information. This example comes from Waupaca County in Wisconsin. Search: Waupaca County Online Mapping.

The user interface in online GIS is generally not very intuitive, and it can sometimes take considerable effort to navigate. Panning and zooming are not as cleanly implemented, making the experience less than satisfactory. Users are typically very motivated when using these systems, and so they are more willing to accept a poor user interface and slow response times in order to get a specific map.

Examples of online GIS include the mapping of flood risks (search: FEMA maps), toxic sites (search: EPA Envirofacts), and sex offenders (search: national sex offender registry). One of the primary applications of online GIS for many cities is making property records available to citizens. The user consults an online database that provides a map of land property boundaries, called a cadastre, along with the taxes paid by each homeowner (see Figure 2.15). In most parts of the world, this information is not available to the public. In North America and parts of Europe and Asia, property taxes are viewed as public information. Making the information freely available is seen as leading to greater fairness in taxation and more trust in government. This online mapping capability is an important element of e-government and is a prime example of how online resources can provide important information to citizens.

Before the late 1990s, GIS systems were reserved for use by a few, trained operators. Only higher-end computers could run the sophisticated software, and GIS software is perhaps the most difficult of all software to learn. Online GIS has brought some of the power of these systems to the general public and this has led to a certain democratization of GIS. The concept of Public Participation GIS developed to underscore the potential of online GIS to involve the public in collectively addressing geographic problems, such as the siting of new schools or businesses. This optimistic vision for the potential of online GIS has largely faded, and a need has arisen for continued effort to bring the tools and procedures of GIS to a broader user audience.

One company dominates GIS. Located in Southern California, Environmental Systems Research Institute (search: ESRI) has grown from its beginnings as a consulting company in the 1970s to a virtual colossus in the volatile software industry. Still privately held, the company sells its Arc-series of software, data, and other services, including training courses in the use of their software. Its major product, ArcGIS, is the most widely used GIS software in the world. Most jobs associated with GIS require background experience with ESRI software as can be seen at websites such as the GIS Jobs Clearinghouse.

Free and open-source software (FOSS) has also become available for online GIS. The most commonly used is QGIS. It has been implemented around the world for a variety of applications.

Another FOSS application, MapServer, is commonly used to present maps through the Internet (see Figure 2.16).

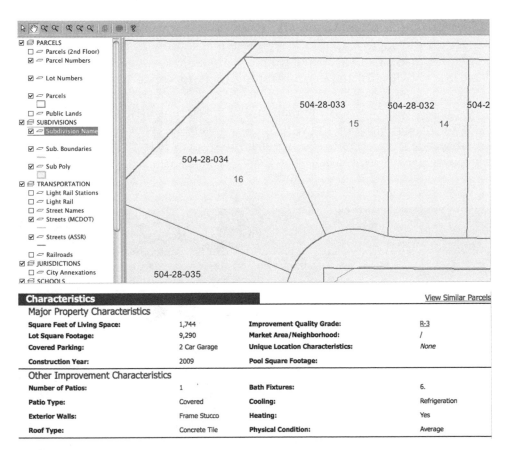

FIGURE 2.15. An online cadastre with tax information on a house. The site displays property boundaries and housing characteristics. The amount of taxes paid is also available for every property in Maricopa County, Arizona. Copyright 2012 AutoDesk; Autodesk screenshot reprinted with permission of Autodesk, Inc.

As with other online GIS implementations, the person who installs MapServer creates the specific user interface. Many decisions made during this process are poor and detract from the usability of the program. In addition, like most older client–server mapping, MapServer sites tend to be slow. Although it is possible to implement a tile-based, Ajax approach, a large number of MapServer sites still use the older client–server method and rely on older servers. With minimal server resources, these sites will likely never achieve the map display speed or general user satisfaction of commercial online mapping systems with their larger data centers and more resources.

2.3.10 Map Mashups

Probably the most important development in mapping during the first decade of the 21st century was the introduction of online mapping tools in the form of the application programming interface (API). APIs are specialized libraries of computer code that are accessible through the Internet. Soon after the introduction of Google Maps, the company made a library of routines available that would allow the creation of custom online maps. Users could map their own points, lines, and areas on a Google map and make these maps freely available to others. The data used for mapping often came from other websites—thus the term *mashup* to indicate the melding of data and mapping tools to create new presentations of information.

An early application was the mapping of apartment listings from Craigslist, a free service for selling goods and services. New businesses were born simply by combining free data from one site and free mapping software in the form of a mapping

FIGURE 2.16. An implementation of MapServer by USGS showing the distribution of the Eastern Newt. From *USGS Amphibian Atlas* 2013; U.S. Geological Survey, Department of the Interior/USGS.

API and *mashing* them together. Websites such as MapsKrieg and HousingMaps are examples of this type of combination (search: MapsKrieg, HousingMaps).

Mashups have made it possible to map data that had never been mapped before. Figure 2.17 shows the distribution of Netflix queues in the Minneapolis-St. Paul area for two movies with clear differences in viewership patterns. At this time, Netflix distributed movies mostly through the mail as DVDs, and the queue would reflect which movies customers had selected for rental.

2.3.11 Animated Maps

Map animations have existed since the early days of film. Although such animations were available, they were difficult to distribute and few people had the opportunity to view them—much like early maps on paper that were copied by hand. Through TV, people have become accustomed to animations showing the movement of clouds on weather forecasts. The Internet is making it possible to distribute many other kinds of animated maps.

A limited form of animation is possible with animated GIFs, a variation of the GIF format. Some animations are stored in a format designed for movies such as MPEG or MOV. The available animations include *terrain flythroughs* in which

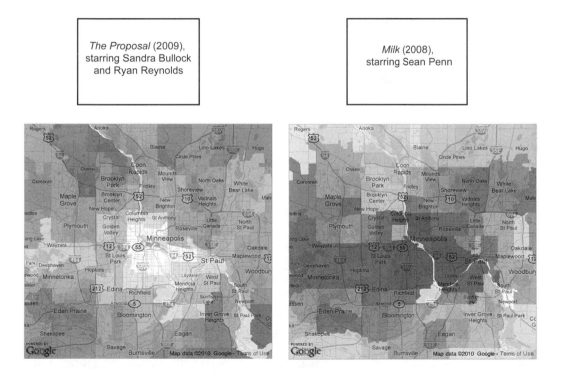

FIGURE 2.17. Two maps of the Minneapolis-St. Paul area in Minnesota showing Netflix rentals for two movies. *The Proposal* (2009), a movie with Sandra Bullock, attracted suburban viewership. *Milk* (2008), about the first openly gay man to be elected to public office in California, was more popular in the downtown parts of both cities. Copyright 2012 Google.

a landscape, usually somewhat mountainous, is viewed as if it were being flown through with an airplane (search: terrain flythrough). Animation has also been applied to show population growth in a region (search: Dynamic Mapping of Urban Regions: growth of the San Francisco/Sacramento region). Here, shadings are applied in a progressive fashion to depict the pattern of population growth. Additionally, animations are available that depict alternative methods of mapping data such as different methods of data classification (search: cartographic animation visualization, temporal and non-temporal). A noteworthy application is the *Animated Atlas of Flight Traffic over North America* that shows air traffic over a 24-hour period in a 49-second time lapse (search: theflightatlas channel youtube). The use and distribution of animated maps is still in an early stage of development.

2.4   Finding Maps

Although millions of static maps are available through the Internet, as well as thousands of interactive mapping sites and map animations, finding the right one is not always easy. This section outlines different options for finding these maps in the cloud.

2.4.1   *Online Map Library*

Map libraries were once the primary storehouse for maps. These libraries, usually associated with the government or a university, house a large number of atlases and individual map sheets. The largest physical map library in the world is the U.S. Library of Congress in Washington, DC.

Online map libraries attempt to make a map collection available through the Internet. An example is the Perry–Castañeda Map Collection at the University of Texas (search: University of Texas Map Library). This website contains a large number of maps produced by the U.S. government. As such, the maps do not have copyright restrictions and can be freely distributed and duplicated. There are also scanned versions of older maps for which copyright restrictions have expired. The advantage of these libraries is that all of the maps are kept on a single server and the links to maps can be maintained. Other university libraries also have online map collections. The United Nations maintains the Dag Hammarskjöld Library for maps (search: Dag Hammarskjöld Library).

Roelof Oddens, the former curator of the map library at the University of Utrecht in the Netherlands, created Oddens' Bookmarks, a database of available online maps. His site began as a list of links and was eventually converted to a searchable database. The use of such a human-crafted database is perhaps the best way to reference online maps. Only a human curator can effectively index and evaluate maps. Maintaining such a database is very time consuming and, with Oddens' retirement in 2005, the database is no longer available.

Although many individual efforts have been made to maintain a collection of online maps, there has been no concerted multinational effort to create a definitive map library or geographic database. The Internet has created the potential of international cooperation, but little real progress has been made in producing an online international map library (Peterson 2007).

2.4.2   Search Engines

The information landscape that most people experience is the one presented by search engines, the only automated tool to traverse the online information landscape. These automated systems explore this ever-growing landscape and create a database of keywords. Of course, there is more to this information landscape than can be found with automated search methods. Search engines lead us to only a fraction of what is available. It is estimated that they have only indexed a third of all web content.

Search engines are based on a program called a web robot or a spider that traverses web pages in an automated fashion. There are two steps in the traversing process: (1) words are entered into a special index; and (2) other links are gathered for later traversing. Web robots need only be instructed to search a single page from which it automatically finds all links from that page and subsequent pages. If a page has links to 100 pages, and each of these pages has links to another 100 distinct pages, a total of a million pages will have been indexed in the first six pages. The web robot captures a small amount of text from each page—less than 2 kb—that is copied to the database of the search engine.

After capturing small amounts of text from billions of pages, the web robot begins to prioritize the links by keyword. This is a weakness of the system because the process is inevitably arbitrary and based on inadequate information. Search engines prioritize in a different manner as well, although search engines copy ranking results from each other. In 2011, Google demonstrated that Bing was copying its search engine results by devising a term and artificially inflating its ranking. Within hours, Bing had the same ranking for the bogus term (Miller 2011). Microsoft had captured search results made with Google through its widely used Explorer browser.

Companies like Google do not provide the exact prioritization algorithms used to rank results because doing so would lead to abuse by those people trying to get their web pages listed near the top in the search results. This creates a circular problem because we need to know how the web robots work in order to provide better matches. If this mechanism were made public, however, the system would be abused by those seeking to manipulate it for their own benefit.

The ranking system used by Google, called PageRank™, relies on a procedure that counts the number of times pages are linked with each other. It also incorporates a weighting factor that analyzes the "importance" of the page that is making the link. Google states that it goes beyond the number of times a term appears on a page and examines all aspects of the page's content, including the content of the pages linking to it, to determine if it is a good match for a query.

Some basic search engine rules seem to be shared by all search engines. Shorter pages are given preference over longer pages with a similar set of keywords. The text in the title of the page is given a higher priority than text within the page. The META tag, listed near the top of the code for the page, is another way of making the web robot assign a high priority to a page. Popular pages that have more links directed to them are listed higher. As a result, new pages, even those registered with the search engine, tend to fall into an "anonymity trap" since they are not linked from other pages, and they cannot be linked from those pages because they cannot be found.

Search engine rankings are extremely competitive. The operators of websites are therefore under pressure to make sure that their sites are listed near the top, within the first 20 matches. Webmasters will spend a considerable amount of time and resources to guarantee that their sites are highly ranked by major search engines. Most search engine companies now accept cash in return for a high placement. Google and others usually indicate which sites are sponsored. The search engine has become a necessary, if sometimes frustrating, part of people's lives.

Sherman and Price (2001) identify five major problems associated with search engines:

1. *Cost of crawling:* Crawling the web is very expensive and time consuming, requiring a major investment in computer and human capital.
2. *Crawlers are dumb:* Crawlers are simple programs that have little ability to determine the quality or appropriateness of a web page, or whether it is a page that changes frequently and should be recrawled on a timely basis.
3. *Poor user skills:* Most searchers rarely take advantage of the advanced limiting and control functions offered by all search engines.
4. *Quick and dirty results:* "Internet Time" requires a fast, if not always thorough, response. A slower, more deliberate, search engine would not gain user acceptance although it might lead to better results.
5. *Bias toward text:* Search engines are highly optimized to index text. It is much more difficult to index images, audio, or other media files.

A long-term solution to the search engine problem is XML and related mark-up languages. Rather than being used to solely format a page, XML describes the elements of the page and their meaning. The search engine can use this information to produce better search results.

2.4.3 *Media Search Engines*

Media search engines index images, sound, or video files, collectively referred to as multimedia (Cartwright et al. 2006). Image indexes are the most common and are offered by all major search engines. Media search engines are an alternative to the text-based approach and are a more promising technology for finding maps.

Billions of images have been indexed, and many of these images are maps. Maps dominate any type of place-name search. For example, a Google image search for "Africa" results in 18 maps in the top 20 links. The corresponding Yahoo image search finds 16 maps in the top 20 images. A majority of these maps are in the JPEG format.

The primary problem with image search engines is that the indexes can only be created from words associated with the images, and not the images themselves. The crawler examines the name of the image file and the accompanying ALT picture tag that is sometimes added when the image is referenced. It also examines words or phrases that are close to the image, or the META tags found at the top of the file. The nature of the website and its provider may also be taken into account.

The image indexing process can easily be misled with a nondescriptive file-name or associated text that does not relate to the image. For example, a search engine would likely be fooled with a page on South America that displayed a map of Africa, especially if the name of the file were "South America." In this case, the map of Africa would almost certainly be indexed as a map of South America.

To improve its image indexing, Google implemented a competitive project to label images called Google Image Labeler. In this project presented as a game, a player was paired with another user, and they were given points for the amount of detail they provided for each image presented over a two-minute period. The purpose was to attach more descriptive keywords to images so that a Google Image search would return better results. If the images presented were maps, the labels given to them would better define these maps in the Google index—depending on how the players responded. Google Image Labeler was discontinued in 2011.

Google has also implemented an automated procedure that classifies images into four categories: Face, Photo Content, Clip Art, and Line Drawings. The Face option is almost perfect at finding pictures of people. Photo Content locates pictures with a more general theme. Clip Art finds smaller files, both in size and dimension. Line drawing images are black and white and mostly composed of lines. Maps can generally be found in all but the Face option.

The attempt to create content-based search engines that index all character-istics of an image, such as its shape and color is still experimental. Thus far, fully automated indexing of images has eluded the designers of search engines.

2.4.4   The Missing Map Search Engine

Whereas search engines have begun the classification of images based on visual characteristics, maps have not yet been indexed separately. An image search engine would benefit from a clear distinction between pictures and maps, a distinction made possible in an automated way by looking at the level of pixel color variability in a file and the presence of text. The map would have large areas with the same color or shading, which would not be the case for ordinary pictures characterized by more variation of pixel values. The recognition of text within images would be possible using the techniques of optical character recognition. The resulting text might be able to separate an ordinary image from a map.

A map author could self-define the map content and insert this into an associ-ated header or metadata file. To make this process easier, a standardized identifica-tion format could be established for all maps so that search engines would be able to determine the characteristics of each map. The content of these headers—called metadata—could then be listed when a match is made. This method would involve considerable cooperation between different web map publishers, and would also require a clearinghouse to check the accuracy of the metadata.

Wikipedia has established a successful model for the submission of encyclope-dic online entries. Wikimedia Commons is the associated portal for all multimedia content, including maps. The complicated submission process for images focuses on establishing copyright clearance. Submissions are denied if the images or maps are not in the public domain, and this limitation extends to a screen capture from any interactive online map provider.

Websites that allow the user to construct a map interactively will partially obviate the map search problem. Rather than searching for an existing map, the user would specify the map he or she wants to make. This is already occurring with street maps that incorporate a high degree of interactivity with the added benefit of being supported by businesses.

2.5   Finding a Map's Address

On the Internet, an address refers to the location of a document on a remote computer. This hypertext transfer protocol (http) address, called a uniform resource locator (URL), has two parts: The first part identifies the computer, and the second part denotes the location of a file on that computer. The URL address:

http://maps.unomaha.edu/

is a specific computer. The address:

http://maps.unomaha.edu/Cloud/index.html

refers to the "index.html" file within the subdirectory "cloud" on this computer.

A two-letter code may be included at the end of the computer's address if the computer is located in a country other than the United States. Some sample codes are: "ca" for Canada, "jp" for Japan, "de" (Deutschland) for Germany, or "uk" for the United Kingdom (search: Internet country codes). Aside from the institution and country code, there is little indication of where the map file actually resides, and so additional information about the site must often be inferred.

2.5.1   *Address of a Static Graphic Map File*

Finding a graphic file is simply a matter of locating a file that has the right suffix—the three letters at the end of the filename (e.g., "GIF," "JPG," or "PNG"). When these letters appear at the end of the uniform resource locator (URL), then the file that is currently displayed is a graphic file. Most often, however, graphic files are embedded within a text-based web page, and the address of the graphic will need to be determined in another way.

The process of finding the address of the graphic file within a web page varies between browsers but usually involves clicking on the picture with the right mouse button (control-click with Macintosh) and choosing either the Properties or Copy Image Address option from the pop-up menu. Pasting this address into the address bar and hitting return displays only the image. This can be accomplished with most graphics that appear on a web page, even those small icons and graphic elements in the header that are part of almost all web pages. Once the address is found, it can be accessed from another HTML file or saved as a separate file. The following are addresses of a JPG and a PDF map from the Perry–Castañeda collection:

http://www.lib.utexas.edu/maps/africa/africa_ref_2007.jpg

http://www.lib.utexas.edu/maps/africa/africa_ref_2007.pdf

2.5.2 Address of an Interactive Map

It is also possible to find the address of an interactive map. Most of these sites incorporate a link option (upper right of the map) that presents the HTML code for insertion into a web page. Figure 2.18 shows the code and the resulting map. Additional options are provided under "Customize and preview embedded map." The code can simply be copied and embedded within an HTML document.

2.6 Summary

The transition to a new medium for maps has been dramatic. Within a few years, the major method of map distribution changed from paper to computer as millions of maps began to be distributed through the Internet on a daily basis. Users adapted quickly to the new interactive maps and to features such as route finding. The pervasive use of computers and the Internet makes it likely that more people are being exposed to more maps than ever before.

The Internet has increased our access to maps. It is possible to find everything from subway maps of distant cities to maps of socioeconomic distributions

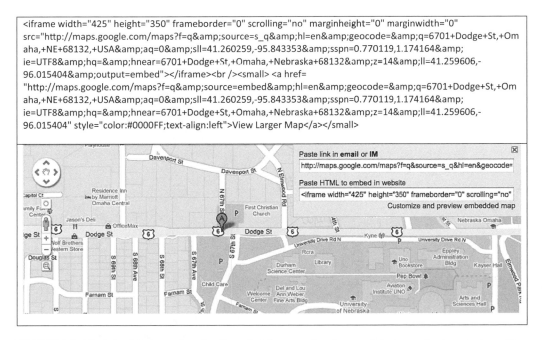

FIGURE 2.18. The Link option in Google Maps makes it possible to embed an interactive map into a web page. The code on the top includes the necessary information to make the map. Copyright 2013 Google.

of countries. Photographs of ancient maps can be found that were made thousands of years ago as well as weather maps that were created just a fraction of a second before they are viewed. In many respects, the maps available through the Internet constitute a huge, if somewhat disorganized, atlas.

The disorganization of the web can make it difficult to find a particular map and be at times frustrating. The use of search engines can be tedious and time consuming. Interactive mapping sites and online map libraries are a solution for basic types of maps. Better methods need to be implemented to organize and access all of the online maps that are available, particularly thematic maps. This will be a continuous and developing process.

The Internet map landscape has evolved since 1993 from scanned paper maps, client–server maps, maps defined in vector format, and maps defined in pre-rendered tiles. The general goal of these developments, particularly tile-based mapping, was to increase the speed of map delivery. Tile-based, Ajax mapping has drastically changed the user experience for such basic actions as zooming and panning. Most map users now expect this form of map interaction, and this expectation partly renders static maps obsolete.

2.7   Exercise

Use the procedures outlined in this chapter to find at least nine different maps on a topic of your choice. Examples of a theme might be population density, wine regions, unemployment, genocide, foreclosure rates, land use, land cover, international aid, or bicycle paths. The series of maps could also be on a particular place, like a city, state, or national park, although captions are easier to write for maps that present thematic information.

Record the URL of the graphic file and save it in a text or word processing file, or email the address to yourself. Most interactive mapping sites provide an option to create a link to the map, either as an http address or as an embed statement. If the address of the map cannot be easily determined, take a screen capture of the map using the PrtScrn procedure in Windows, the snipping tool in newer versions of Windows, or the Grab program on the Macintosh. Save the graphic file in PNG format. The preview program on the Macintosh can convert the TIF file from Grab to a PNG file. The Paint program in Windows can be used to crop the image and save the file in a particular format.

2.8   Questions

1. What are some advantages of paper maps?

2. Why is it so difficult to make maps by computer?

3. Describe the label placement problem with maps.

4. What is the most common image file format for maps as determined by an image search engine? (*Tip:* Advanced Search allows a search by file type.)

5. How did Google Maps change the presentation of online maps, and why is it

no longer possible to count the number of individual maps that are distributed through the Internet?

6. When did Google Maps overtake MapQuest as the main provider of interactive maps?

7. Will services like Google Maps remain free? What is the benefit for companies that offer this service?

8. Describe how you can distinguish between vector and raster elements in a PDF file.

9. What is the best scenario for maintaining an updated map given the tremendous investment of resources needed in doing so?

10. Describe common vector and raster graphic file formats.

11. What file formats are best suited for printing maps and why?

12. What is tile-based mapping, and how did it come to dominate the distribution of interactive maps?

13. Describe how the data storage costs for a tile-based mapping system can be calculated.

14. Most interactive mapping systems provide maps at 20 levels of detail. For the 21st level of detail, how many tiles would be needed, and what would be the storage requirements at 15 kb per tile?

15. Compare Google Maps and Google Earth. Which service would offer a greater revenue stream for the company?

2.9   References

Cartwright, William, Michael Peterson, and Georg Gartner (2006) *Multimedia Cartography*, 2nd ed. Berlin: Springer.

Dougherty, Heather (2009, Apr. 14) Google Maps Surpasses MapQuest in Visits. *Hitwise*. [http://weblogs.hitwise.com/heather-dougherty/2009/04/google_maps_surpasses_mapquest.html]

Godse, A. P. (2008) *Computer Graphics*. Technical Publications Pune: Pune, India.

Miller, Claire Cain (2011, Feb. 1) Google to Microsoft: Search "Gotcha." *New York Times*.

Peterson, Michael P. (2007) *International Perspectives on Maps and the Internet*. Berlin: Springer.

Sample, John T., and Elias Ioup (2010) *Tile-Based Geospatial Information Systems: Principles and Practices*. New York: Springer.

Sherman, C., and Gary Price (2001) *The Invisible Web: Uncovering Information Sources Search Engines Can't See*. Medford, NJ: Cyberage Books.

Suzuki, K., and Yoshiki Wakabayashi (2005) Cultural Differences of Spatial Descriptions in Tourist Guidebooks, in C. Freksa et al. (Eds.), *Spatial Cognition IV, LNAI 3343*, pp. 147–164. Berlin Heidelberg: Springer-Verlag.

USGS National Amphibian Atlas (2013) Eastern Newt (Notophthalmus viridescens). Version Number 2.2 USGS Patuxent Wildlife Research Center, Laurel, MD [www.pwrc.usgs.gov/naa]

CHAPTER 3

The Meaning of Mapping

You want the truth? You can't handle the truth!

—JACK NICHOLSON IN *A FEW GOOD MEN*

3.1   Introduction

Truth may be as hard to handle as reality is to represent, especially when working with an object as large as the world. The only way to show the world "truthfully" would be as a globe at a 1 to 1 scale—that is, with no reduction in size or change in shape. The comedian Steven Wright comments on the absurdity of the 1:1 map with the line delivered in his characteristic monotone voice:

> I have a map of the United States. . . . Actual size. It says, "Scale: 1 mile = 1 mile." I spent last summer folding it. I also have a full-size map of the world. I hardly ever unroll it." (search: map scale of 1:1 quotes)

A 1:1 globe has also been envisioned that is somehow attached to the Earth and follows behind as a second sphere. Besides being impossible to make and store, any 1:1 map or 1:1 globe wouldn't be any more useful than simply looking at the world itself.

As scaled depictions, maps are by necessity abstractions of reality. The abstraction takes many forms. Some features are left out because there is no room to show them while others are exaggerated in size so that they will be made visible on the map. Roads, for example, would not be visible on most maps if they were not represented as being wider than they are in reality. Taken together, maps are a collection of lies, and yet we find them useful. According to Picasso, "art is a lie which makes us realize the truth." The same can be said of maps. Maps are meaningful to us because they present a generalization of reality, not reality itself.

The true meaning of maps goes beyond their abstraction. Maps are an important source of information from which people form their impressions about places and distributions. Each map is a view of the Earth that affects the way people think about the world. Maps are always made from a specific point of view and may be constructed to influence us in a particular way. Our thoughts about the space in which we live and especially the areas beyond our direct perception are influenced by these representations, and the way we think about our environment influences the way we act within it. Ideas about the use and purpose of maps are particularly relevant as we change how maps are distributed. These various conceptions of maps paint a picture of what maps are and what they mean to us.

The role of maps in a technological world is particularly important. Advancements in the sciences, in the exploration for natural resources, and in other areas of study are the result of a continued emphasis on the analysis of data in visual form. What we derive from these displays is information—information that can be of incalculable value. It is this information that ultimately gives meaning to mapping. In this chapter, we examine the map as a mental creation, the map as a form of communication, the relationship between maps and society, and ways in which maps are used not only as forms of representation but as tools for analysis.

3.2   Maps in the Mind

Before the first map was ever drawn, it had to exist in the mind of an individual. We can say then that the very first maps were those that existed as a mental construct. The Japanese cartographer, Takashi Morita, has conceived of the very first map in the sand as resulting from a mental map with the illustration shown in Figure 3.1.

Internal representations of the environment are referred to as mental or cognitive maps. Examples of mental maps would be the maplike representations of

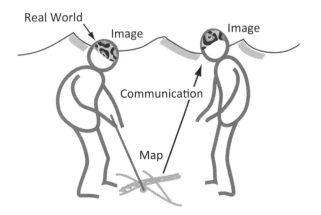

FIGURE 3.1. A hypothetical illustration of the drawing of an early map. One individual is drawing his mental map in the sand with a stick. A second individual creates a mental map by looking at the map in the sand. After illustration by Takashi Morita, used with permission.

the shape of continents or the internal representations that we have of the neighborhood in which we live. These "internal maps" help us not only in navigating but also in structuring information about the world beyond our direct perception. Maps help us to organize information about the environment.

3.2.1 Acquiring Mental Maps

We acquire information for the formation of mental maps in two basic ways. The first source is direct environmental perception and occurs from such actions as simply walking or driving through the environment. This movement results in a mental map of our *activity space*. The movements that we make, even when moving from one room to another, are dependent on the mental representations of space acquired through direct experience interacting with the space itself. Initially, the internal maps are based on landmarks. With time, they become geocentric as if viewing reality from above.

The creation of such mental maps is based on the ability to see. People who have been blind since birth use different strategies to code the spatial environment (Kitchin and Freundschuh 2000, p. 224). Blind people tend to code spatial relationships egocentrically, where the individual is in the center, rather than geocentrically because egocentric coding works best for them (Kitchin and Freundschuh 2000, p. 234). For example, their experience of moving between rooms in a building is like walking along an open path. There may be no sense of being in individual areas.

The second source for internalized maps are representations of the environment, and the most useful of these representations are maps. The mental map that we have of the outline of a country, for example, can only be obtained from a map. Although boundaries for certain countries are defined by oceans and would be visible from space, the borders for most countries cannot be seen from above. The only way that we can derive a mental map for the outline of most countries is from a map. The way we identify countries on maps is then based on mental maps that we have formed (see Figure 3.2).

The form and function of mental maps have been objects of inquiry for many years. A common method of studying these internal representations is to ask people to draw their mental maps on paper. One observation from this testing is that people are rarely satisfied with the map they have drawn. In fact, they can subsequently identify specific errors with their own drawings. Thus people have much more accurate mental maps than they are able to reproduce on paper.

Another way to demonstrate the accuracy of mental maps is to introduce errors into a map. For example, most people are able to identify the three errors in the outline of Africa in Figure 3.3. The only way that this outline could have been learned is from a map.

Our minds are a storehouse of mental maps that we might better refer to as a mental atlas. We have internalized maps of street layouts, bus routes, and the corridors of the buildings where we work or go to school. Internalized maps of the outlines of states, river networks, and the location of cities relative to each other have by necessity been gathered by examining maps. Most of these mental maps

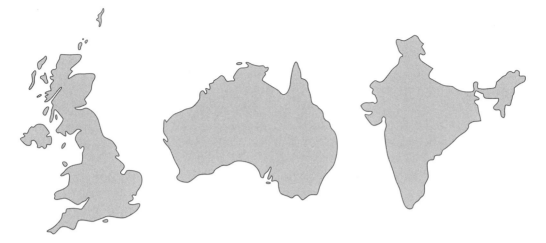

FIGURE 3.2. The outlines of three different countries are represented. Most people are able to identify them as the UK, Australia, and India.

have been added without conscious effort or awareness. Psychology has directed a great deal of study to the understanding of how we process these mental images.

3.2.2 Maps on the Right Side of the Brain

Initial evidence for where mental maps may actually reside in the brain can be traced to early studies in psychology. As early as the 1920s, the psychologist Stroop noted the problem with saying the name of a color when that color is used to write the name of another color. For example, when red is used for the letters that spell "green," we struggle to say the word "red" and not "green" (search: Stroop Effect). Stroop showed that people had difficulty saying the name of the color because we

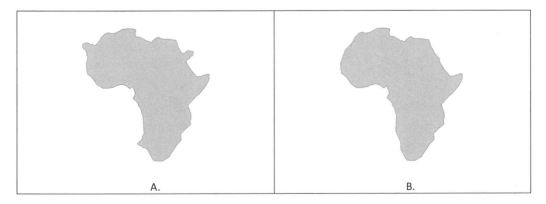

FIGURE 3.3. Map A has three errors in the outline. Map B is correct. Most people are able to instantly identify the errors, indicating that people use maps to create very accurate mental maps.

cannot help but read the word. This has been interpreted as a struggle between different parts of the brain used in color recognition and speech.

Roger Sperry, who won a Nobel Prize in 1981, stated that the brain can be viewed as two separate realms of conscious awareness; as two separate sensing, perceiving, thinking, and remembering systems (search: Roger Sperry left right brain). He argued that the right side processes intuitive, imaginative, and artistic operations while the left hemisphere works primarily with language, analysis, and reason (see Figure 3.4). Sperry discovered this bilateralization of the human brain by examining people in whom the connection between the two sides of the brain had been severed.

Sperry's work on the brain involved patients who had a severe form of epilepsy that could not be controlled through drugs. Severing the two sides of the brain by cutting the corpus callosum—a "cable of connections" between the right and left hemispheres—controlled the frequency and severity of seizures. The patients seemed completely normal after the operation with no visible side effects. Through experiments with these patients, Sperry found that learning, perception, and memory were completely independent on each side of the brain.

This independence could be demonstrated by placing an object in the split-brain patient's right hand without permitting the patient to see it. The right hand sends its information to the left hemisphere, which is regarded as the only side capable of producing speech. In this case, the patient could easily describe what she was holding. However, an object placed in the left hand that sends its information

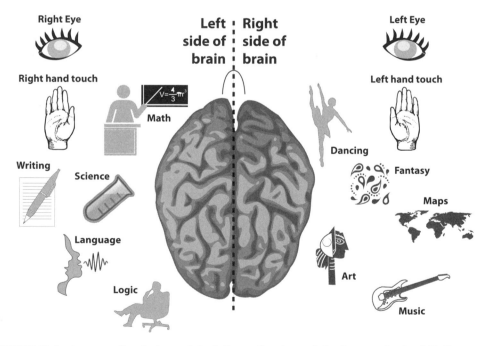

FIGURE 3.4. A generalized view of the bilateralization of the human brain. While mental activity associated with these activities may be predominantly on the side depicted, few mental processes are entirely localized to one side of the brain.

to the right hemisphere, which has no role in speech, produced no description. Similarly, if the patient were shown a picture to only the right eye, which sends its information to the left hemisphere, she could verbally describe what she saw. The patient could not describe what was shown only to the left eye because the right hemisphere has no role in speech.

In a further experiment, an object was placed in a patient's left hand, which was hidden by a screen, and asked what he was holding. The patient would respond, "I don't know." If he was then shown pictures of different objects and told to point to what he was holding, he could do that very accurately. The right hemisphere seemed to know what it was perceiving, but it could not produce speech to tell about it.

Finally, an experiment was developed with chimeric faces where two halves of different faces are placed together (see Figure 3.5). A number of such faces were shown to the patients so that the right eye saw only the right face and the left eye saw only the left face. If asked to describe the face they saw, the patients would always describe the right face. If asked to point to the face that they saw, they would point to the left face. In one experiment, the procedure was interrupted in the middle of describing the right face. At this point, the patients were asked to point to the face that they saw. They would invariably point to the opposite picture. Asked to explain why they described a face of a young girl but pointed to a face of a man, they were embarrassed and explained that they weren't paying attention.

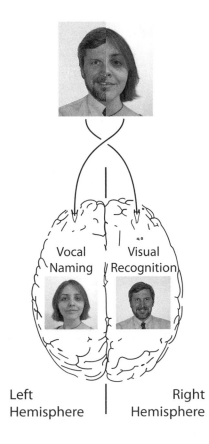

FIGURE 3.5. In Sperry's chimeric face experiment, people who had the connection cut between the two hemispheres of the brain were shown two different faces in the right and left visual fields. If asked to describe the face, they would describe the face in the right field. When asked to point to the correct face, they would point to the face on the left.

In a sense, this was true because the right hemisphere was not paying attention to what the left hemisphere was describing. When interrupted just as they were about to point to the correct person and describe the face, they would provide the same contradictory response.

While modern studies of the brain using positron emission tomography (PET) scans and nuclear magnetic imaging (NMR) show significant bilateral brain activity (search: bilateral brain activity), the split-brain concept is a subject of much debate in psychology. It is pointed out that the corpus callosum is intact for most individuals and is constantly exchanging information between the two hemispheres. Also, not all people have language controlled by the left hemisphere. While the left hemisphere does control language in the great majority of right-handers, left-handers have a more even distribution of language control on both sides of the brain (Zillmer et al. 2008, p. 164).

The split-brain notion has also been popularized in unfortunate ways with assertions that every conceivable human behavior can be categorized as "left brain" or "right brain." In popular culture, the left-brain "person" is seen as extremely analytical and "cold." By contrast, the right-brain person is never good at mathematics and is seen to be more empathetic and artistic. These are false characterizations that detract from the scientific work in this area.

3.2.3 *Educating the Two Sides of the Brain*

A variety of observations have been made since the notion of functional differences between the two sides of the brain was introduced. Education, it is argued, builds the left hemisphere by concentrating on reading, writing, and mathematics. It does not adequately engage the functions associated with the right hemisphere. This argument has been used as a successful strategy in art education.

In a book entitled *Drawing on the Right Side of the Brain* (1979), Dr. Betty Edwards observes that older students who could still draw well were using mental images, presumably in the right side of the brain, and those who could not were attempting to draw from the left side. Edwards's first step in teaching students to draw from the right side of the brain is to have students re-create a picture that is upside down. The object of this exercise is to "turn off" the dominant, symbolic left hemisphere, which interferes with drawing, and "turn on" the subordinate right mode, which functions best for drawing.

Edwards presents this method in her book through a series of drawing exercises designed to help shift from the dominant verbal/logical thinking to a more global, intuitive mode of thought, with the ultimate goal being to teach people how to "see." The right brain, it is argued, has the capacity to perceive and process the visual information utilized in the skill of drawing. Edwards argues that as computers take over more and more "left-brain" tasks, it is particularly important that we recognize the need for training the visual mode of thinking—the types of tasks that computers cannot do well.

Another assertion made by Edwards is that Western culture emphasizes the development of the left side of the brain—language, mathematics, and logic—at the expense of the right. Many children, for example, enter the school system being artistically creative but within a few years have lost the ability to draw or perform well with any functions traditionally associated with the right hemisphere. Similarly, education with maps is not emphasized, and many children do not develop the ability to work with them.

3.2.4   Maps in the Visual Cortex

More recent studies in the psychology and physiology of visual information processing have focused on the primary visual cortex, centrally located in the back right of the brain (see Figure 3.6). The cerebral cortex is fed by fibers, called axons, that transmit messages from the eye with electrical impulses. The number of brain cells involved in vision is phenomenal. One million axons connect to about 100 million cells in the cerebral cortex. It is theorized that the images "displayed" in the cerebral cortex can be created not only from vision but also from information stored in the brain. The cerebral cortex might by thought of as a "universal imaging area" that is capable of displaying images directly from the eye or visual images constructed from the brain.

Evidence for the dual-use function of the imaging part of the brain is apparent when people are asked to draw a mental map or explain a route between two places. When faced with such a task, it is common for individuals to look upward to defocus their eyes. This action is thought to deactivate the visual stimulus in order to "project" a map internally for viewing.

A variety of studies in psychology have attempted to define the properties of mental images. Psychologists have theorized that imagery uses representations and processes that are ordinarily associated with visual perception. The visual mental image is used for all types of mental tasks and seems to have map-like qualities. In one compelling study, subjects were asked to visualize a rabbit next to an elephant and a rabbit next to a fly (see Figure 3.7). When asked questions about the rabbit, it was found that subjects could respond faster when the rabbit was visualized next to a fly (search: Kosslyn mental maps and images; Kosslyn 1975). The interpretation was that this larger scale rabbit was more detailed, much like a larger scaled

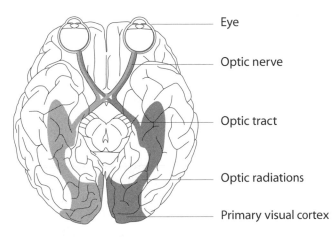

Eye

Optic nerve

Optic tract

Optic radiations

Primary visual cortex

FIGURE 3.6. Depiction of the visual cortex in the brain. It is believed that the visual cortex constructs images from the eyes and internally from information in the brain.

FIGURE 3.7. Kosslyn's experiment involved evaluating the relative detail in the visual mental image of the rabbit in two imagined scenes. It was found that subjects could respond to questions about a rabbit imaged next to a fly much more quickly because that rabbit constituted more of the visual field of the mental image and was therefore more detailed.

map. This shows that mental images are not simply pictures in the brain but are put together with more features if there is room to show them—much like a map.

There is also evidence that people differ in their ability to form "visual mental images." Such differences are difficult to verify because it is impossible to determine the exact form and function of these images. Many psychologists seem to agree that males are generally more adept at mental imagery and spatial problem solving than females (Sigelman and Rider 2009, p. 344). One observation in navigation is that females rely more on landmarks while males use more of an overall mental map (Kimura 2002). Other studies have shown that there is no difference in map readings skills between males and females (Montello et al. 1999). If there is a difference, it is likely more the result of environment than biology. For example, boys may get more attention from adults for using maps than girls.

The quality of our mental maps is strongly related to how we use them. Humans seem to have an amazing capacity to acquire highly complex spatial representation of a large city (Maguire et al. 2006, p. 1099). In a study comparing taxi and bus drivers in London using nuclear magnetic imaging (NMR), it was found that the brains of taxi drivers had a greater gray matter volume in the posterior hippocampus, a part of the brain that plays an important role in consolidating information from short-term to long-term memory and spatial navigation. The study showed that, like a muscle, the brain grows with greater usage.

Many questions remain concerning the relationship between maps and mental images—how they are used and how they combine with linguistic symbolism in thought processes. The location of where things occur in the brain is probably not that important, even for studies in psychology. Two facts are accepted: (1) some parts of the brain specialize in certain tasks; and (2) people differ in their spatial abilities, with possible differences in such abilities between males and females. These differences can be traced to both heredity and the environment (e.g., education). Certainly, if we are to work effectively with maps, it is important that we develop those parts of our mental faculties that deal with spatial abilities. Some have argued that interactive computer displays, particularly computer games, are contributing to better spatial abilities in children (search: effect of video game practice on spatial skills).

3.3   The Purpose of Maps

As depicted in Figure 3.8, four broad areas can be identified that give purpose to mapping.

3.3.1   Maps as Communication

The first maps were based on direct experience acquired by walking through the environment. Instruments of measurement were later used to increase their accuracy. Eventually, aerial photography and other forms of remote sensing were employed to create views from above that could be analyzed directly or transformed into maps. The use of computer databases was then applied to store and analyze "data" about the environment in what became known in the 1980s as geographic information systems (GIS). Although increasing amounts of technology have been applied to the mapmaking process, the purpose of maps remains to understand and communicate information about the spatial world.

Externalizing mental maps and mapping the world beyond our direct experience were major accomplishments for humans. Mapping involves both the abstraction of information and its display on a medium that others can examine, thus facilitating communication. The ability to communicate information about both the physical and human environment is the primary purpose of maps.

One of the more recent developments in cartography has been the study of maps as a distinct form of communication. The general goal was to improve the communication potential of maps. This study occurred at the same time that the production of maps by computer was taking hold. This dual emphasis in cartography merged to become what was called multimedia cartography (Cartwright et al.

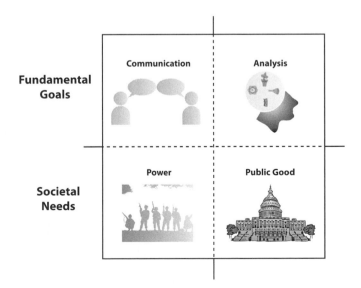

FIGURE 3.8. The purpose of maps. Communication and analysis represent the fundamental goals, while power and public good represent societal needs.

2006) or "visualization" (Hearnshaw and Unwin 1994) in the late 1980s and early 1990s. Here, the computer was being used to help improve communication with maps.

3.3.2 Maps as Power

It is important to understand the use of maps by society. To many people, maps may simply be graphical illustrations that provide an overall view of an area. But to people in the military, in government, and in business, maps are vital tools. For these people, maps are instruments that can be used to both gain and maintain power (Wood 1992).

The vital need for maps is especially apparent in time of war, and their utility to help wage battle was recognized long ago. Advancements in the technology of mapping are closely associated with periods of war, when governments invest enormous amounts of money in the making and updating of maps. Pilots need accurate and up-to-date maps to fly to unfamiliar areas. The Navy needs bathymetric maps (maps of the ocean floor) to effectively navigate ships and submarines. Armies need maps to move people and equipment in the most effective and direct manner. Computerized maps are embedded within missiles to help find their targets with remarkable accuracy.

Federal, state, and municipal governments use maps in various ways—to exploit resources, assess taxes, and maintain records of land ownership. Assessing taxes is one of the major reasons for cadastral mapping. These tax maps will typically include the size of houses, patios, swimming pools, and other taxable items. Utilities maintain maps on the electrical, water, sewer, natural gas, and telephone networks. Knowing the location of natural gas pipelines, for example, is critical. Explosions of gas pipelines caused by digging have been known to destroy entire city blocks. The importance of these maps means that governments and utility companies must make a considerable investment in their creation and maintenance. Most cities maintain a hot line to help avoid accidents with the underground infrastructure.

Businesses also use maps to obtain a competitive advantage over other businesses. Maps of population, income, and other socioeconomic variables are used to site business establishments, to attract potential customers, or for direct mailing. Insurance companies use maps to determine insurance rates for houses, cars, and other types of property. The utility of maps for these purposes has been increasingly recognized with the computerization of the underlying data, leading ultimately to the creation of information systems of geographic data.

3.3.3 Maps for Analysis

Until the 1700s, all maps were for general reference, primarily showing oceans, seas, continents, and other points of interest. Another type of map developed in the latter part of the century. These maps concentrated on a specific theme or distribution and were directed more at analyzing a particular geographic feature.

In 1855, the physician John Snow created a map that depicted cases of cholera in a part of London, England (see Figure 3.9). On a map of the Broad Street area,

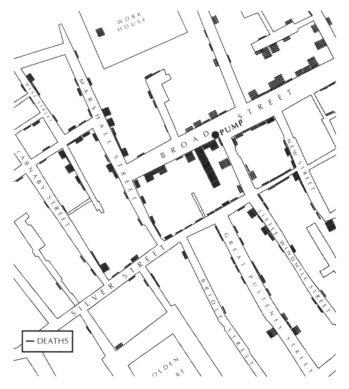

FIGURE 3.9. A re-creation of Snow's 1854 cholera death map of London. Each bar represents a death from cholera. By mapping deaths by residence, Snow was able to show that the water in the Broad Street pump was responsible for the outbreak. Snow's solution to the epidemic was to remove the pump's handle.

Snow placed a dot at the location of each death from cholera and a cross for each water pump. As there was no running water in homes, residents of the city used public pumps to obtain water. The map revealed that the incidence of cholera was mainly among persons near a pump located on Broad Street. Identifying the source of the disease helped stop its spread. Dr. Snow's map was an early example of how maps can be used to analyze a distribution rather than simply depicting the location of places.

Maps that depict a particular distribution, such as population, income, or the incidence of cholera, are referred to as *thematic maps*. These kinds of maps stand in contrast to *general reference maps* that show where things are located relative to each other. The difference between the two forms of mapping has been labeled "in place, about space" (Petchenik 1979). General reference maps, including everything from a globe to a map of a city, put objects in place. Thematic maps that concentrate on specific distributions—like the cholera outbreak in London—are designed to convey a spatial pattern. These patterns about space help us to understand why things occur in space the way they do. The ability to interpret all aspects of the map comes from a solid background in geography and related disciplines.

Mapping for the purpose of analyzing a distribution has led to a form of cartography that is still fascinating researchers in a variety of areas. Most recently, a field called scientific visualization applied the tools of mapping in much the same way with the help of computers. Maps are used in the process not only to display phenomena in space but also to analyze the resulting distribution.

Most maps that are produced today are thematic maps. They may show weather patterns such as temperature, rainfall, and the location of severe weather. They are used to show the distribution of earthquakes, volcanoes, and fault lines. They show population, prosperity, poverty, and patterns of language and religion. Each thematic map provides a piece of the puzzle in understanding the world in which we live. Reading these maps involves analyzing the distribution and attempting to understand why the patterns exist. These interpretations are often based on the mental maps of other maps that we have seen.

3.3.4 *Maps as a Public Good*

The Internet has brought to light the differences in the way governments distribute maps and map-related data. National and state mapping agencies in many countries charge the public for these products. In other countries, principally the United States and Canada, such information is made available at no cost through the Internet. Government-produced maps are viewed more as a "public good," a product of society that has been paid by taxes so that all taxpayers should be provided access.

Free access to maps and geographic data is not common in most of the world. Many countries believe that charging for such information will raise funds to support data-gathering and mapping operations. It has been found, however, that charging for such information usually does not raise even the funding required to pay the salaries of those few individuals who are actually selling the maps and data. As a result, tax income is used to pay the employees to make the maps available for sale.

Another argument for free access to such information is related to the question of double taxation. National mapping agencies are funded by tax dollars. It is argued that to charge for the products of a national mapping agency would represent another form of taxation. In those countries where taxes can only be assessed by specific units of government, for example, a congressional or legislative body, such added charges would be unconstitutional. It is against the law to charge for government-produced maps because this would represent an unauthorized form of taxation.

In addition to highlighting differences between government mapping agencies, the Internet has also begun to make government mapping agencies more responsive to public needs. The United States Geological Survey not only supplies topographic maps but also has websites that show the current locations of earthquakes, volcanoes, drought, floods, and hurricanes through its natural hazards support system. The United States Department of Agriculture displays updated forest fire maps. The National Oceanic and Atmospheric Agency Administration (NOAA) supplies satellite weather images. All of these agencies supply maps and data at no cost. In addition, many agencies make the information available in a useful form. Governments play an important role in providing a common set of map resources for the public good.

3.4 Summary

Maps would not exist if not for the human ability to both generalize and symbolize the world. In one sense, we are all cartographers because we convert our surroundings into functional mental maps that help us navigate. We also use maps to form mental maps of the world beyond our direct perception. How humans developed the ability to map the world, and how information is acquired from maps are issues that are still central to cartography.

What is clear is that maps are an important source of information from which people form their impressions about places and distributions. Although we still do not completely understand how these abstract representations are internalized or how they influence our actions and the way we think about the world, we can say that maps are an amazing, if evolving, source of information.

3.5 Questions

1. Why is the abstraction of reality necessary.

2. Describe a possible scenario for the first map that incorporates interaction.

3. What are the two basic ways that information is acquired for mental maps?

4. Explain some differences in functions that have been found between the two sides of the brain and how these differences relate to maps and map use.

5. Describe the Stroop Effect. What is a possible explanation?

6. What did the experiments with split-brain patients demonstrate about language?

7. Does education equally address functions typically associated with the right and left sides of the brain?

8. What does Kosslyn's experiment with images of a rabbit, elephant, and fly indicate about visual mental images?

9. How can maps be improved as a communication device?

10. What does "maps as power" mean?

11. Describe the presentation/analysis distinction with maps. Do you think it is a real distinction, or is analysis a part of all map use?

12. How do maps serve a public good?

3.6 References

Cartwright, William, M. Peterson, and G. Gartner (2006) *Multimedia Cartography* 2nd ed. Berlin: Springer.

Edwards, Betty (1979) *Drawing on the Right Side of the Brain.* New York: Penguin Putnam.

Hearnshaw, Hilary M., and D. Unwin (Eds.) (1994) *Visualization in Geographical Information Systems.* New York: John Wiley & Sons.

Kimura, Doreen (2002, May 13) Sex Differences in the Brain: Men and Women Display Patterns of Behavioral and Cognitive Differences That Reflect Varying Hormonal Influences on Brain Development. *Scientific American.*

Kitchin, Rob, and S. Freundschuh (2000) *Cognitive Mapping: Past, Present and Future.* London: Routledge.

Kosslyn, S. M. (1975) Information representation in visual images. *Cognitive Psychology* 7: 341–370.

Maguire, Eleanor, Katherine Woollett, and Hugo J. Spiers (2006) London Taxi Drivers and Bus Drivers: A Structural MRI and Neuropsychological Analysis. *Hippocampus* 16: 1091–1101.

Montello, Daniel R., Kristin L. Lovelace, R. G. Golledge, and C. M. Self (1999) Sex-related differences and similarities in geographic and environmental spatial abilities. *Annals of the Association of American Geographers* 89(3): 515.

Petchenik, B. B. (1979) From Place to Space: The Psychological Achievement of Thematic Mapping. *The American Cartographer*, pp. 5–12.

Sherman, C., and Gary Price (2001) *The Invisible Web: Uncovering Information Sources Search Engines Can't See.* Medford, NJ: Cyberage Books.

Sigelman, C., and E. Rider (2009) *Life-Span Human Development.* Belmont, CA: Wadsworth Cengage.

Wood, D. (1992) *The Power of Maps.* New York: Guilford.

Zillmer, Eric, M. Spiers, and W. Culbertson (2008) *Principles of Neuropsychology.* Belmont, CA: Thomson Wadsworth.

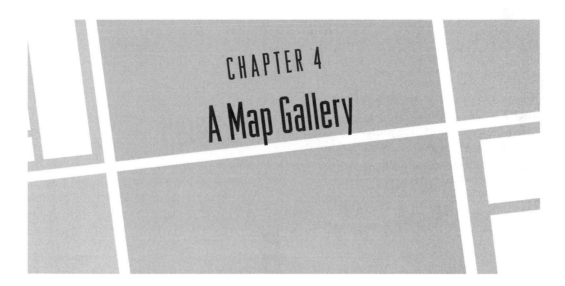

CHAPTER 4

A Map Gallery

Living is like tearing through a museum. Not until later do you
really start absorbing what you saw, thinking about it, looking it
up in a book, and remembering—because you can't take it in all
at once.

—AUDREY HEPBURN

4.1 Introduction

A gallery is a collection of selected objects with a specific theme. Whether consisting of works of art, historical artifacts, or maps, a gallery brings meaning and a sense of understanding to a particular collection. The caption, the description placed prominently beside the object, is a vital element of the gallery. Many museum visitors move between the captions, glancing only briefly at the objects they describe.

An Internet map gallery is a web page that presents and describes a selection of maps that are available through the Internet. The purpose is to show how a series of maps can be brought together in a single collection. The maps for this gallery will not be printed on paper, nor will the files be transferred to your computer. Rather, a link will simply be established to the file, and the maps will be displayed with an accompanying caption. Making links to material on different computers is a major advantage of the Internet. The resultant map gallery can be an international collection of maps because links can be made to maps located on computers in different parts of the world.

Presenting maps in an online gallery involves the use of the hypertext markup language (HTML), the main language of web pages. HTML is the script that defines the layout of the page (Willard 2009). It consists of simple text codes, surrounded by the "<" and ">" delimiters, that specify how the document will appear in the browser. Certain HTML tags create links to documents or display a graphic file. While there are many tools to help with the formatting of these pages (and these will be introduced in this chapter), it is good to have a basic understanding of HTML because it represents a container for other types of code that will be introduced in later chapters.

The coding of the gallery is relatively simple. An ordinary text editor can be used to enter the file (Notepad on Windows or TextEdit with Macintosh, with appropriate settings). Unlike word processors, these programs are intended for the entry of unformatted text. Once the files are created, they may be opened with a browser such as Explorer, Firefox, or Chrome. A working knowledge of HTML is particularly important when we examine JavaScript, application programming interfaces (APIs), and the PHP (personal home page) programming language that are all embedded within HTML code.

4.2   Single-Page Map Gallery

HTML is very similar to early word processing programs like WordStar of the late 1970s and the 1980s that required the user to insert formatting codes within the text (MacDonald 2006). To indicate a paragraph break, for example, the user needed to enter a special code in the document, such as ".p". HTML goes beyond simple text formatting by incorporating hyperlinks and graphics. The ability to link to graphics and other files makes HTML a powerful hypermedia language.

All HTML files begin with the <html> tag and end with the same code prefaced with a slash, for example, </html>. The slash in front of the end code indicates that the HTML coding is finished. Technically, all HTML tags have a beginning and ending, with the end tag indicated with "/". For example, the <h1> command is used to begin header text—larger text used for titles—and the </h1> code stops the header text format. The following code would display a single line of text in the largest header text size:

```
<html>
<h1> My Map Gallery </h1> <!--This writes "My Map Gallery."-->
</html>
```

These three lines of code can be typed into a text editor (e.g., NotePad or TextEdit) and saved as a file called: Gallery.html. After the file has been saved and closed, it can be opened using the Open File command in the File menu of your browser (see Figure 4.1). An easier option to open the file is to double-click on the file icon. The easiest way to display the HTML file is to drag and drop the HTML file on top of an open browser window such as Explorer, Firefox, or Chrome.

The header text sizes go from h1, the largest, to h6, the smallest. The <h2> and <h3> can now be used to add some more text to this file. The result is displayed in Figure 4.2.

```
<html>
<h1> My Map Gallery </h1>
```

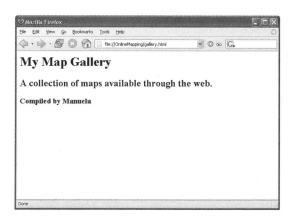

FIGURE 4.1. An open file dialog in the Internet Explorer browser program. This dialog is almost never used because the file can be more easily opened by double-clicking on the icon of a file or dragging and dropping the icon of the file on an open browser window. Copyright 2012 Microsoft.

```
<h2> A collection of maps available through the web. </h2>
<h3> Compiled by your name </h3>
</html>
```

Next, we can add a horizontal ruler line with the header ruler <hr> command and the title of the first map. Note how this tag does not have an ending such as </hr>. In the more rigorously defined XHTML format, all tags need to be ended. A shorthand for a code like <hr> that does not have a natural ending is <hr />.

```
<html>
<h1> My Map Gallery </h1>
<h2> A collection of maps available through the web. </h2>
<h3> Produced by your name </h3>
<hr />
<h2> Births to Mothers under the Age of 20 in the US </h2>
</html>
```

FIGURE 4.2. Initial map gallery page with title lines and name using <h1>, <h2>, and <h3> header text.

We can now display a map with the img tag. All file names that are referenced with img must end with GIF, JPG (or JPEG), or PNG, as these are the common file types for the different browsers. The img "src" option is used to specify the file's URL address. In the example below, notice the image file designator .gif at the end of the URL. The img command also includes a number of options that can be used to alter the size of the image or change its placement on the page. Like hr, img does not have a standard ending tag. In strict XHTML formatting, it is written as <img src=map.png />

```
<html>
<h1> My Map Gallery </h1>
<h2> A collection of maps found on the web. </h2>
<h3> Produced by your name </h3>
<hr />
<h2> Births to Mothers under the Age of 20 in the US </h2>
<img src="http://maps.unomaha.edu/Cloud/Chapter4/MapExample1.gif"
width="500" height="389">
</html>
```

Altering the size of the image will almost certainly be necessary because many of the maps available through the web are very large and won't fit onto the screen. Maps represented as pixels do not resize well because the line-work and lettering are adversely affected. Resizing a larger image may be done to create a slightly smaller image, but a small image should never be made larger. Increasing the size of a map in raster format makes the map look fuzzy.

Reducing the size of an image is done by specifying new pixel dimensions for either or both the height and the width (i.e., HEIGHT=# and WIDTH=#). If values are used for both fields, they need to be in the same proportion as the original values; otherwise the map will appear distorted, with one dimension altered more than the other. The original size of a saved graphic file is shown in the Properties or View Image Info window after right-mouse clicking on the image. New values can be calculated by changing one dimension and then cross-multiplying to find the other. For example, if the original image has a width of 900 and a height of 700 pixels, and you choose 500 for the new width, then $(700 \times 500)/900$ would give the proportional height of 388.88. All pixel dimensions need to be expressed as whole numbers, so the new proportional dimensions would be 500×389.

When displaying a smaller version of a map, it is always good to make a link to the larger version. This is done by embedding the img statement within an anchor using the <a href= tag. The command attaches a link from the smaller image to the larger image, resulting in the display of the larger image when the small image is clicked.

```
<a href="http://maps.unomaha.edu/Cloud/Chapter4/MapExample1.gif">
    <img src=http://maps.unomaha.edu/Cloud/Chapter4/
    MapExample1.gif width="500" height="389"> </a>
```

The reduction in display size is not reflected in the size of the file being transmitted; it is simply being displayed at a smaller size. If the file is still the original size, the download speed will be very slow. Rather than resizing the file at the destination, it would make sense to reduce the image size using a program such as Adobe Photoshop™. An option called *Save for Web and Devices* produces very small file sizes. PNG (portable network graphic) would be the preferred file format.

In the gallery.htm file, the paragraph that describes the map follows after the img tag. The <p> is used to create a space and move the text down to the next line. Like a caption for a piece of artwork in a museum, the caption for the map should explain what the map depicts and the information it is trying to convey.

A separating line surrounded by paragraph breaks (<p>
<p>) is then added. These three formatting tags are all on the same line because there is no need to separate formatting codes on separate lines, other than for readability. The next map is then displayed with the <img> tag followed by its caption, the next separating line, followed by the next map. This pattern repeats itself for all of the maps in the map gallery.

```
<html>
<h1> My Map Gallery </h1>
<h2> A collection of maps found on the web. </h2>
<h3> Produced by [your name] </h3>
<hr>
<h2> Map of _____ </h2>
<img src="http://maps.unomaha.edu/Cloud/Chapter4/MapExample1.
gif">
<p>This map shows the distribution in the 48 United States of births
to mothers under the age of 20 years old in 1982. Shadings are used
to show the five categories of data. The states with the higher number
of births to mothers under 20 are in the southern part of the United
States.</p>
<p> <hr> <p>
</html>
```

The <embed> tag is used to display graphic files that are not in the GIF, JPEG, or IMG format. Examples include Adobe's PDF, Flash, SVG, and QuickTime. The format is identical to that of the <img> option:

```
<embed src="http://maps.unomaha.edu/Cloud/Chapter4/MapExample4.pdf"
    width="500" height="389">
```

Formats such as PDF, Flash, and QuickTime usually require a separate plug-in to display the image properly. Before the file can be displayed, the plug-in needs to be downloaded and installed. Displaying images through a plug-in takes longer because the plug-in program needs to be activated before the image can be displayed.

The SVG graphic file format is the preferred format for Wikimedia Commons, the graphics repository for Wikipedia. But support for the display of SVG files has been problematic. Adobe stopped supporting its plug-in for the format in 2009. As an open standard, browsers such as Firefox have implemented some degree of native support, bypassing the need for a plug-in. SVG files can be displayed through the <img> tag in browsers such as Opera, Safari, and Chrome. With other browsers, SVG files can only be displayed with the <embed> or <object> tags. Microsoft

Internet Explorer, still the most widely used browser, has been slow to provide support for this format.

4.3 HTML Editors

HTML editors make it possible to avoid learning the HTML coding tags. The two disadvantages of HTML editors is that they normally don't implement all of the possible HTML formatting options and the files they create are usually larger than HTML files that are "hard-coded" with the use of an ordinary text editor. These disadvantages are minor for most applications, and it is useful to learn one of these programs. If used properly, these programs can help in learning HTML.

A good HTML editor should make it possible to manipulate the display of text and graphic elements, including their size and placement. More advanced HTML editors allow links to be added to parts of a graphic. Figure 4.3 depicts the toolbar for the HTML editor associated with a free HTML editor called Nvu (Heng 2010). The text layout functions are located across the bottom of the menu. The upper part contains a series of file management functions, including the "Insert Link" option that creates a link to another file and the "Insert Image" option that embeds a graphic file into the web document. The "Insert Table" places a table into the document with editable fields.

Tables are one of the more useful formatting elements of HTML. Cells within tables can be made larger or smaller and may contain text or images. Lines that separate the cells can be made invisible, and tables like this are often used to help place elements on a page. Hard-coding tables in HTML is complicated, and HTML editors can help to perform this task. Figure 4.4 depicts a table with nine cells and the corresponding HTML code.

Figure 4.5 includes the format dialog for manipulating images with an HTML editor. The top part (a) displays the filename of the image. The size can be changed with the Dimensions settings in (b). The placement of the text around the image is controlled with the pop-up dialog in (c). Text can be placed in the upper, middle,

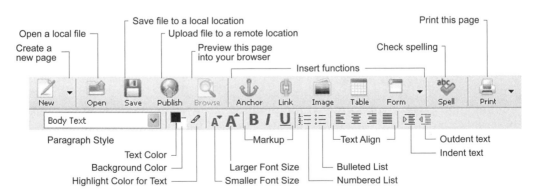

FIGURE 4.3. Toolbar for the open source program Nvu. Text layout functions are located across the bottom of the menu bar. The top line includes the hyperlink functions for files and images, as well as the table function and a spell checker. Heng 2010.

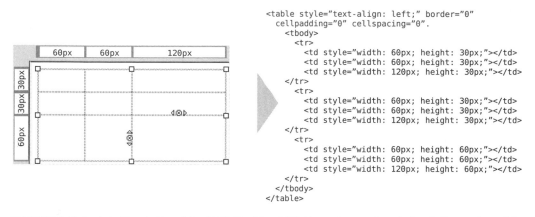

```
<table style="text-align: left;" border="0"
  cellpadding="0" cellspacing="0".
    <tbody>
      <tr>
        <td style="width: 60px; height: 30px;"></td>
        <td style="width: 60px; height: 30px;"></td>
        <td style="width: 120px; height: 30px;"></td>
      </tr>
      <tr>
        <td style="width: 60px; height: 30px;"></td>
        <td style="width: 60px; height: 30px;"></td>
        <td style="width: 120px; height: 30px;"></td>
      </tr>
      <tr>
        <td style="width: 60px; height: 60px;"></td>
        <td style="width: 60px; height: 60px;"></td>
        <td style="width: 120px; height: 60px;"></td>
      </tr>
    </tbody>
</table>
```

FIGURE 4.4. A table defined in HTML. The HTML code on the right defines the table shown on the left. Each <tr> defines a new table row. Each <td> code defines a cell. The width and height modifiers adjust the size of each cell.

or bottom area of the image, or around the image on both sides. These are standard formatting functions that are implemented by all HTML editors.

4.4 Multipage Gallery

Rather than displaying the gallery as one web page, each map in the gallery could be placed on a separate page. Using a table with one row and two columns on each page, the caption could be displayed either to the right or the left of the map. The first page in the series would only include the title of the gallery and a link to the first map. The page that displays the first map could have a link that displays the second map, and so on. But, instead of implementing this kind of sequential access between the individual files, each web page in this map gallery could have links to

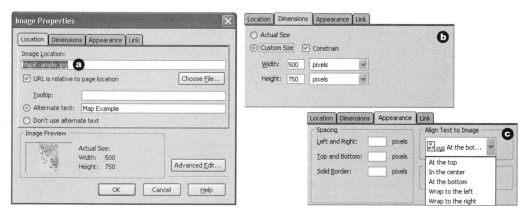

FIGURE 4.5. The Format dialog for manipulating images in Nvu. The selection of an image file is shown in (a). The resizing of an image is shown in (b). Controlling the placement of text around the image is shown in (c).

all of the other maps in the gallery. These links could be done on one line with a vertical bar between each link. Clicking on each link would direct the user to that page, as shown in Figure 4.6.

The coding of this would be done with a series of anchor tags that would reference a series of HTML files with names like map1.htm, map2.htm, and map3.htm. Each of these pages would have the same lines of code to allow the user to go to any page of the gallery, as shown in Figure 4.7.

All of the individual pages would be identical except for the map that is displayed and the caption, and some indication of which page is being viewed. Each page would have a title, a two-column table, and the links to all of the other pages (see Figure 4.8). To indicate which page is currently being displayed, the text for that link could be made bold by adding the <b> and </b> tag on either side of the text. In this example, the <p align="center"> tag would center the links on the page.

4.5   Describing the Map

It is impossible to put into words precisely what maps convey. This is exactly why maps are such a vital form of communication. Yet, for the sake of communicating with others, we need to express the message of the map. Attempting to translate the graphic of a map into words is an important part of map use and involves an understanding of both the mapping process and the geographic features that are represented. A proverb states that a picture is worth a thousand words. This is also true for maps if one has the background to properly interpret what the map shows. Gallery captions are typically on the order of 70 to 100 words.

Much of the information for the caption will come from the title and legend of the map. Some maps are designed primarily to emphasize the location of objects relative to each other, like cities (general reference maps), while others show broad patterns, such as the pattern of temperatures on a weather map (thematic maps). The data shown may be of a physical characteristic of Earth or something associated with human population. The information in the map caption should describe, as much as possible, the "who, what, when, where and how" of the map. It is generally easier to write a caption for a map that depicts a spatial pattern rather than a general reference map that only shows the location of features.

```
<a href=title.htm> Title Page </a> |
<a href=map1.htm> Map1 </a> |
<a href=map2.htm> Map2 </a> |
<a href=map3.htm> Map3 </a> |
<a href=map4.htm> Map4 </a> |
<a href=map5.htm> Map5 </a> |
<a href=map6.htm> Map6 </a> |
<a href=map7.htm> Map7 </a> |
<a href=map8.htm> Map8 </a> |
<a href=map9.htm> Map9 </a> |
```

FIGURE 4.6. The HTML code creates a series of links to separate HTML pages.

```
<html>
<h2> Title of Map #5 </h2>
<table width="1000" height="597" border="1">
  <tr>
    <td width="734" height="393"><img src=http://maps.unomaha.edu/
Cloud_Mapping/Chapter3/MapExample1.gif>
</td>
    <td width="250">Caption</td>
  </tr>
</table>
<p align="center">
<a href=title.htm> Title Page </a> |
  <a href=map1.htm> Map1 </a> |
  <a href=map2.htm> Map2 </a> |
  <a href=map3.htm> Map3 </a> |
  <a href=map4.htm> Map4 </a> |
  <a href=map5.htm> <b>Map5 </b> </a> | <!--This makes the link bold-->
  <a href=map6.htm> Map6 </a> |
  <a href=map7.htm> Map7 </a> |
  <a href=map8.htm> Map8 </a> |
  <a href=map9.htm> Map9 </a> |
</html>
```

FIGURE 4.7. An HTML file that links to nine different HTML files, each displaying a map and having links to the other nine maps.

Major League Baseball Fan Map

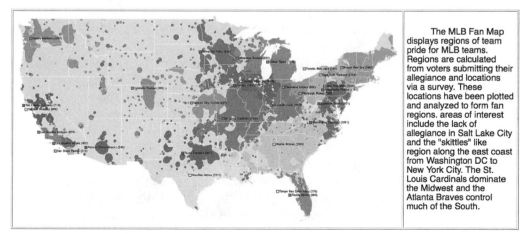

FIGURE 4.8. A single page from a multipage map gallery. The links at the bottom access the overall assignments page, the title page, and other maps in the gallery. Map courtesy of commoncensus.org.

Analyzing the map for content requires some experience with maps and a background in fields like geography, history, biology, and environmental science. Map analysis necessitates both an understanding of how maps are made and how the data are symbolized. Critically analyzing the design of the map requires examining the graphic attributes of the map based, in part, on aesthetic criteria. In other words, does the map look good?

Maps should be enjoyable to examine. The overall appearance is a determining factor in how we judge a map. It has been found that well-designed maps are viewed for a longer time and users have greater confidence in what they show. Conclusions about the graphic quality of a map are based on such attributes as the selection of colors, symbolization, text legibility, and layout. A map user should be able to observe and comment on these qualities.

4.6   Website Hosting

The best way for others to see your web page is by placing it on a web server. Such a server can be an ordinary desktop computer or even a laptop that is connected to the Internet and has the proper software to respond to web requests. A common web server software package is Apache, distributed under an open-source license on all major computing platforms.

Operating a web server is time consuming, particularly protecting it from attack by viruses and other malicious software. Viruses are especially malicious with servers, turning them into instruments for their own purposes. Another option is to use the web server offered by an educational institution or one of the many online web server providers. Usually, these commercial providers offer a free option for a limited amount of storage space. Remote servers can be easily operated through a web page. Even at US $5 or $10 a month, these services are more cost effective than operating a server and paying for all of the associated connection and maintenance costs.

Website hosting services vary in the number of services they offer. One service that will be important in later chapters is support for an online database. A database program called MySQL (based on the structured query language) will be used to store data for mapping. It is important that the web-hosting service use MySQL version 5 or newer. Access to MySQL may also be provided for free for a limited number of databases. All of the assignments in this book can be hosted through a commercial website-hosting service for no charge.

There are literally thousands of different online service providers. These companies provide a certain amount of disk storage space, access to the necessary server software, and a web address to make your online services available to others. Figure 4.9 depicts the sign-up page for the 000webhost.com provider and the resulting account information. An email address is all that is needed to sign up for the service. The Account Information page in Figure 4.9 shows that the web address that was assigned by the service provider is http://geographyprof.hostei.com (or http://64.120.177.162) and that 1500 mb of free storage is available.

The web-hosting company also provides a web page interface for uploading and editing files. First, an account is created and associated with an email address, as shown in Figure 4.9.

» Account Information	
Domain	geographyprof.hostei.com
Username	a8040697
Password	******
Disk Usage	0.2 / 1500.0 MB
Bandwidth	100000 MB (100GB)
Home Root	/home/a8040697
Server Name	**server33.000webhost.com**
IP Address	64.120.177.162
Apache ver.	2.2.13 (Unix)
PHP version	5.2.*
MySQL ver.	5.0.81-community
Activated On	2011-03-10 16:23
Status	Active

FIGURE 4.9. The left panel shows the sign-up page for http://000webhost.com. Account information is shown on the right. Choosing a subdomain for the web address is free. A charge is incurred for specifying a domain with a specific name, such as http://www.peterson.com. But having the site choose a free domain like http://geographyprof.hostei.com is free. Copyright 2013 First-Class Web Hosting.

Included in most online hosting services is a graphical interface to all of the services that are offered. This is called the control panel, or cPanel (see Figure 4.10). Separate tools handle email, the editing of files, the scheduling of tasks, and account management. All of these tools represent open-source projects that are written and maintained by a small legion of programmers. File Manager is the major tool for managing files and building a website. MySQL and phpMyAdmin will be used later for building a database. Most online hosting services use the same cPanel to access server resources.

Figure 4.11 shows the FileManager window with access to tools for uploading and creating new files and directories (subfolders). Tools are also available to upload, move, delete, and rename files. The file listing shows the name, type, and size of the file, while the Owner, Group, and Perms fields depict security settings. Mod Time indicates when the file was last modified. The files can be edited directly from this window by clicking "Edit" at the end of each filename.

The public_html or htdocs folder is the directory from which all web files are served. If an HTML file is to be presented through a web page, it must reside in this folder. Usually, this folder contains a file called index.htm (or index.php) that is the first page accessed when the site is referenced. For example, if an address such as:

http://geographyprof.hostei.com/Cloud/

is entered into a browser, the browser will look for a file called index.htm in a directory (folder) called Cloud that is itself located in the public_html directory. That means that the following two addresses would display the same file:

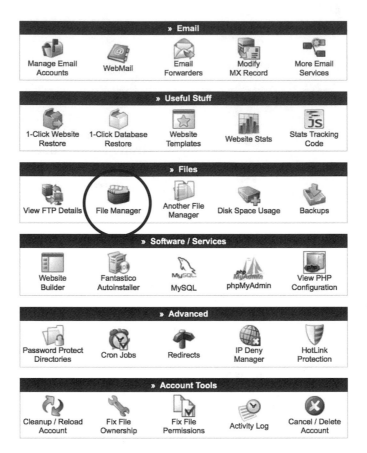

FIGURE 4.10. A standard web-hosting control panel, called cPanel, gives access to the different hosting tools. File Manager is the major program for uploading and editing files. Copyright 2013 First-Class Web Hosting.

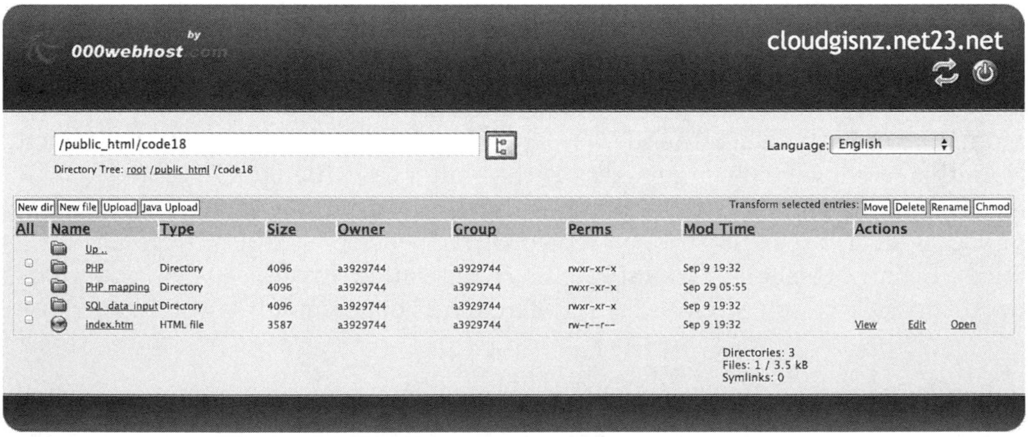

FIGURE 4.11. The File Manager window from an online hosting service. This service allows files to be created and edited. All files to be served through the web must be in the public_html. directory/folder. Copyright 2013 First-Class Web Hosting.

http://geographyprof.hostei.com/Cloud/

http://geographyprof.hostei.com/Cloud/index.htm

Normally, this `index.htm` file serves as an entry point to the website and will have links to all other files in the directory.

The `index.htm` file will have a relatively simple structure—a title line followed by links to all of the assignments. It would also be useful if this file would have a picture of the website owner and links to the websites for all students in the course, as shown in Figure 4.12. Figure 4.13 shows a part of the HTML code for this `index.htm` file. The picture is inserted using the `<img src=filename>` tag. The links to the student pages are separated by two vertical lines ("||"). The code for the index file can be obtained by selecting View Source on this web page:

http://maps.unomaha.edu/Cloud/template/

4.7 Summary

HTML is the building block of the web. It is the language that makes it possible to present information through web pages. A number of programs facilitate making pages without any knowledge of HTML. The importance of HTML is that it is a container for other tools. As we will find in the following chapters, the ability to incorporate user-defined maps in a web page is based on some knowledge of HTML coding.

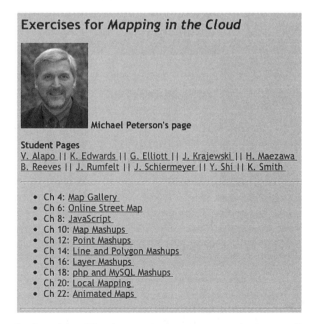

FIGURE 4.12. An `index.htm` file that includes a picture, links to all other students, and links to the assignments.

```
<html>
<head></head>
<body bgcolor="#CCCCFF">
<h2> Exercises for <i>Mapping in the Cloud</i> </h2>
<img src=peterson.jpg height=150><b> Michael Peterson's page </b><p>
<p>
<b> Student Pages </b><br>
<a href=http://victoriaA.site88.net> V. Alapo </a> || <a href=http://
mapsarefuntoo.web44.net> K. Edwards </a> ||   <a
....................
<br> <hr>
<ul>
<li>Ch 4: <a href=http://maps.unomaha.edu/onlinemapping/code04.zip> Map
Gallery </a><br>
<li>Ch 6: <a href=code06.zip> Online Street Map</a><br>
<li>Ch 8: <a href=http://maps.unomaha.edu/onlinemapping/code08.zip>
JavaScript </a><br>
....................
</a><br>
</ul>
<hr>
</body>
</html>
```

FIGURE 4.13. Code for the index.htm file that displays a picture using the img tag, links to the pages for the other students, and to the assignments.

Writing the HTML file is only the first step. The file, or files, must also be uploaded to a web server so that they can be made available to others. Although it is possible to create a web server using any computer, it is best to use the services of a web-hosting site. Many of these online services offer a limited amount of storage space at no cost. Tools like File Manager facilitate the uploading and editing of HTML files.

4.8   Exercise

Make a web page of the nine maps found for the assignment in Chapter 2 using both the single page and multipage map gallery code from this chapter. Upload the map galleries to your web server and create an index.htm file that links to both galleries.

4.9   Questions

1. In HTML, how can text size be changed?

2. What does the <HR> tag do?

3. Describe the img tag. How is the size of the image adjusted?

4. Why is it not good practice to increase the size of images with the img tag?

5. What types of image formats can be displayed with the img tag?

6. How is the embed tag used?

7. Show how a hyperlink is associated with a text string and with an image.

8. Compare a standard web-hosting site like 000webhost.com to Amazon Web Services.

9. What is the public_html directory?

10. What is the File Manager?

4.10   References

Heng, Christopher (2010) How to Design and Publish your Website with NVU. *Nvu TheSiteWizard.com* [http://www.thesitewizard.com/gettingstarted/nvu1.shtml]

MacDonald, Matthew (2006) *Creating a Web Site: The Missing Manual*. Sebastopol, CA: O'Reilly.

Musciano, Chuck, and Bill Kennedy (2007) *HTML & XHTML: The Definitive Guide*. Sebastopol, CA: O'Reilly.

Willard, Wendy (2009) *HTML: A Beginner's Guide*. Berkeley, CA: Osborne/McGraw-Hill.

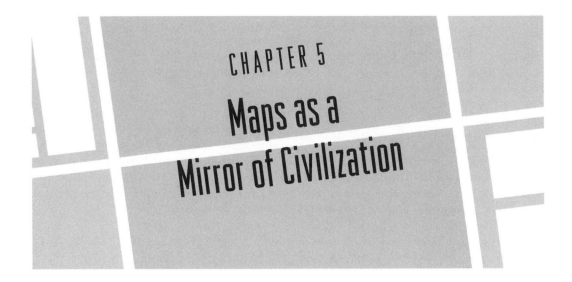

CHAPTER 5

Maps as a Mirror of Civilization

The characteristic of scientific progress is our knowing that we did not know.

—GASTON BACHELARD

5.1 Introduction

Maps hold a special place in human understanding. Thrower (1972, p. 1) describes maps as a "sensitive indicator of the changing thoughts of Man," and argues that no other work better reflects the human struggle to comprehend the world. As a mirror of civilization, maps convey as much about the ideas and the priorities of the people who make them as they do about the environments they depict.

Early maps were sometimes more conjecture than fact. Some were used to reflect religious beliefs. By the 1500s, maps became closely linked to navigation and developed into a commercial product in the great map houses of European port cities. Here the knowledge of the world was gathered and put in graphic form. The importance of maps for navigation gave way to military applications by the 1700s. Early mapping was done without the ability to see the world from above. Aerial photography, and later remote sensing, drastically changed our ability to map the world.

5.2 Early Cartography

Although it is generally agreed that maps predate writing, there is no way of knowing when the first map was made. The association has been made between the human ability to draw pictures that have been found on cave walls (Davis 1986) and the ability to draw maps. The geocentric view required to create maps of an area may have taken much longer to develop. For example, it has been found that children navigate with more egocentric and landmark-based mental maps of their surroundings and are generally unable to create a more geocentric view until a much later age. Artists also could not create pictures in correct three-dimensional perspective until the 1400s. Similarly, it is likely that the human capacity to represent the world in a geocentric way developed many years after humans began to draw.

The maps of Babylonia from about 2300 BC used stylized symbols that were impressed or scratched on clay tablets. The larger scale maps showed such features as canals, city walls, houses, and terraces. An intermediate scale map, with an orientation of east at the top, included water bodies, settlements, and mountains. The small-scale world map shows a flat Earth with Babylon at the center (Thrower 1972, p. 13). The method of symbolization is not that different from maps of today (see Figure 5.1).

Very few maps have survived from the time of Mesopotamia up to about 1000 years ago. Some argue that the human ability to create such displays had not developed until then or that the knowledge to make maps had not been widely disseminated. A more accepted notion is that the material on which maps were created was too perishable to survive. Indeed, it seems that as societies become more advanced, the medium on which they choose to convey information becomes less substantial. The first known maps were made on clay tablets. Later a fibrous material, such as paper, was used. Today, most maps are stored on computer media that can be easily destroyed with a slight amount of magnetism or heat. In addition, most modern maps require the use of specialized equipment to make the maps viewable. It is doubtful that any current map, particularly a map from the Internet, will survive or be viewable in its present form for more than a decade or two.

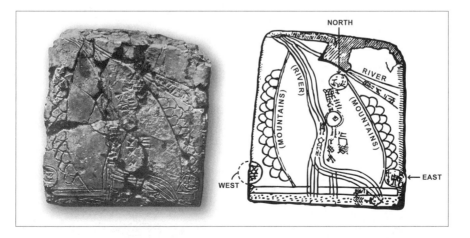

FIGURE 5.1. Ancient map from Ga-Sur in Nuzi [Yorghan Tepe] from 2300 BC, near the city of Kirkuk, 200 miles north of the site of Babylon in present-day Iraq. The map is small at only 7.2 cm × 6.8 cm. The tablet depicts two mountain ranges with a river flowing down the middle as seen in the drawing on the right. Clark and Black 2005. Image on left courtesy James Seybold.

Not many maps have survived from the Roman period. The most famous is a Roman road map known as the Peutinger Table, passed-on by the librarian of the emperor Maximilian of Austria to Konrad Peutinger in the German city of Augsburg in 1508 (see Figure 5.2). The manuscript can be dated to the 12th or 13th century, but it is clear that it is a copy of a much older original. The map represented the Roman road system on a 20-ft-long, 1-ft-wide scroll. The result was that distances were distorted to make the map fit on the 1-ft-wide dimension. The map is sometimes compared to modern-day subway maps that drastically distort reality but effectively show travel routes.

While few maps have survived from ancient times, scientific knowledge about the size, shape, and representation of the Earth can be traced to the Greeks more than 2000 years ago, particularly Eratosthenes and Ptolemy. Eratosthenes lived between 276 and 194 BC and was the chief librarian at the Great Library in Alexandria, Egypt; he is credited with making a very accurate calculation of the Earth's circumference. His method to calculate the Earth's size was based on the observation that, on a certain day of the year—the Spring Equinox, sunlight perfectly illuminated the bottom of a well near Aswan, Egypt, located near the Tropic of Cancer at 23½° north latitude (see Figure 5.3). Measuring an angle cast by a pole in the more northerly city of Alexandria, he was able to associate this angle, through the laws of geometry, to the arc of the circle between Aswan and Alexandria. This angle of 7.5° was calculated to be 1/50 of the circumference of the Earth (360/7.5 = 50). The circumference of the Earth was determined by multiplying this distance by 50.

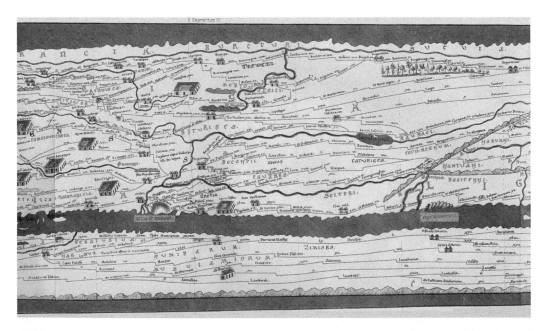

FIGURE 5.2. Example of a Roman road map. The Peutinger Table, discovered by Konrad Peutinger, represented the Roman road system on a 20-ft long (6.75 m), 1-ft-wide (0.34 m) scroll. The result was that distances were distorted to make the map fit on the 1-ft-wide (0.34 m) dimension. Image courtesy James Seybold.

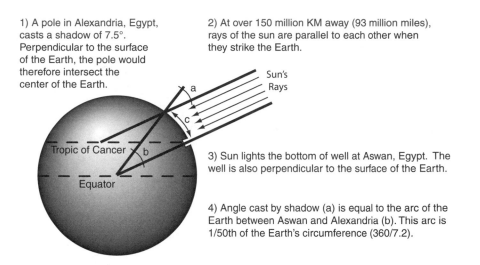

1) A pole in Alexandria, Egypt, casts a shadow of 7.5°. Perpendicular to the surface of the Earth, the pole would therefore intersect the center of the Earth.

2) At over 150 million KM away (93 million miles), rays of the sun are parallel to each other when they strike the Earth.

Sun's Rays

3) Sun lights the bottom of well at Aswan, Egypt. The well is also perpendicular to the surface of the Earth.

4) Angle cast by shadow (a) is equal to the arc of the Earth between Aswan and Alexandria (b). This arc is 1/50th of the Earth's circumference (360/7.2).

5) The ground distance (c) between Alexandria and Aswan is 5000 stadia. At 607 feet per stadia, the circumference of the Earth would be about 28,740 miles (50x5000x607/5280). The actual circumference is 24,900 miles or 40,075 KM.

FIGURE 5.3. Eratosthenes' calculation of the Earth's circumference, considered one of the great accomplishments of early Greek civilization.

Over 300 years later, Claudius Ptolemy (AD 90–c. AD 168), also a Greek, assembled the scientific accomplishments of the Greek and Roman empires, and worked at the same Great Library in Alexandria. Ptolemy's treatise, *Geographia*, was a compendium of knowledge about the Earth, its size, and methods of representation. He is also credited with devising a system of latitude and longitude, making map projections, and introducing a way to make large-scale sectional maps.

Knowledge of the Earth as embodied in Ptolemy's *Geographia* was not improved upon substantially for nearly 1000 years. Indeed, the great scientific accomplishments of the Greeks were lost during the Middle and Dark Ages, a time when maps in Europe were used primarily to reflect religious doctrine. The T in O map, for example, having Jerusalem at its center and dividing the Earth into three parts for the three sons of Noah, was widely reproduced by hand in monasteries beginning in the 1000s, and then later through printing beginning in the latter part of the 1400s. About 100 hand-drawn versions of the T in O map still survive (McCarthy 2006, p. 490), including the famous Hereford Mappa Mundi (see Figure 5.4).

In contrast to the T in O maps, some very accurate maps were being constructed by the end of the 1200s in seaport cities of the Mediterranean, primarily Italy and the Catalan Islands off the east coast of Spain. Called Portolan (Italian for sailing manual) or Catalan charts, these ornate navigational maps were designed to be used with a compass. The maps depicted the coastal areas from northern Europe to southern Africa in great detail. A distinguishing characteristic of these maps that depicted in great detail the coastal areas from northern Europe to southern Africa is a "web" of lines that radiate outward from a series of compass roses (see Figure 5.5). Used by sailors involved in trade through the 1600s, the maps have no lines of latitude or longitude and appear to depict the Earth as flat.

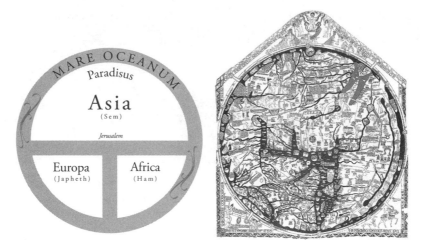

FIGURE 5.4. Stylized version of T in O map on left showing the world divided into three parts, each for one of the three sons of Noah (Sem, Japheth, and Ham). The Hereford world map on the right from around 1275, also referred to as Mappa Mundi, is the most famous example of this type of map. Image on right from http://en.wikipedia.org/wiki/File:Hereford_Mappa_Mundi_1300.jpg.

Ptolemy's *Geographia* was regarded as a lost treasure when it was discovered, first by Islamic cartographers and later in Europe. Ptolemy became known as a great authority on all aspects of cartography more than a millennium after his death. With the development of printing in the 1400s, his book was widely distributed. Unfortunately, Ptolemy did not include the calculations of Eratosthenes in *Geographia*. Instead, Ptolemy relied on a much smaller calculation of the Earth's circumference attributed to Posidonius, thus misleading early explorers. In particular, *Geographia* influenced Columbus's ideas about the size and shape of the Earth and contributed to his misguided search for a western route to India and his inadvertent discovery of the New World. Columbus was so convinced of this smaller size of the Earth that he argued until his death that he had found islands off the coast of India.

The Renaissance culminated at the end of the 1400s with a series of major developments, the most significant of which was the invention of printing. Mastered in Western civilization in 1467, printing had a dramatic effect on the production and distribution of maps. Initially carved in reverse in wood, printing made it possible for the first time to produce a large number of maps that were identical to the original. Before this time, maps had to be reproduced by hand—most often done by monks in monasteries. The first printed map was a T in O map that was also mass-produced in monasteries, thus extending the life of this map well into the 1600s.

The early printing of maps has an interesting analogy to the distribution of maps through the Internet. Maps were first printed in large quantities by monks in monasteries who quickly adapted to this method of reproduction. Isolated in monasteries, monks had little knowledge of the world. In the same sense, the initial maps on the Internet were placed there by people who quickly learned how to adapt to this new technology, not by those who knew about the making of maps. A new medium seems to initially attract people who can adapt to its particular technology, and not necessarily those who can contribute significantly to its content.

The Age of Discovery—a period when explorers would map out the new world and refine the techniques of navigation—followed the Renaissance. The 1500s are marked by the work of Mercator, Ortelius, and others that was done in the great map houses of the port cities in Europe. Located near the ports and often in close proximity to other establishments that provided services to sailors, including houses of prostitution, these shops became the center of knowledge about the new world. The Mercator projection (see Figure 5.6), though severely distorting the size of northern and southern latitudes, was a useful tool for navigation when used in conjunction with a compass.

The 1600s and 1700s are characterized by the greater use of mathematics in cartography. Working under the French Academy of Sciences (search: French Academy of Sciences), researchers would determine that the Earth was an irregular sphere, called a spheroid, and would mathematically describe its shape. Mathematicians in France and Germany would tackle the map projection problem—now expressed as a mathematical conversion of the round Earth to a flat piece of paper. Initially, it was hoped that a solution to the conversion could be found without introducing distortion in the flat map. Later, when it was proved that a mathematical solution did not exist, methods were simply devised to reduce the amount of distortion. The Frenchman, Nicholas Tissot, went on to use ellipses to represent the amount of area and angular distortion at each point on a flat map (search: Tissot indicatrix).

Many early maps from other parts of the world have been lost, so we have little idea what they may have looked like or what they represented. The discovery in the

FIGURE 5.5. Portolan and Catalan charts, appearing toward the end of the 1200s, were based on direct observation by means of the mariner's compass. They portrayed the coastlines and ports for sailors. Lines of latitude and longitude are not included. From http://en.wikipedia.org/wiki/File:Jorge_Aguiar_1492_MR.jpg.

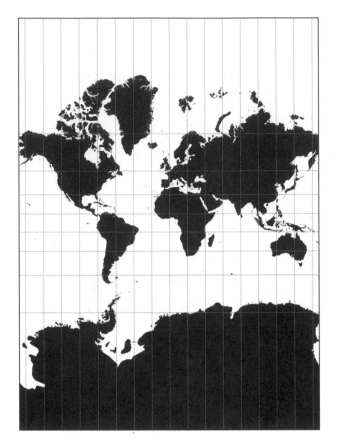

FIGURE 5.6. The Mercator projection is a cylindrical map projection. Its parallels (latitude) and meridians (longitude) are straight lines, and the map can be effectively used for navigation with a compass over short distances. To maintain the correct relative sizes, the lines of either latitude or longitude must come closer together with increasing distance from the equator. Notice in the Mercator projection how the lines of latitude actually become further apart.

1890s of maps made by Marshall Islanders in the Pacific Ocean provided a window to a totally different form of mapping. The Marshall Island maps were constructed using sticks and shells, lashed together with cord (see Figure 5.7). Not only was a different medium used but the maps represented features from the Earth that are typically not mapped, even today. Most of the sticks depicted sea swells that were used by the Marshall Islanders for navigation (Wise 1976). Sea swells are a type of wave that are regular and persistent in their flow; they are more easily felt than seen.

Additionally, only one person in a group of Marshall Islanders could read the map, and this knowledge was passed down from father to son. This made it very difficult for Westerners to discover exactly how they were used for the father could not divulge the information to anyone but his son. In his account, the German Navy captain Winkler describes how alcohol and gifts were used to extract information about the maps from a chief of the Marshall Islanders. Another interesting finding concerned the actual use of the map. It was determined that the "map reader" could direct the boat while lying in a prone position in the bow of the boat and could do so at night, even while intoxicated (Winkler 1899). The navigator did not rely on visual cues from the environment.

5.3 Early Navigation

Cartography has always been closely linked to navigation and its associated measuring devices. The compass has been in use for perhaps 1000 years. Latitude was measured fairly accurately as early as the 1400s, but it was not until the early 1800s that longitude could be precisely determined.

The compass, consisting of a magnetic needle freely suspended so that it aligns with the Earth's magnetic north and south, has been used for navigation since at least the 1100s and perhaps earlier. With the help of a map, it was possible to reach a destination by laying a course in a certain direction. To determine the current position, a dead-reckoning system of navigation developed that used point of departure, the course as shown by compass, the speed and distance traveled, and the time elapsed.

The compass became so popular for navigation that maps, such as the Portolan and Catalan charts, incorporated compass directional lines. While the maps could be used with a compass, lines of constant direction—called rhumb lines—do not follow the great circle and therefore do not represent the shortest distance between two points (see Figure 5.8). Moreover, the compass itself did not help the sailors

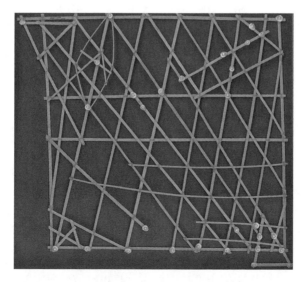

FIGURE 5.7. A Marshall Island stick chart with bamboo sticks and shells to represent islands. The map uses a different medium and shows different elements of the environment than maps of Western culture. It is an indication that early maps may have used different methods of symbolization and mapped different features than maps of today. Image courtesy of Library of Congress. From www.loc.gov/item/2010586182.

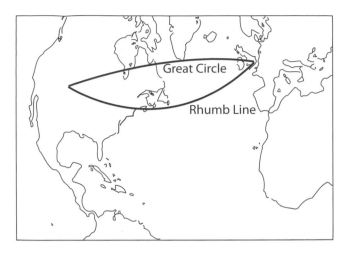

FIGURE 5.8. Comparison between the great circle and a rhumb line between Denver and London. The great circle dissects the Earth into two equal parts and represents the shortest distance between two points. The rhumb line is the line of constant direction that would be followed with a compass. The projection used here causes neither line to be straight.

determine their position en route but only helped them to follow a course between two points.

5.3.1 Determining Latitude

The *cross-staff* was used to find latitude by the early 1400s. It was constructed from two pieces of wood with the cross at right angles to, and sliding on, the staff. A sight was fixed at the end of the staff. Holes were bored at the ends of the 26-in (66 cm) cross. The instrument was sighted in the direction of a heavenly body until the star appeared through the upper hole and the horizon through the lower. The altitude was then read on a scale marked on the staff. The device was not very accurate. Columbus navigated with a cross-staff, a compass, and a table of the Sun's declination on his voyages to the New World in the 1490s. The most Columbus could determine was the direction of travel and approximately how far north or south he was.

The *astrolabe* was another instrument used to determine latitude. It consisted of a disk of wood or metal with a circumference marked off in degrees. Pivoted at the center of the disk was a pointer called an *alidade*. Angular distances could be determined by sighting with the alidade and taking readings of its position on the graduated circle. More elaborate astrolabes included a star map and a zodiacal circle. Calculations of longitude were also attempted with the device. It was used on voyages of discovery beginning in the 1500s and was still used with the advent of the sextant in the 1700s. An ornate astrolabe from Nuremburg, Germany, was manufactured in the workshop of Georg Hartmann in 1537 (search: Hartmann astrolabe). Like the cross-staff, it was also not very precise.

Using adjustable mirrors to measure the exact angle of the stars, Moon, or

Sun above the horizon, the *sextant* further refined the determination of latitude. Invented independently in England and America in 1731, it is based on the principle that a reflected ray of light leaves a plane surface at the same angle at which it strikes the plane. The horizon and the reflected image of a celestial body such as the Sun are aligned on the "index arm," and the altitude is determined from the reflection on this arm (see Figure 5.9). Latitude is then determined with reference to navigational tables.

5.3.2 The Search for Longitude

Despite all of the progress made in finding latitude, the determination of longitude still remained elusive. In 1675, the Royal Observatory was established in Greenwich, England, to help solve the problem of finding longitude. John Flamsteed was appointed as its first astronomer, and the 28-year-old clergyman was instructed to "apply himself with the most exact care and diligence to rectifying the tables of the motions of the heavens, and the places of the fixed stars, so as to find out the so much-desired longitude of places for perfecting the art of navigation" (Bailey 1835, p. 107).

Flamsteed and the astronomers who followed him at Greenwich measured the position of stars with transit telescopes. These devices were aligned in a north–south direction and were used to record the position of the stars as they came into view at various times during the night along this north–south transit line. The intent was to provide sailors with the star positions as they were viewed at Greenwich so that their position could be compared to their location in the sky as viewed from other parts of the world. Meticulous readings were taken at the observatory

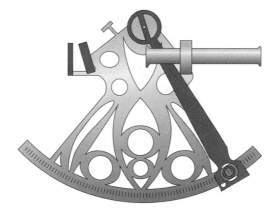

FIGURE 5.9. The sextant consists of a triangular frame, the bottom of which is a graduated arc of 60°. A telescope is attached horizontally to the plane of the frame. A small mirror is mounted perpendicular to the frame at the top of an index bar that swings along the arc. A half-transparent, half-mirror called the horizon glass in front of the telescope reflects the image of the Sun or other celestial body from the index mirror to the mirror half of the horizon glass into the telescope. When the horizon is seen through the transparent half of the horizon glass, with the reflected image of the celestial body lined up with it, the altitude of the Sun or star can be read from the index arm of the arc. The latitude can then be determined with reference to navigational tables.

for many years but astronomers did not want to release their data until the readings could be repeatedly verified.

Once the system of navigation by the stars was introduced on the ships it proved to be both extremely complicated and cumbersome, as there was no instrument that could make the necessary planetary sightings and measurements. The first *Nautical Almanac and Astronomical Ephemeris* was nevertheless published in 1767 listing the position of celestial bodies for each day of the year. The ephemeris data is still being published as the *Astronomical Almanac*, a joint American and British publication, and contains such information as the daily ascension and declination of the Sun, Moon, planets, and other celestial bodies.

In 1707, before the publication of the *Almanac* and more than 30 years after the Royal Observatory had been established to solve the problem of determining longitude, the British suffered a disastrous loss of a fleet of ships off the Isle of Scilly to the southwest of Britain that resulted in the death of nearly 2000. Dissatisfied with the progress of the observatory in finding a practical method to determine longitude, the British formed a separate Board of Longitude that, in 1715, offered £20,000 (equivalent to 1 million dollars today) to anybody who could find a way to determine a ship's longitude within 30 nautical miles.

In the end, it was not the astronomers at the Royal Observatory who solved the longitude problem but a cabinetmaker by the name of John Harrison. Born in Yorkshire, England, in 1693, the son of a woodworker, Harrison apprenticed as a carpenter and cabinetmaker and had little formal education (Sobel 1995, p. 9). Working essentially alone on his development of the chronometer, Harrison challenged the scientific and academic establishment. He built his first clock entirely out of wood at the age of 20. He was commissioned to build several clocks thereafter, one of which—the so-called Pelham clock, built of wood in 1722—is still in operation (Duffy 2000, p. 37).

In 1730 Harrison went to London to convince the Board of Longitude that he could build a clock that would be accurate enough to determine longitude. The Board was filled with astronomers and mathematicians who were not sympathetic to Harrison, and so they did not allow him to present his ideas. One of its members, however, Sir Edmund Halley, known for Halley's Comet, was impressed with Harrison and introduced him to the eminent clockmaker, George Graham. Graham became his benefactor, loaning him money to be repaid "at no great haste, and at no interest" (Sobel 1995, p. 77).

Harrison completed a device he named the H1 in 1735, the first of five maritime chronometers he would build. Made of brass, it weighed 70 pounds and was nearly 4 ft high, wide, and deep. The clock was tested at sea, and though it met the requirements for the prize, Harrison believed he could build a better clock and merely asked for £500 to continue his work. Harrison was still not satisfied with the H2, completed two years later, and again merely asked for more money to continue work on an improved design. It was 17 years before the H3 was completed. Innovative in design, it incorporated many new inventions and was half the size of the H1 at only 55 pounds. Satisfied with his accomplishment, and now 60 years of age, Harrison believed the H3 deserved the Longitude Prize.

In the intervening years, however, another device called the Quadrant had been developed that made navigation by the stars more practical, if enough planets and stars could be accurately plotted. Members of the Board of Longitude were now actively engaged in finding the astronomical solution to navigation and thereby claiming the prize for themselves. Because the Board was not inclined to let Harrison prove the utility of his chronometer, no ship could be found to carry it until three years later. By then, Harrison had completed the H4 (search: Harrison H4).

The H4 looked like a big pocket watch, around six inches in diameter. Completely enclosed, it was perfect for seagoing navigation. It is presumed that much of the 17 years used in the development of the H3 was actually used in making the H4. It was tested on a Royal Navy ship to the West Indies, and it was shown to have an error of only 1¼ minutes. A 30-minute accuracy was sufficient for the prize. The Longitude Board, however, did not believe the results and commissioned another test, this one to Barbados. The H4 passed again.

The newly appointed Astronomer Royal at Greenwich, Neville Maskelyne, who wanted to claim the prize based on his own astronomical work, forced Harrison to disassemble his clock in front of other watchmakers to prove it was not some kind of "trick" (Sobel 1995, p. 122). Then, another clockmaker was given the designs and asked to reproduce the clock—a task that took two years. Harrison was then ordered to produce two copies of the H4 without having either the original clock or the designs.

In 1772, Harrison's first copy was finished and, at the age of 76, he was incapable of producing a second. He appealed to King George III for help. The king ordered the Board of Longitude to meet and explain why the Longitude Prize had not been awarded to Harrison. Lacking any explanation, the Board of Longitude finally gave him £14,315, not the £20,000 award, and withheld bestowing the actual Longitude Prize. The Parliament later awarded him another £8750.

The terms for the prize were reset in such a way that it was impossible for either astronomers or clockmakers to meet the requirements. It was eventually withdrawn 50 years later, still unclaimed. Before Harrison died in 1776, Maskelyne published Harrison's clock design, effectively putting his ideas in the hands of other clockmakers who seized the opportunity to become wealthy from the new maritime chronometer that Harrison spent his life developing. Navigation by chronometer was used for the next 150 years, and the pocket watch—and later the wristwatch based on Harrison's designs—became a major commercial success.

Navigating was an early impetus for mapping. The extended search for longitude contributed to a prolonged interest in maps and the growth of commercial cartography. The Longitude Prize was an example of early government involvement in mapping.

5.4 Maps from Photos

Developments in printing and further refinements to the technology of mapping marked the 1800s in cartography, while aerial photography dominated cartography during the early part of the 1900s. The earliest photographs from above were made from balloons in the 1850s, but it wasn't until the 1920s that aerial photography by airplane became widespread (see Figure 5.10). What followed was a major transition in mapmaking through the science of photogrammetry. Literally, measuring

FIGURE 5.10. Section of a mosaic of photos along the Atlantic shore near Atlantic City, New Jersey. The photos were taken by the Army Air Service in 1919. From http://oceanservice. noaa.gov/news/features/oct10/mapmaking.html.

(metry) from a photo (photogram), this new area of study dominated cartography throughout the middle part of the century.

5.4.1 Photogrammetry

While it is possible to make measurements from individual aerial photographs, including the measurement of shadows to determine the height of buildings, the major accomplishment of photogrammetry was the use of stereo information from adjacent photos. Captured with 60% forward overlap and 40% on each side, pairs of aerial photos could be examined in unison (see Figure 5.11). In addition to viewing the landscape in three dimensions using stereo glasses, the relative displacement of objects between two photos, called parallax, could be quantified. This displacement is used to calculate elevation.

The use of aerial photographs to make maps progressed through three distinct stages (Konecny 1985). The first stage was analog photogrammetry. This phase, lasting until about 1960, culminated in the development and widespread use of the stereoplotter, a device designed for topographic mapping (see Figure 5.12). The stereoplotter is used to make maps from airphotos. This instrument is used to extract precise elevation information (i.e., point elevations and contours) from overlapping images. The most widely used of these devices is the Kelsh Optical Projection Stereoplotter introduced by Harry T. Kelsh in 1945. The device allows a trained operator to trace contour lines based on the elevation information in the stereophotos. An entire topographic map can be created in a week or two, depending on the complexity of the surface features. Accurate elevation measurements are derived by adjusting a floating point on a device called a platen, an elevated platform that is moved around on a table. By placing the floating point to the perceived surface of the stereo model, accurate point elevations or contours are determined.

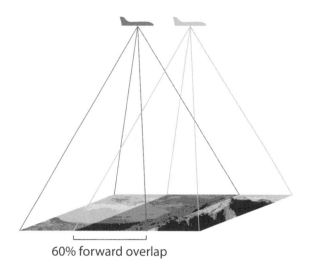

60% forward overlap

FIGURE 5.11. In acquiring stereo aerial photography, 60% forward overlap is needed to provide sufficient stereo information. The adjacent flight line results in a 40% side overlap.

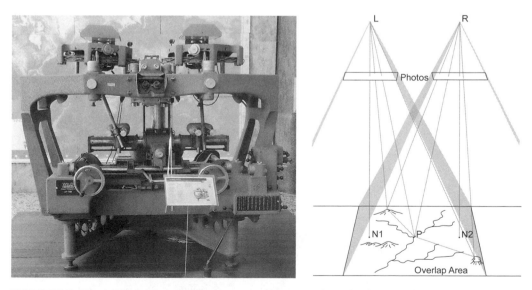

FIGURE 5.12. A Wild A7 Autograph stereoplotter on the left is an analog instrument used to make topographic maps. Two stereo photos are inserted at top. The diagram on the right explains the stereo mapping process. Ground point P, midway between N1 and N2, is the stereoscopic perspective center.

The second stage in the development of photogrammetry was based on the analytical stereoplotter, developed in 1957 in Canada (see Figure 5.13). In contrast to the analog stereoplotter, the analytical stereoplotter uses a more mechanical and computer-controlled system of deriving elevation information. A computer was used to move the instrument around the stereomodel and to digitally transform coordinates between the image and the map. The analytical stereoplotter solves the parameters of the photogrammetric equations and determines the adjustments required to form a stereo model while correcting for all known sources of error. The third stage, digital photogrammetry, added the use of digital photos and incorporated a computer specially configured to perform rigorous photogrammetric tasks.

Elevations for deriving contours and orthophoto correction are now obtained through a laser technology called LIDAR (light detection and ranging). Mounted in a small jet, the device determines the distance to the ground for small areas by measuring the amount of time it takes for a low-energy laser signal to return after reflecting from the surface of the Earth. The narrow beam of the laser is ideal for mapping applications, with many elevation points derived in a raster pattern over a small area. Figure 5.14 shows a LIDAR image over a relatively flat area.

5.4.2 Orthophotographs

The direct use of photographs that have been rectified through the removal of distortion is very common for a variety of applications. Such images are typically used as a background image to a map and to update the map with new features. These modified photographs, called orthophotos, are produced through a process called orthorectification, which can occur in two ways. One approach uses ground control points (GCPs) that are associated with visible physical features in a landscape and have a surveyed position. For a city, these points may be manhole covers that

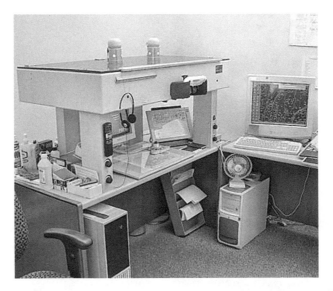

FIGURE 5.13. Analytical stereoplotter. The device is used to automatically capture elevation points from stereo photographs. Manual assistance is possible in difficult terrain.

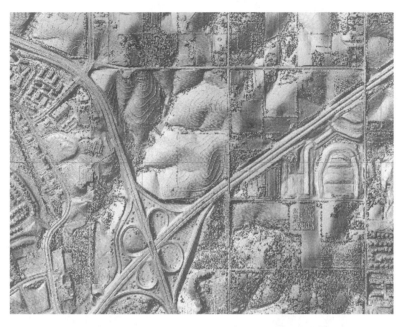

FIGURE 5.14. Example of a LIDAR image depicting elevation differences in shaded relief for an area north of Lincoln, Nebraska. The image includes an interstate interchange with terraced fields to the north. Even houses and trees are visible on the image because they have a different elevation than the background. Photo courtesy of Scott Robinson.

are painted with an X before the air photos are taken. The image is then re-sampled based on the relationship between the GCPs and the x and y photo coordinates.

In the second approach, elevation points are collected through photogrammetric methods, and these are used for establishing the mathematical relationship between the control points and the corresponding points on the image. In both cases, the image is warped so that it conforms to real-world coordinates. Displacements in the photo caused by terrain relief and optical distortion are removed. Slopes are ignored so that the distance measured on a flat field is the same as on a steep incline. The photo is essentially transformed to have the geometric qualities of a map. A digital orthophoto quadrangle (DOQ) is a particular example of an orthophoto that matches the size of a United States Geological Survey 1:24000 quadrangle map (see Figure 5.15).

5.5   Remote Sensing

The field of remote sensing began with the introduction of satellite imagery in the 1960s. Many of the developments in aerial photography and remote sensing were fueled by the Cold War and associated spying operations, particularly of missile installations. The combination of the computer and digital images led to major developments in digital image processing, including photorectification and multispectral pattern recognition.

FIGURE 5.15. Digital orthoquadrangle (DOQ) of Washington, D.C. The scanned image has been rectified with ground control points for both distortion caused by the camera lens and differences in elevation. Source: U.S. Geological Survey (USGS), Department of the Interior.

Remote sensing expanded drastically in the 1970s as extensive digital satellite imagery of the Earth began to be collected. Satellites in a Sun-synchronous orbit captured images on a continuous basis, imaging the same area more than 20 times a year. In addition, electronic scanners had been developed that could capture images in multiple parts of the spectrum. Rather than just taking one image of the Earth, these scanners capture separate images in the blue, green, red, and infrared parts of the spectrum—sometimes even hundreds of separate pictures, or bands, from the same area of the Earth.

While multispectral images provided a new perspective on the Earth, the massive quantities of imagery made it necessary to find automated methods of analysis. This came with development of multispectral pattern recognition that facilitated the recognition of features based on their spectral reflectance. Figure 5.16 depicts the spectral responses for alfalfa, potato, canola, and oat plants that would make these crop types distinguishable under ideal conditions. Differences in reflectance patterns between features at different times of the growing cycle and under different moisture conditions meant that the procedure was flawed; many problems remain in completely automating the recognition of features through multispectral pattern recognition. Nevertheless, the technique can quickly map large areas and is commonly used to create land-use and land-cover maps.

5.6 Summary

Both navigation and aerial photography have influenced the development of cartography. Exploration led to the development of commercial cartography. Furthered by governmental interests, the commercial aspect of cartography has continued to the present. Even in the age of GPS, navigation continues to be a major use for maps.

Images from above, whether taken by airplane or satellite, provided a new perspective on the Earth and a new source of map information. Stereo images made it possible to derive elevation, now acquired mainly through laser technology in the form of LIDAR. To determine land cover and land use, scanners are used to acquire multiple images from different parts of the spectrum that can then be classified based on differing spectral response characteristics of objects.

There has always been a close association between cartography and technology. Gathering, storing, processing, and displaying information about the Earth is not an easy matter leading to a continual search for better methods. In many ways, this aspect of the history of cartography continues today—perhaps at an even more rapid pace.

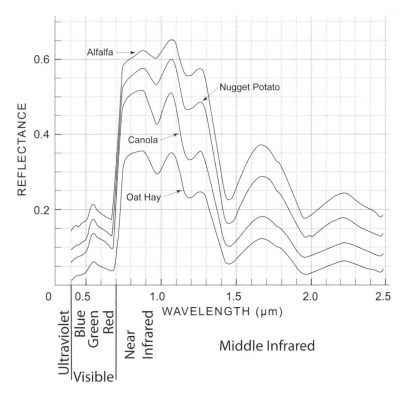

FIGURE 5.16. Spectral response pattern for different crop types over the visible, near, and middle infrared parts of the spectrum. Variations in spectral responses between features is the basis of multispectral pattern recognition.

5.7   Questions

1. What is a disadvantage of computer maps compared to the ancient maps of Mesopotamia?

2. How are Roman road maps similar to contemporary subway maps?

3. Describe the major assumptions made by Eratosthenes in his early calculation of the Earth's circumference.

4. Who was Ptolemy, and why was his book, *Geographia*, so important 1200 years after his death?

5. What inspired the T in O map, and what does it indicate about the didactic value of maps?

6. How were Portolan and Catalan charts significant to the early development of cartography?

7. What do the Marshall Island Sea Charts, not discovered until the latter part of the 1800s, indicate about ancient maps?

8. What is the distinction between the great circle and the rhumb line?

9. For navigation, how was latitude and longitude ultimately determined to a sufficient degree of accuracy, and when?

10. Most flights from New Zealand to London fly through Australia and Dubai. Would this be the shortest distance to fly between these locations?

11. Why is latitude easier to determine than longitude?

12. How was longitude ultimately determined?

13. What were the major contributions of photogrammetry to mapping?

14. How is the stereoplotter used?

15. Describe LIDAR and its applications.

16. How is an orthophotograph created, and why are they so important to GIS?

17. Describe the process of multispectral image classification.

5.8   References

Baily, Francis (1835) An Account of the Rev. John Flamsteed, the First Astronomer Royal. *The Quarterly Review*. London: W. Clowns and Sons.

Clark, John, and Jeremy Black (2005) *100 Maps: The Science, Art and Politics of Cartography Throughout History*. New York: Sterling Publishing.

Davis, W. (1986) The origins of image making. *Current Anthropology* 27(3): 193–215.

Duffy, Trent (2000) *The Clock*. New Canaan, CT: CommonPlace.

Konecny, G. (1985) The International Society for Photogrammetry and Remote Sensing—75 Years Old, or 75 Years Young, Keynote Address, *Photogrammetric Engineering and Remote Sensing, 51*(7): 919–933.

McCarthy, Thomas (Ed.) (2006) *Exploring the Middle Ages*. Tarrytown, NY: Marshall Cavendish.

Sobel, Dava (1995) *Longitude: The True Story of a Lone Genius Who Solved the Greatest Scientific Problem of His Time.* New York: Walker Publishing.

Thrower, Norman J. W. (1972) *Maps and Man: An Examination of Cartography in Relation to Culture and Civilization.* Englewood Cliffs, NJ: Prentice-Hall.

Winkler, Captain (1899) On sea charts formerly used in the Marshall Islands, with notices on the navigation of these islanders in general. *Annual Report of the Board of Regents of the Smithsonian Institution.* (Translated from Marine-Rundschau, Berlin, 1898, pp. 1418–1439.)

Wise, Donald (1976, June) Primitive Cartography in the Marshall Islands. *Cartographica: The International Journal for Geographic Information and Geovisualization, 13*(1).

CHAPTER 6
The Online Street Map

A digital map of a country that is revised constantly will be more useful than any printed atlas, even one updated every year.

—RICHARD WOODWARD IN A *NEW YORK TIMES* REVIEW OF THE *OXFORD ATLAS OF THE WORLD*

6.1 Introduction

No other type of online map has been as important or as significant to our daily lives as the interactive street map. While the major sites like MapQuest, Google Maps, Bing Maps, and Yahoo Maps provide smaller scale maps that do not depict city streets, it is the large-scale versions with streets that make these sites distinctive and useful. Without maps at this scale, it would be impossible to show the locations and addresses and provide directions between places. These online services have transformed how people find where they are, how they travel, and how they view the world.

We have previously examined the tile-based mapping system with its massive data storage requirements. In this chapter, we take a detailed look at these online map services. In particular, we examine the various views offered, how the data are collected and assembled, the underlying map data, and its representation.

6.2 Map Views

In an attempt to attract more users, online map providers are in heated competition to provide better, faster, and ever more powerful services. All major mapping sites provide the street/road map and a so-called satellite view. Some offer an additional terrain map and digital globe or "Earth" view. These views represent the base maps on which other information is placed. The various offerings can thus be separated into those that are stand-alone—map, satellite, terrain, and digital globe—and those that are transparent overlays. These overlays would have little or no meaning if displayed by themselves. Figure 6.1 lists the different views and overlays produced by the major map providers.

6.2.1 Map

The basic street map is the most functional of all the views provided by these services. The map is provided at over 20 levels of detail (LOD). Each map at a particular LOD is created from an underlying vector database consisting of points, lines, and areas. Once all the elements of the map have been properly placed, it is converted into a matrix of pixels. In the process, anti-aliasing is performed by automatically shading pixels around sharp edges to soften any stair-step effect. After the map is placed in a matrix that can be millions of pixels on each side, it is divided into tiles—usually 256×256 pixels (see Figure 6.2). This process is repeated for every LOD. The size of the matrix at the 20th zoom level would be 256,000,000 by 256,000,000 pixels, for a total of 65,536,000,000,000,000 pixels (6.5536×10^{16})—nearly 66 gazillion.

6.2.2 Satellite

Selecting the satellite view provides access to multiple images with varying resolutions (ground area per pixel). Most higher resolution imagery is from lower-altitude aerial photography. In these cases, the "satellite" description is somewhat misleading. Oblique air photo images may also be offered providing the user with a bird's eye view. These images may also be acquired by aircraft.

Making such high-resolution images available through the Internet was originally done as an experiment initiated by Microsoft™ (Barclay et al. 1998). The company wanted to test the "scalability" of the Windows NT operating system, SQL database, and server software. Scalability refers to how the system responds to

Stand-alone Views

Map - basic map view
Satellite- images and air photos
Bird's Eye (45°) - oblique photos
Terrain - shaded relief
Earth - digital globe
Street View / Streetside - panoramas

Scale-dependent Views/Overlays
Bicycling - bicycle paths

Overlays

Traffic - current traffic conditions
Photos - location of photographs
Weather - Current conditions for selected cities
Labels - text labels in English or other language
Transit - public transportation options
Webcams - location of webcams
Videos - location of videos
Wikipedia - Wikipedia pages on locations
Gas prices - updated gasoline prices
Real Estate - property for sale

FIGURE 6.1. Stand-alone and overlay views offered by Google Maps. Overlays consist of transparent layers that are placed on top of one of the stand-alone views. At more detailed scales, the bicycling layer becomes a stand-alone view under the Map option.

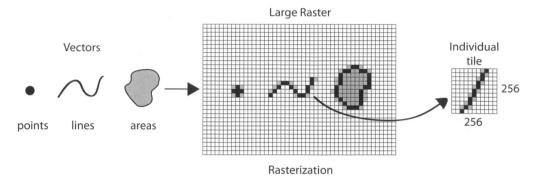

FIGURE 6.2. The process of making map tiles. A database of points, lines, and areas is rasterized into a large matrix that is subsequently divided into 256 × 256 square tiles.

increases in usage. To find a data set for testing, it asked the United States Geological Survey (USGS) for its largest data set. In response, the USGS supplied all of its public domain orthophoto quadrangles and scanned topographic maps, the latter known as digital raster graphics (DRG). Most of the database consisted of higher-resolution imagery of urban areas. The site also included imagery from other parts of the world provided by the Russian Federal Space Agency. Starting out in 1997 with 5 terabytes of image data compressed into a 1-terabyte database, TerraServer was the largest online database at the time, equal to all HTML pages in existence. An associated gazetteer included locations of 1.1 million places in the world to help the user find a particular location. Although outdated and based on older client/ server technology that does not incorporate interactive panning or zooming, the site continues to operate at http://msrmaps.com (see Figure 6.3).

Contrary to what many mistakenly believe, the images provided by online mapping services are never current. The imagery could be anywhere from a few months to a few years old. The major providers put their satellite view together in similar ways. Images for Bing and Yahoo are provided from the National Aeronautics and Space Administration (NASA), the USGS, and a variety of companies, including I-cubed, Harris, Earthstar Geographics, and Pictometry. In some cases, they might only rely on three different images for all 20 levels of detail. To create the smaller-scale images, larger-scale images are generalized by averaging pixel values (see Figure 6.4).

Google stitches together imagery from multiple sources, sometimes combining satellite images and air photos at a certain LOD. Suppliers include free government sources such as the USGS, the USDA Farm Service Agency, and commercial entities, including Terrametrics, Digital Globe, and GeoEye. Many cities have donated their orthophotographs to Google, resulting in extremely detailed images of those urban areas. These high-resolution orthophotos are acquired by cities on a regular basis at considerable expense to assist with their GIS database development. Google has also partnered with GeoEye, the world's largest space-imaging corporation, giving exclusive online use of imagery from the GeoEye-1 satellite. Although the satellite provides black and white images at 41 cm (16 in) resolution per pixel, this imagery

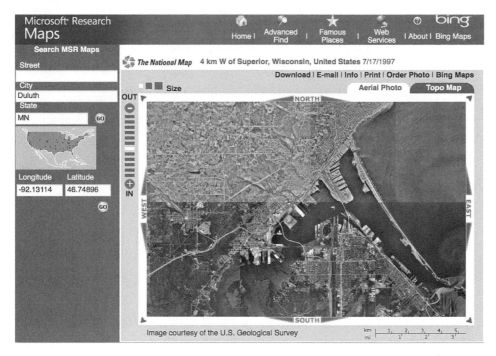

FIGURE 6.3. The original Terraserver is now operated under MSRmaps.com. Terraserver is an early website that displayed imagery provided by the U.S. and Russian governments, and scanned versions of U.S. topographic maps. Copyright 2013 Microsoft Corporation and its suppliers.

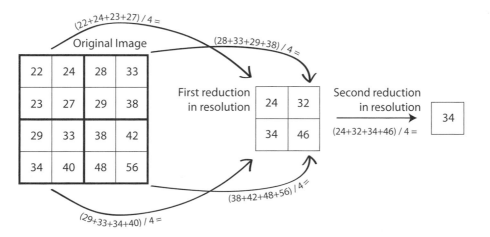

FIGURE 6.4. Reduction in image resolution. The numbers represent gray scale values. A high-resolution image can be manipulated for use at a smaller scale by averaging pixel values for the 2×2 squares.

is only available to the U.S. National Geospatial-Intelligence Agency. Google is provided imagery at 50 cm (20 in) per pixel but relies mostly on lower-resolution color images that are 1.6 m (5.4 ft) per pixel.

Security concerns have led to some imagery being intentionally obscured or pixelated as, for example, with nuclear power plants in the United States (see Figure 6.5). The power plants are still visible but at a reduced resolution with obvious pixelation. The intentional degradation of imagery has probably brought even more attention to the places being depicted. A Wikipedia page includes a list of all such degraded sites.

6.2.3 Terrain

In addition to the map and satellite views, Google also offers a terrain map that shows elevation using a shaded-relief mapping method. Slopes facing to the southeast receive the darkest shading, while those facing the northwest are "illuminated" (see Figure 6.6). The terrain map is only available at 15 levels of detail and is not frequently updated. Larger scale versions include contour lines in addition to the shaded relief.

Not long ago, shaded relief mapping could only be done by hand with an airbrush. The hand-held airbrush uses compressed air to transfer small droplets of ink onto paper. Moving the device repeatedly over the same area would result in a darker shading. Attempts to automate the process began in the 1970s (Brassel 1973). The automated process starts by calculating the slope and aspect for each pixel based on a digital elevation model, a matrix of elevation values. Slope is the

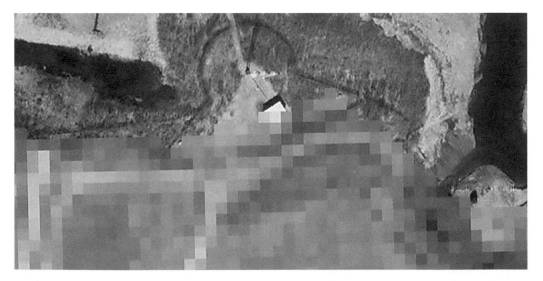

FIGURE 6.5. A Google satellite view on the boundary of the Seabrook Station nuclear power plant in New Hampshire, approximately 40 miles north of Boston. Part of the image has been intentionally degraded out of concerns for national security. This image degradation has brought even more attention to these locations. The Wikipedia page "Satellite map images with missing or unclear data" includes a list of many of these sites. Copyright 2013 Google.

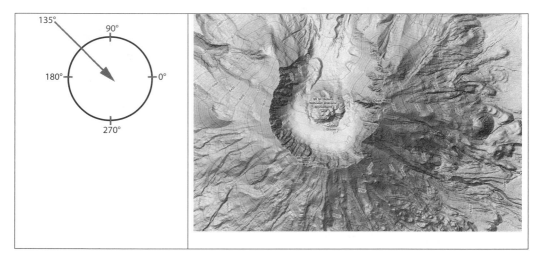

FIGURE 6.6. A Google Terrain map of Mount St. Helens in Washington State. A 3D effect is achieved by shading southeast-facing slopes darker and northwest-facing slopes lighter. The map is constructed from U.S. government digital elevation data. Copyright 2013 Google.

amount of change in elevation over a certain distance, and aspect is the direction of that slope as expressed in degrees between 0° and 360°. A gray value is then calculated for each pixel based on the direction and amount of slope. For reasons that are not entirely clear, shading the southeast slopes with a darker shading results in the best impression of elevation.

6.2.4 Overlays

Overlays are map layers that are placed over a map, satellite, terrain, or digital globe view to provide some type of additional information. Overlays can be done in two different ways. The first is as a series of points or flags that are placed on top of the map. Google, for example, uses the upside-down teardrop as a standard point symbol to indicate a location.

The second type of overlay is a series of tiles that have the same size and dimension as the base tiles, but these overlay tiles are made transparent so that you can see the tile underneath. Whatever part of the tile is opaque becomes superimposed on the underlying map (see Figure 6.7). Overlaying transparent tiles in this way is faster than overlaying individual points or lines. Most views are transparent tile overlays with a particular theme such as traffic, webcams, or photos.

6.2.5 Street View

Street View provides panoramic views along streets and was added as an option to Google Maps in 2007. Similar services have since been introduced by other map providers but with much less coverage. Initially available only for certain cities, Street View has since expanded to all major roads, major streets in populated areas of the United States, and at least 30 countries. The implementation is an ongoing

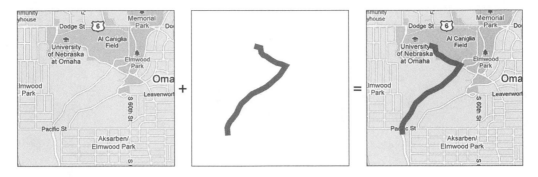

FIGURE 6.7. The tile overlay method. A map tile on the left is overlaid with a transparent PNG file with an opaque line. Combining the two tiles produces the display on the right. The traffic, transit, and most other types of maps provided by online services are the result of such an overlay. Copyright 2012 Google.

process, and some streets have already been imaged multiple times. The coverage is inconsistent, with some areas having more streets included than others. Dropping the Pegman icon on the map along an imaged street accesses the series of panoramas for that area. Movements can then be made forward or backward along the street or road, but the system does not yet allow for uninterrupted movement. It is impossible, for example, to create the effect of smoothly driving down a street.

Street View images are acquired with a special multiple-camera device that is mounted on top of a car (see Figure 6.8; Image search: Google Street View car). The device also includes a GPS to capture the location of the images and a laser scanner to measure distance. All images, locations, and distances are captured on the hard

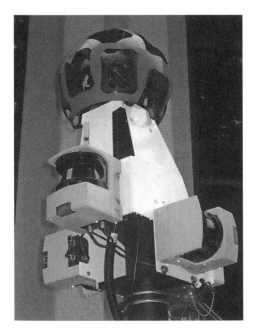

FIGURE 6.8. A Google Street View imaging device. A series of wide-angle cameras take pictures at specific intervals that are subsequently stitched together for a panoramic view. The device also includes a GPS to determine the location of the pictures and laser range scanners for distance calculations. Image courtesy of Kowloonese.

drive of a computer in the car. In postprocessing, the images are "sewn" together to create the panoramic views. The position provided by the GPS helps locate the panorama image along a street in Google Maps and Google Earth views.

Soon after the introduction of Street View, privacy concerns began being voiced. In response to these concerns, the product manager for Google Maps defended Street View by saying "All of these images are taken on public streets. It's exactly what you could see walking down the street" (Weismann 2007). Initially, faces of people and license plates on cars were visible on the images. It was noted that this was done without user consent. A Canadian initiative in 2007 prompted Google to obscure faces and license plate numbers, a process that began the following year.

The blurring of faces and license plate numbers is now done automatically for all Street View images. Requests can also be made to Google through an online form to blur an entire person, house, or car. Some have voiced concerns that this blurring has degraded the quality of the entire Street View imagery. A variety of Street View images have also been found that show such things as public urination, nudity, street fights, and apparent drug deals, and a number of websites have cataloged such images (search: top Street View images). Users can also ask for the removal of such inappropriate content.

Street View has been even more controversial in countries outside of North America. In Japan, for example, the height of the cameras was lowered in 2009 after concerns were expressed that the images were taken above the height of fences and walls intended to maintain privacy (search: Street View under fire in Japan). In England in April of 2009, a Street View vehicle was forced out of a prosperous neighborhood by residents. Citing recent burglaries, one homeowner said, "If our houses are plastered all over Google, it's an invitation for more criminals to strike" (Kennedy 2009). In Austria, an elderly gentleman attacked a driver of a Google Street View vehicle with a pickaxe (*Austrian Times* 2010).

German Consumer Protection Minister Ilse Aigner referred to Street View as a "photo offensive" that is "nothing less than a million-fold violation of the private sphere." She went on to say that there was "not a secret service in existence that would collect photos so unabashedly" (*Der Spiegel* 2010). The city-state of Hamburg demanded that the company not archive raw images of people, cars, and property. While Google pixelates the license plates and faces, it keeps the raw data. Germans, however, can ask the company to delete even these raw images.

In October 2010, in an investigation initiated by European privacy officials, it was determined that Google's Street View device was also capturing Wi-Fi signals as they passed through neighborhoods, a violation of privacy laws in some countries (Halliday 2010). It was claimed that Google had captured email passwords and entire emails using receivers that were concealed in the Street View vehicles. After numerous protests, Google ended Wi-Fi data collection by Street View vehicles (Miller 2010). Wi-Fi signals are used to help identify location with mobile devices by matching a particular configuration of signals.

The pace at which Street View images are captured has slowed in countries that have formally objected to this form of data collection. Google has imaged parts of Europe, Japan, Taiwan, Hong Kong, Australia, South Africa, and Brazil. It is also improving the resolution of the cameras being used. Google has managed to install its bulky camera on the back of a bicycle to take images along bicycle paths. The amount of time it takes to drive or bike all of the possible streets, roads, and bike paths means that updates will not be completed for many years.

6.3   Map Data

Making detailed street maps available through the Internet allowed the public to clearly see errors in the maps. Almost everyone could look at the map of their neighborhood and see that it was inaccurate or out of date. Creating accurate, large-scale maps has become a major concern, with the underlying goal of improving the accuracy of maps to garner the confidence of the map user.

6.3.1   Government and Private-Sector Street Maps

The origins of most digital maps are associated with early government initiatives. In the United States, the major government entities involved were the USGS and the U.S. Census Bureau which, in the 1980s, cooperated in the creation of the TIGER database (topologically integrated geographic encoding and referencing system). TIGER is a combination of the rural road information from the USGS and the urban street data from the Census Bureau. Released freely with the 1990 census, TIGER was the most accurate map of the entire country.

Several companies began offering car navigation services based on government map data. An early car navigation system developed by ETAK prior to the introduction of GPS (global positioning system) used dead reckoning to determine the current position. Given a starting point and using a database of streets stored in the car, the system would determine the direction being traveled with a digital compass and the distance traveled along each street with sensors on two wheels. Even with this technology, the car would become easily lost. A complex procedure then needed to be performed to reorient the system. The company stopped selling its navigation product after a short time.

ETAK had made a considerable investment in its database of streets, adding more detail to the government-supplied map. Attributes such as the direction of travel for one-way streets and the location of business had been added to the map data. The company reinvented itself as a supplier of street maps.

Digital map suppliers are in a major competition to have the best, most up-to-date, and most useful database of streets in the world. The major current suppliers of street data are TeleAtlas (formerly ETAK) and NavTeq—now owned by the mobile phone company, Nokia. They have supplied data to all of the major online map providers. Both companies employ a small army of drivers that traverse the streets in GPS-equipped cars. While initially dependent on these companies, Google has since developed its own map database.

6.3.2   Crowdsourcing Maps

The idea of asking map users to update and correct the online map was voiced from the very beginning, even before websites like Wikipedia confirmed that crowdsourcing was a viable model for gathering information. Google Map Maker was started in 2008 to allow users to draw features such as roads directly onto the map, but the system was initially only implemented for certain countries where Google did not have adequate map data.

All contributions to Map Maker are checked by more experienced moderators. The problem is that these moderators may not have any place-specific knowledge and therefore cannot adequately monitor user-supplied additions. To evaluate the quality of the moderation, the author of this book inserted a nonexistent "Star Cinema" on the outskirts of Multan, Pakistan, in 2008. The addition was approved by a moderator and remained on the Google Map of Multan (see Figure 6.9). The cinema also became part of the online yellow pages and a list of cinemas for the city (search: Star Cinema, Multan, Pakistan). The experiment would seem to indicate that moderation of map content does not work as well as moderation of content on sites like Wikipedia where there is a larger "expert" audience.

OpenStreetMap (OSM) is a major online map database that was entirely developed through collaborative means. Begun in 2004, OSM gathers data from multiple sources, including GPS data from users and donated aerial photography. Registered users can both upload GPS track logs and edit vector data using the editing tools provided. There are over 200,000 registered users. OSM maps are royalty free.

The amount of street data in OSM has expanded dramatically. The two maps in Figure 6.10 show the road network around São Paulo, Brazil, in 2007 and 2011. Much of the added data was acquired from GPS traces that were submitted by volunteers. The national mapping agency of Brazil, controlled by the military, has not made its map data freely available.

Providing users with accurate and up-to-date maps will remain a continuing problem. Although significant progress has been made in crowd-sourcing map information, this may not represent the ultimate solution in making an accurate map of the world—although it is difficult to conceive of a workable alternative.

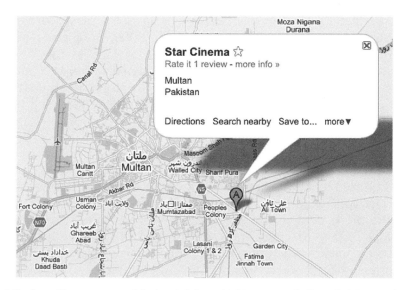

FIGURE 6.9. Star Cinema was added to Multan, Pakistan, with Google Map Maker in 2008. The addition was approved by a moderator, but the theater does not exist. In Google Maps, search for "Star Cinema, Multan, Pakistan." Copyright 2013 Google.

| 2007 | 2011 | Sample GPS Trace |

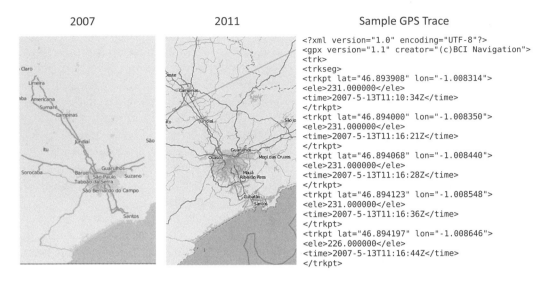

```
<?xml version="1.0" encoding="UTF-8"?>
<gpx version="1.1" creator="(c)BCI Navigation">
<trk>
<trkseg>
<trkpt lat="46.893908" lon="-1.008314">
<ele>231.000000</ele>
<time>2007-5-13T11:10:34Z</time>
</trkpt>
<trkpt lat="46.894000" lon="-1.008350">
<ele>231.000000</ele>
<time>2007-5-13T11:16:21Z</time>
</trkpt>
<trkpt lat="46.894068" lon="-1.008440">
<ele>231.000000</ele>
<time>2007-5-13T11:16:28Z</time>
</trkpt>
<trkpt lat="46.894123" lon="-1.008548">
<ele>231.000000</ele>
<time>2007-5-13T11:16:36Z</time>
</trkpt>
<trkpt lat="46.894197" lon="-1.008646">
<ele>226.000000</ele>
<time>2007-5-13T11:16:44Z</time>
</trkpt>
```

FIGURE 6.10. The São Paulo area of Brazil depicted in OpenStreetMap in 2007 and 2011. The listing on the right shows an example GPS trace submitted by a user. The Brazilian government has not made its data freely available, leading map users to create their own online map. Copyright 2013 OpenStreetMap contributors CC-BY-SA.

6.4 Rendering

The process of making a map from a digital database is called rendering. It is the most complicated aspect of the mapmaking process because it involves making a legible and well-designed map that looks like it could have been made by a trained cartographer. The major companies are vying to make a map that is the most accepted by the public, the map that people will see as the most "correct." Companies continue to update the look of their maps. In 2009, Google re-rendered all of its maps using a new map design (see Figure 6.11).

Rendering involves taking the point, line, and area data from the map database and converting it to a usable map. Since the data are stored in three-dimensional coordinates of latitude and longitude, the first step in the rendering process is projecting the data into a two-dimensional form.

6.4.1 Projection

The projection used by all major online map providers is a slight variation of the Mercator projection called the Web Mercator (see Figure 6.12). Gerhard Mercator developed the original Mercator projection for use by sailors in 1569. The map shows lines of constant direction, or loxodromes, as straight lines. It could therefore be used with a compass. The projection produces severe distortion in the northern and southern latitudes. This distortion is often referred to as the "Greenland Problem" because Greenland, positioned at a high latitude, is depicted larger than Africa when in actuality Africa is 14 times larger than Greenland. Australia appears much smaller than Greenland but is nine times bigger.

Although one would expect the scale bar to change by selecting a different level of detail, with the Mercator projection it also changes by simply moving north or south. Figure 6.13 shows the scale bar for various latitudes. The scale varies so much that it is impossible to make an accurate distance calculation other than exactly along a line of latitude.

The Mercator projection shows angles correctly, at least over small areas, resulting in the more correct depiction of shape. The distortion of distance and area that is apparent at smaller scales is not noticeable on street-level maps that most people use. In addition, the rectangular projection works well for computer monitors. There are other rectangular projections that depict the size of landmasses correctly. The extent of the area distortion is apparent with Google Maps by switching between the map and digital globe (Earth) views. Land areas are shown in correct relative proportion with the digital globe but it does not show the entire world in a single view.

It is unlikely that map services will cease to use the Mercator projection, and so we are left to work around its limitations. For example, the smaller scale depictions that show most of the world as depicted in Figure 6.12 should simply not be used, except when illustrating the level of distortion. In addition, any comparison between two maps at the same level of detail between two different latitudes

FIGURE 6.11. In 2009, Google introduced a new look to their maps. The new look required rerendering all map tiles. Copyright 2013 Google.

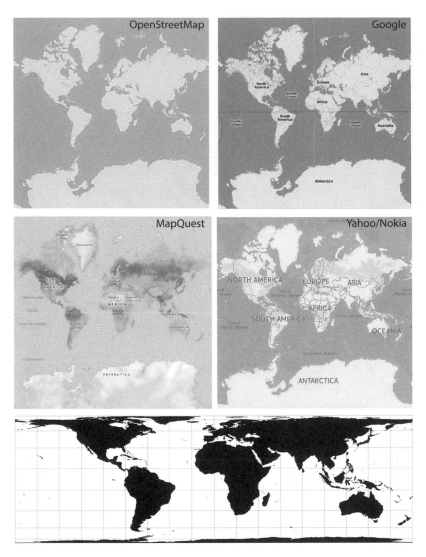

FIGURE 6.12. All major online map providers use the spherical Mercator projection. The projection severely distorts the area of landmasses in the higher latitudes. The equal-area cylindrical projection on the bottom shows areas accurately. Copyright 2013 OpenStreetMap contributors CC-BY-SA.

should be avoided. Any size comparison between maps at different latitudes is also problematic.

6.4.2 Symbolization and Labeling

While often starting with the exact same database of points, lines, and areas, each major map provider symbolizes its maps differently. Figure 6.14 shows four renderings of the same neighborhood from the major providers—Google, Bing, Yahoo,

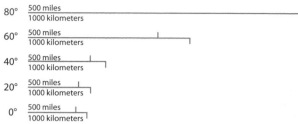

FIGURE 6.13. Scale varies continually by latitude on a Mercator projection. The five scale bars depicted here are from 0°, 20°, 40°, 60°, and 80° in north or south latitude.

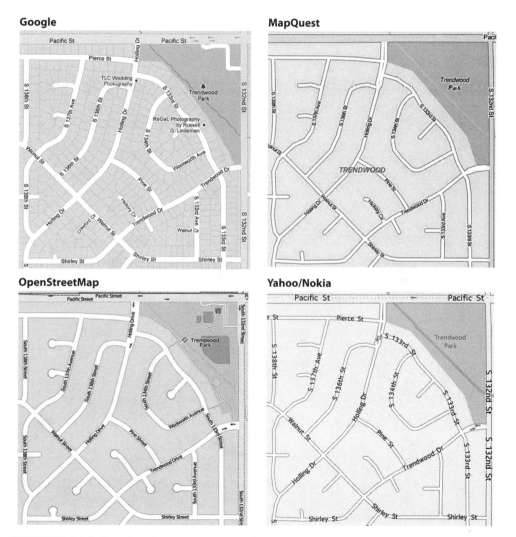

FIGURE 6.14. Rendered maps from Google Maps, MapQuest, OpenStreetMap, and Yahoo/Nokia. MapQuest and Yahoo/Nokia are based on data from NavTeq. Google has integrated parcel data showing land ownership boundaries from data provided by the city. Copyright 2013 (clockwise from top left): Google, MapQuest, Yahoo/Nokia, OpenStreetMap contributors CC-BY-SA.

and MapQuest. Notice the differences in the representation and labeling of the streets.

In contrast to the commercial online map providers, OpenStreetMap allows users to download the underlying vector data. This has made it possible for other organizations to render their own tiles. CloudMade has become the largest provider

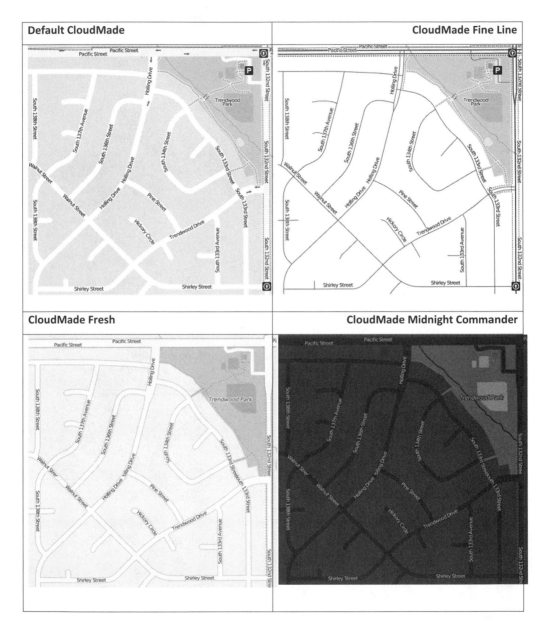

FIGURE 6.15. Rendered maps from CloudMade based on vector data from OpenStreet-Map. CloudMade provides a large variety of differently rendered maps and charges based on the number of tiles that are downloaded. Copyright 2013 CloudMade—Map data 2013 OpenStreetMap contributors CC-BY-SA.

of these alternative mapping tiles. They provide a number of different map designs, with each given a name such as Fine Line, Fresh, Midnight Commander, No-Names, and Pale Dawn (see Figure 6.15). The company provides the first 500,000 tiles for free and then charges US $25 for 1 million downloaded tiles.

6.4.3 Cultural Differences in Mapping

One of the more interesting aspects of the map-rendering process is designing the map to conform to cultural norms. Figure 6.16 shows the difference between the Google map of San Francisco and London. The London map has been made to look more like a map designed in Europe with colors that match the paper maps from the area. Online map providers realize that in order to make their maps acceptable, they must adapt to the look of maps in that part of the world.

6.4.4 Map-Labeling Decisions

Besides the complicated process of label placement, decisions need to be made about which labels will actually be placed on the map. The so-called *Baltimore Phenomenon* is a good illustration of the compromises that need to be made when labeling a map—decisions that are caused by a competition for limited map space.

The Baltimore metropolitan area has a population of over 2.7 million. It is by far the largest city in the state of Maryland, and yet it is not depicted on many maps of the United States because of its close proximity to Washington, DC, and

FIGURE 6.16. Map providers render their maps to look like the paper maps they are replacing. The map of London has been symbolized to look more like a map from Europe. Copyright 2013 Google.

the competing problem of labeling the oddly shaped state of Maryland. Figure 6.17 depicts the east coast of the United States at a relatively large scale. Baltimore, located close to the lettering for Maryland, is missing. By contrast, there is a large amount of space in the middle of the United States on a map at the identical scale to show the locations of cities that are only a fraction of the size of Baltimore, such as Norfolk, Nebraska, with a population of just 24,000.

6.5 Summary

Map users quickly embraced the online interactive street map. The map form has evolved dramatically since its introduction by MapQuest in 1996. It has revolutionized the way people use maps and the way they navigate the world.

From the standpoint of business and long-term sustainability, the introduction of Google Maps in 2005 is perhaps even more important. Google showed that the map represented not only a tool for navigation but a way to search for information. All features in the world that have a spatial component could be placed on the map and found in this way. As a result, the map became another way to display the results of a search engine. This may represent a more sustainable model for the continued free access to maps.

The number of map views, both the stand-alone and overlays, will certainly increase. All manner of information will be overlaid on these maps, and they will become a collective repository of information about the world.

FIGURE 6.17. Baltimore, located between Washington, DC, and Philadelphia, has a population of 2.7 million. It is not included on the Google Map on the left because of a competition for limited map space and the need to label the state of Maryland. In contrast, many cities are depicted in the map from the identical Google version of the less populated Midwest. The largest city is Omaha with nearly 2 million less people than Baltimore. Cities like Norfolk, Nebraska, with a population of only 24,000 people are included on the map. Copyright 2013 Google.

6.6 Exercise

Everyone is an expert on the area around where they live. Produce a gallery of maps of the area where you live from all the online map service providers and comment on both the accuracy of the maps and the method of representation. Use the templates from Chapter 4.

6.7 Questions

1. What is the distinction between a stand-alone view and an overlay with online mapping systems?

2. What does Google's bicycle map indicate about the future development of maps through the Internet?

3. How are points, lines, and areas defined as vectors rendered into a raster representation at 20 levels of detail?

4. What types of images are used to assemble the Satellite view?

5. Describe the spatial and spectral resolution of images from the GeoEye satellite.

6. Explain how Google's terrain view map is constructed.

7. How are Street View images collected and processed?

8. Why would people believe that Street View images are an invasion of privacy?

9. What does the addition of the Star Cinema in Multan, Pakistan, indicate about collaborative mapping?

10. What is a GPS trace?

11. What does rendering refer to in online mapping?

12. Describe some differences in symbolization and labeling between the major online map sites. Which do you think is the best?

13. What is the Baltimore Phenomenon with maps?

14. What efforts are online map suppliers making to adapt to local map styles?

6.8 References

Austrian Times (2010, Aug. 4) "Google Street View" Driver Escapes Axe Attack.

Barclay, Tom, et al. (1998) Microsoft Terraserver. *MSDN Library*. [http://msdn.microsoft.com/en-us/library/aa226316%28v=sql.70%29.aspx]

BBC News (2009, May 14) Street View Under Fire in Japan.

Brassel, K. (1973) *Modelle und Versuche zur automatischen Schraglichtschattierung*, Buchdruckerei E. Brassel: Klosters, Switzerland.

Der Spiegel (2010, Feb. 8) Million-Fold Violation of the Private Sphere: German Minister Takes on Street View.

Halliday, Josh (2010, Oct. 20) Google Street View Broke Canada's Privacy Law with Wi-Fi Capture. *The Guardian.*

Kennedy, Maev (2009, Apr. 3) Coy Village Tells Google Street View "Spy" to Beat a Retreat. *Guardian.*

Miller, Claire Cain (2010, Oct. 27) A Reassured F.T.C. Ends Google Street View Inquiry. *New York Times.*

Weismann, Robert (2007, Dec. 11) Get Ready for Your Close-Up. *Boston Globe.*

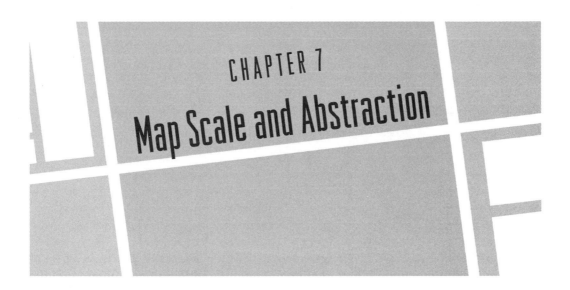

CHAPTER 7
Map Scale and Abstraction

Art is a lie which makes us realize the truth.

—Pablo Picasso

7.1 Introduction

Cartographic abstraction includes the processes of selection, generalization, and symbolization. Generalization involves an entire series of operations including classification, simplification, and exaggeration. All of these aspects of *reality transformation* control the mapmaking process. Cartographic abstraction depicts the world in a way that is more useful than looking at the world itself. We begin by examining map scale, an aspect of selection and the single most important element controlling the amount of cartographic abstraction.

7.2 Map Scale

Map scale is a ratio of map distance to ground distance. It can be represented in three forms: (1) a representative fraction (RF), as in 1:24,000; (2) a verbal scale, either in the English or metric measurement systems, where the units need to be verbalized, as in 1 in:10 mi or 1 cm:10 km; or (3) a graphic or bar scale, where a line is associated with a distance (see Figure 7.1).

For conversion purposes, the representative fraction is the most useful. It simply states that one unit of measure on the map represents a certain number of that same unit on the surface of the Earth. This would mean that on a map with a 1:50,000 RF, 1 cm on the map is 50,000 cm on the surface of the Earth. It also means that 1 in on the map is 50,000 in on the ground. We could also say that the

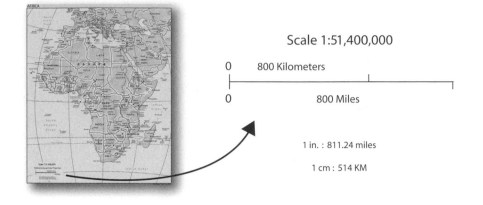

FIGURE 7.1. Example of a representative fraction (RF) and a graphic scale from a map of Africa. A verbal scale is not provided but can be calculated based on the RF. Because the map has been altered in size, only the bar scale is still correct. Map of Africa courtesy of University of Texas Libraries.

length of a pencil on the map would take 50,000 identical pencils placed end to end on the Earth. The scale is an RF, as long as the unit of measure is the same on both sides of the expression, even if written in crude form as 10:100 or 5.4:1078.6.

With the verbal scale, one needs to "put into words" or specify exactly what the units represent because they differ, as in 1 cm:10 km. The most commonly used units in the metric system are centimeters, meters, and kilometers. There are 100 cm in a meter and 1000 m in a kilometer. As a result, there are 100,000 cm in a kilometer. The 1 cm:10 km verbal scale is identical to an RF of 1:1,000,000 (10 × 100,000).

English measurements, based on units that date to medieval England, include inches, feet, and miles. There are 12 inches in a foot and 5280 feet in a mile. This means that there are 63,360 inches in a mile (12 · 5280). This system of measurement has been abandoned in almost all countries, even in those that were once a part of the British Empire. The United States, Myanmar (Burma), and Liberia are the only countries where it is still in official use. For many years, the country of Myanmar intentionally isolated itself from the rest of world.

Conversions between metric and English map scales would use the representative fraction. This means that a scale of 1 cm:10 km would be converted first to an RF and subsequently to the English verbal. A 1 cm:10 km verbal scale is the same as an RF of 1:1,000,000. This RF can be interpreted as 1 inch on the map is to 1 million inches on the ground. Dividing 1 million inches by 63,360—the number of inches in a mile—results in an English verbal scale of 1 in:15.78 mi.

As the simplest of all scales, the graphic scale is a line with an associated distance. It is the least accurate of all scales because its interpretation is based on measurement and it is often only an approximation of the distance that is being represented. The advantage of the graphic scale is that it changes in size as the map changes size, with any enlargement or reduction of the map also reflected in the length of the bar. By contrast, the numbers in the numeric scales, both RF and verbal, remain the same after any enlargement or reduction and therefore become meaningless once the map has changed in size. Because it is impossible to maintain the size of maps presented through the Internet, the graphic scale has become the major way of indicating scale on a map.

With proper measurement, the graphic scale can be used to determine the numeric scale of the map. For example, if the line for the graphic scale represents 10 miles and it is measured to be 1.5 inches, then the verbal scale of the map is 1.5 in:10 mi. This type of expression is referred to as a crude expression of scale because the numerator is not equal to one. It can be converted easily by dividing both sides by the numerator. The result is: 1 in:6.667 mi. To convert this verbal scale to an RF, we would multiply 6.667 inches by 63,360. The representative fraction for this map would be 1:422,421.

A similar conversion can be done in the metric system. If the graphic scale measures 2.3 cm and represents 10 km, then the crude verbal metric scale would be 2.3 cm:10 km. The proper verbal scale would be 1 cm:4.34783 km (10/2.3). To convert to a representative fraction, we multiply 4.34783 by 100,000 to determine the number of centimeters on the ground. Rather than typing these numbers into a calculator, the multiplication can be done by moving the decimal place five places to the right, giving us an RF of 1:434,783. Figure 7.2 summarizes the most common numbers used in scale calculations.

7.2.1 Scale and Maintaining the Size of the Map

The two numeric scales—RF and verbal—are dependent on the map itself remaining a consistent size. They work best when maps are printed on a plastic material that does not change in size. The numeric scales are mostly used with paper maps, although even paper changes in size as a result of differences in humidity. A paper map increases slightly in size during the humid summer months and becomes smaller again during the drier winter months. The change is small but measurable and would influence the calculation of distance.

The small changes in the size of the paper map as a result of differences in humidity are insignificant compared to how the map changes in size when

Important numbers for calculations of map scale

Metric System	English System
100 - cm in a meter	12 - inches in a foot
1000 - meters in a km	5280 - feet in a mile
100,000 - cm in a km	63,360 - inches in a mile
100 hectares = 1 sq. km	640 acres = 1 sq. mile

2.54 cm = 1 inch
1.609344 km = 1 mile
1 hectare = 2.47105381 acres

FIGURE 7.2. Numbers used in calculations of map scale and related conversions.

displayed by computer. For example, a map shown by projector on a screen is not the same size as the map on the monitor of the computer. The size of the map is also altered by such variables as the resolution setting of the monitor and its dimensions. Therefore, the numeric scales are misleading and essentially useless with maps distributed by computer, and they are rarely used. One should treat with suspicion any RF or verbal scale on a computer map. The only expression of scale that is still valid with maps distributed through the Internet is the graphic scale.

7.2.2   Small-Scale and Large-Scale Maps

Terms such as small scale and large scale are often confusing. To understand these terms, consider scale as a fraction with a numerator and a denominator. If viewed as a fraction, both 1/25,000 and 1/100,000 are small numbers, but it should be apparent that 1/100,000 is much smaller than 1/25,000. Therefore, 1:100,000 represents a smaller scale map than 1:25,000.

Another way to understand the difference is that a map covering a relatively large area has a small scale, while one covering a relatively small area is a large-scale map. Since the terms are relative, there is no sharp division between the two. Thus, a map of scale 1:100,000 is a large-scale map when compared to 1:1,000,000 but is small scale when compared with one of 1:25,000.

As scale decreases, the amount of area coverage increases but the amount of detail decreases. Cartographers reduce detail by selectively removing features from a larger scale depiction. The amount of detail shown depends on several factors including the amount of space that is available and the intended use of the map. The selection of features is a central component of the cartographic generalization process.

7.2.3   Distance Measurements

Typical calculations with scale involve finding the ground distance for a certain distance on a map. As an example, a standard topographic map of the United States has a scale of 1:24,000. If two points measured on the map are 3.2 cm apart, the ground distance between them is 76,800 cm ($3.2 \times 24,000$). Dividing by 100,000 (the number of cm in a km) gives us 0.768 km.

To determine the miles on the ground, we convert the 3.2 cm to inches by dividing by the conversion factor of 2.54 cm to the inch. The map distance is therefore 1.2598 inches. Multiplying this number by the scale of the map, 24,000, gives us a ground distance of 30,236 inches ($1.26 \times 24,000$). Dividing by 63,360 gives us 0.4772 miles. A summary of these calculations is presented in Figure 7.3.

Multiplication with the RF can lead to very large numbers, often into the billions. To avoid this problem, it is sometimes best to convert immediately to the verbal scale equivalents. The conversion to the metric verbal between cm and km is always easier because it simply involves moving five decimal places. For a 1:24,000 map, the metric verbal scale is 1 cm:0.24 km (24,000/100,000). To find the English verbal scale involves dividing by the number of inches in a mile. The corresponding English verbal scale is 1 in:0.3787 mi (24,000/63,360). It is important to note

Metric System

1) 3.2 cm x 24,000 = 76,800 cm

2) 76,800/100,000 = 0.768 km

3.2 cm

English System

1) Convert distances to inches:
3.2 / 2.54 = 1.26 in

2) 1.26 in x 24,000 = 30,236 in

3) 30,236 / 63,360 = 0.4772 mi

Check if 0.4772 mi x 1.61
(the conversion factor
between miles and KM)
is about 0.768 km.

Scale of map: 1:24,000

FIGURE 7.3. Calculations to determine the length of a line on a 1:24,000 map.

that calculations using the verbal scale may be inaccurate because of rounding errors.

To calculate the ground distance in kilometers for 3.2 cm on a 1:24,000 map, one simply multiplies 3.2 by 0.24 to get 0.768 km. After finding that 3.2 cm is equal to 1.2598 in using the 2.54 conversion factor, the distance on the ground would be 1.2598 × 0.3787 or 0.4770 mi. Notice how this number is different from the 0.4772 calculated using the RF approach. This discrepancy is caused by the rounding of both the centimeter to inch conversion (1.2598) and the inches to mile scale conversion of 0.3787.

The measurement of distance on maps generated by computer is often assisted by a distance-measuring tool. The user clicks on two positions, and the distance is automatically calculated. More commonly, the computer provides the distance between two locations along a certain route. This distance is not the actual distance as the "crow flies" between the two locations, but the distance that we travel along roads. This distance cannot be used for calculations of map scale.

7.2.4 Converting Scale

Converting between the three different representations of scale—RF, verbal, and graphic—is an important skill for anyone working with maps. The conversion between RF and verbal involves multiplication or division by the two conversion factors, either 100,000 for the metric system or 63,360 for the English system. This would mean that a 1 cm:2.2 km verbal scale has an RF of 1:220,000 (2.2 × 100,000). The English verbal scale as calculated from the RF is 1 in:3.472 mi (220,000/63,360). A map with an RF of 1:63,360 (a map series that was once produced by the U.S. government) has an English verbal scale of 1 in:1 mi. The alternate verbal scale is 1 cm:0.6336 km.

Converting between metric and English measures is sometimes necessary. When going between miles and kilometers, or vice versa, it is always best to first change the units to inches and centimeters. While it is good to remember that one mile is approximately 1.6 km, the actual number is 1.609344. Using the 1.6 conversion factor in a calculation could lead to large errors. The conversion between inches and centimeters has fewer decimals and is exactly 2.54 cm to 1 in.

The same type of conversion can be made between English and metric scales. Beginning with a scale of 1 in:10 mi, we find the RF by multiplying 10 × 63,360. The resulting RF would be 1:633,600. Simply moving the decimal place by five places—the equivalent of dividing by 100,000—results in a metric verbal scale of 1 cm:6.336 km.

One may also be presented with a so-called mixed-verbal scale such as 1 in:10 km. In this case it is best to convert the inches to centimeters, so the scale can be restated as 2.54 cm:10 km. This crude metric verbal scale becomes proper by dividing both sides by 2.54. This gives us 1 cm:3.93701 km. Moving the decimal to the right by five places gives us an RF of 1:393,701.

Verbal scales may also be expressed with meters, feet, and yards. A scale of 1 ft:21,000 m would translate to a crude RF of 30.48:2,100,000 by converting both 12 in and the 21,000 m to centimeters (12 in × 2.54; 21,000 × 100). The proper RF would be 1:68,898, producing a slightly smaller scale map than 1 inch to 1 mile.

Conversions to and from a graphic scale proceed in a similar way. If we wanted a graphic scale for a 1:24,000 map that represents 3 miles, we would first determine the English verbal scale. Dividing 24,000/63,360 gives us 1 in:0.378 mi. So, we know that if we wanted to make a graphic scale that represents 0.378 mi, it would be 1 in long. For a graphic scale that represents 3 mi, we divide 3 by 0.378. This gives us a graphic scale that is 7.94 in long to represent the 3 mi distance.

Alternatively, we might have a graphic scale that is 3.4 in and represents 10 mi. This is identical to the verbal scale of 1 in:2.94 mi (10/3.4). To determine a corresponding graphic scale for the metric system, we would first find that the RF is equal to 1:186,353 (2.94 × 63,360) and that the metric verbal scale is 1 cm:1.86353 km. A metric graphic scale that is 3.4 in long would be 8.64 cm (3.4 × 2.54). The number of kilometers represented by 8.64 cm would be about 16.0704 km (1.86 × 8.64). A graphic scale to represent exactly 16 km would be determined by dividing 16 by 1.8635. The resulting 8.59 cm line would show 16 km, and the 8.64 cm line would be 10 mi. The difference between the two of only 0.06 cm would be impossible to represent with a graphic scale.

7.2.5 Area Scale

All calculations to this point have involved straight-line distances. The area scale is calculated by squaring both sides of the regular scale. A map with an RF of 1:24,000 is thus converted to an areal RF of 1:576,000,000. This means that 1 square unit on a 1:24,000 map represents 576 million square units on the surface of the Earth. This could be interpreted with centimeters or inches, as in 1 sq cm:576,000,000 sq cm or 1 sq in:576,000,000 sq in.

Because of the large numbers that result from squaring, it is usually best to use the areal verbal scale. The 1:24,000 map has a verbal scale of 1 in:0.378 mi. The corresponding areal verbal scale is 1 sq in:0.1429 sq mi (0.378 × 0.378). The corresponding metric verbal scale is 1 sq cm:0.0576 sq km (0.24 × 0.24). Taking the square root of both sides of an area scale would convert it back to a regular scale. For example, the square root of 0.0576 is 0.24.

If a field on a 1:24,000 map measures 4.2 cm by 6.8 cm, it would be 28.56 sq cm on the map (4.2 × 6.8). We know that 1 sq cm on a map of this scale is 0.0576 sq km. Multiplying the two would mean that the field is 1.645 sq km (28.56 × 0.0576) on the surface of the Earth. A measure of area in the metric system that is widely used around the world is the hectare, equivalent to 10,000 sq m or 0.01 sq km. Since 1 sq km is 100 hectares, the number of hectares in our field is 164.5 (1.645 * 100).

Converting the dimensions of the field to inches, we get 1.65 in (4.2/2.54) by 2.68 in (6.8/2.54), for a total of 4.42 sq in. We know that 1 sq in is 0.1429 sq mi for a 1:24,000 scale map. This means that the field is 0.63 sq mi (4.42 × 0.1429). Since there are 640 acres in a square mile, the field is 404 acres. There are 2.471044 acres in a hectare, the approximate ratio between 404 acres and 164.5 hectares. These area calculations are summarized in Figure 7.4.

7.3   Cartographic Abstraction

Once the scale of the map is chosen, the complicated process of cartographic abstraction begins. Formerly a mostly artistic process, map making has been formalized with specific steps. Collectively termed cartographic abstraction, the process can be divided into three interrelated phases: selection, generalization, and symbolization. Generalization involves a whole series of operations that are dependent on the map scale, including classification, simplification, and exaggeration. We begin with selection.

Metric System

1) Area on map:
3.2 x 6.8 = 28.56 sq. cm

2) An RF of 1:24,000 equals a metric verbal scale of
1 cm : 0.24 km (24,000/100,000)

3) Squaring both sides of this verbal scale results in a metric verbal areal scale of
1 sq. cm : 0.0576 sq. km (0.24²)

4) Since 1 sq. cm equals 0.0576 sq. km on this map, 28.56 sq. cm is 1.645 sq. km. (0.0576 x 28.56)

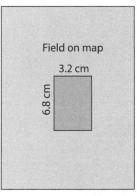

Field on map

3.2 cm

6.8 cm

Scale of map: 1:24,000

English System

1) Convert distances to inches:
3.2 / 2.54 = 1.26 in.
6.8 / 2.54 = 2.68 in.

2) Area on map:
1.26 x 2.68 = 3.37 sq. in.

3) An RF of 1:24,000 equals a English verbal scale of
1 in : 0.378 mi (24,000 / 63,360)

4) Squaring both sides of this verbal scale results in an English verbal areal scale of
1 sq. in : 0.1434 sq. mi. (0.378²)

5) Since 1 sq. in. equals 0.1434 sq. mi. on this map, 3.37 sq. in. is 0.483258 sq. mi. (0.1434 x 3.37)

FIGURE 7.4. Calculations used to determine the area of a field on a 1:24,000 map.

7.3.1 *Selection*

Selection includes everything from the selection of scale, to the selection of what to map, what data, and the projection to be used. These decisions are the basis of all mapping. Through the example of the Baltimore Phenomenon, we have already examined the complex decision-making process for knowing what to include on the map. It is also important that good decisions be made concerning the type of data and projection that is used.

In terms of data, a basic distinction is made between qualitative and quantitative data. Qualitative data refers to features like those between land and water. This type of data is also referred to as nominal, from the Latin *nomen*, meaning name. General reference maps depict qualitative or nominal data because they distinguish between cities, roads, land, and water. The land-cover map in Figure 7.5 is an example of a thematic map that shows qualitative data.

By contrast, a quantitative map depicts some type of numerical data, such as population. Quantitative data exist in three forms: ordinal, interval, and ratio. Ordinal represents categories such as low, medium, and high. A numeric order is implied, but numbers do not need to be used. Interval data uses numbers, but they are simply overlaid onto a specific range.

Temperature is a good example of interval data because the zero value in both the Fahrenheit and Celsius scales do not represent true zero. While 0 Celsius is the freezing point of water, 0 Fahrenheit has no real meaning (to the German physicist, Fahrenheit, it was the freezing point of a certain mixture of saltwater). Most of the world uses the Celsius scale, while only the United States and the small Central American country of Belize use Fahrenheit as the primary system of temperature measurement.

Ratio data has a true zero value and would apply to anything dealing with the counting of objects or people. For temperature, ratio data would be the Kelvin scale. One way of explaining the difference between interval and ratio data is based on making ratio comparisons. For example, if we are using Kelvin temperatures, we can say that 20°K is twice as warm as 10°K. In contrast, with Celsius or Fahrenheit, we cannot say that 20° is twice as warm as 10°. Ratio comparisons can also be made with people. For example, 20 people is twice as many as 10. Any number that is itself a ratio of two numbers, such as population density, can also be compared in this way.

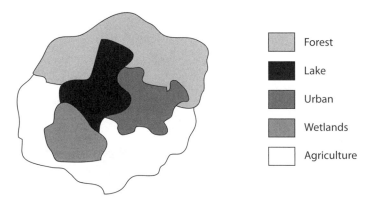

FIGURE 7.5. A land-cover map showing qualitative or nominal classes.

Choosing the map projection is another step in the selection process, and cartographers have spent a great deal of effort trying to properly transfer the spherical Earth to a flat surface. Ptolemy addressed the projection problem almost 2000 years ago. Later, projections of the Earth were actually made using a light source within a translucent globe. Cartographers would draw the lines of latitude and longitude and the outlines of continents that were *projected* onto a wall or another flat surface.

Mathematicians subsequently examined the projection problem, first in France and then in Germany. To them, the accurate transformation of the spherical Earth to a flat piece of paper was a mathematical puzzle. In the beginning, it was theorized that this transformation was mathematically possible without introducing distortion. Although this was proven to be impossible, it was shown later that certain properties of the globe could be preserved on a flat surface. One of these properties was the representation of angles, at least over small areas. The most well known of these projections is the Mercator.

Projections like the Mercator are called conformal, and we have already examined the severe area distortion of this category of projections. Another class of projections is the equal area, and these projections, in contrast, severely distort angles. Conformal and equal-area maps are said to be mutually exclusive. A map that shows angles correctly cannot show areas correctly, and vice versa. The conformal Mercator and the equal-area Mollweide projections are compared in Figure 7.6. The

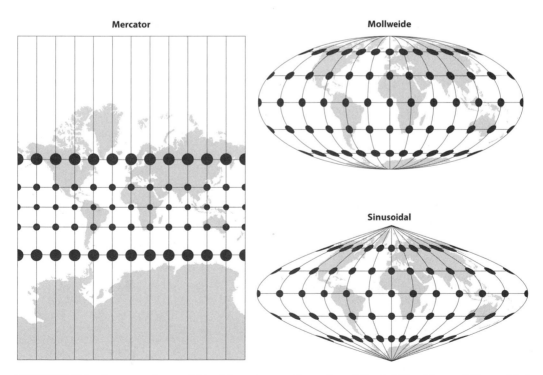

FIGURE 7.6. A comparison of the Mercator conformal and the Mollweide and Sinusoidal equal-area projections. The ellipses represent the degree of area and angular distortion. The amount of area distortion on the Mercator increases with latitude. Angular distortion on the Mollweide and Sinusoidal is greatest where the intersection of the lines of latitudes and longitude vary the greatest from the perpendicular.

two dimensions of the ellipses represent the amount of area and angular distortion. This method for depicting the symbolization of distortion was developed by the French mathematician, Tissot, in the 1850s and is called the Tissot Indicatrix.

One advantage of the Mercator is that it is ideally suited for the rectangular display of the computer monitor. The tiling of the map is also more straightforward when lines of latitude and longitude are straight. The amount of area distortion is extreme, and the projection should not be used to display data at the world level, although this is quite often done. Both the Mollweide and Sinusoidal, like all equal-area projections, distort angles and therefore the shapes of continents. This effect is particularly noticeable on the map with Alaska, northeastern Russia, and Australia.

A number of compromise projections have been developed that attempt to minimize both angular and areal distortion. Of these, the Robinson projection is commonly used (see Figure 7.7).

Thousands of map projections have been developed, but only a few have any notable characteristics. The azimuthal equidistant projection (see Figure 7.8) shows distance correctly but only from the middle point outward. Distance is generally distorted on all projections except along certain lines of latitude or longitude. The gnomonic projection shows the great circle as a straight line. On all other projections, this shortest distance between two points is shown as a curved line.

Finally, the Goode's Homolosine projection combines the Mollweide and the Sinusoidal projections and then "interrupts" the ocean areas to depict land areas with no areal distortion and a minimal amount of angular distortion (see Figure 7.9). The projections are joined at the 41st parallel, visible by a discontinuity on the side of the map.

7.3.2 Classification

Classification refers to a whole series of operations that take place to generalize the map. Classification involves the creation of borders as there are few natural

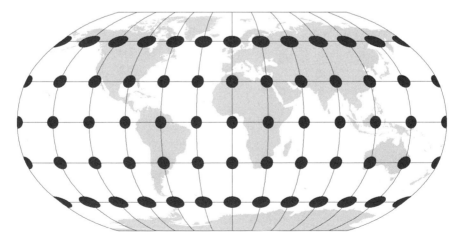

FIGURE 7.7. The Robinson is a compromise projection. It distorts both angles and areas but attempts to minimize each.

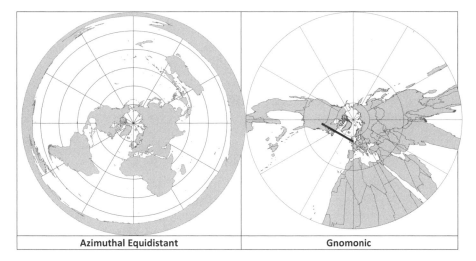

| Azimuthal Equidistant | Gnomonic |

FIGURE 7.8. The azimuthal equidistant projection on the left shows distances correctly, but only from the center point (North Pole) outward. The gnomonic projection on the right shows the great circle as a straight line, such as the line depicted between Chicago and London. All other projections show the great circle as a curved line.

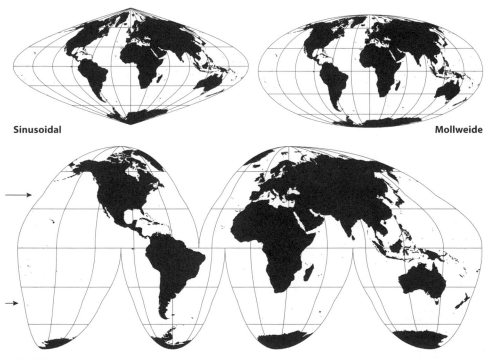

Sinusoidal

Mollweide

Goode's Homolosine

FIGURE 7.9. The Sinusoidal and Mollweide projections on top are combined to form the Goode's Homolosine projection on the bottom, with ocean areas interrupted to decrease angular distortion. A discontinuity in the combined projection can be seen at 41° degrees north and south of the equator and is indicated with arrows along the left side of the map.

borders in the environment. Even the border between land and the oceans is not a definitive line because it changes as a result of tides. Sea level represents an average elevation created by humans. Although some borders between countries are based on natural features, most are of human origin.

A particular example of an arbitrary classification is the border between the United States and Canada that runs through the city of Derby Line, Vermont, and Rock Island, Quebec (see Figure 7.10). The border weaves between the two towns, a result of the work by a surveyor in the 1700s who had been drinking. The line was supposed to be drawn along the 45th parallel, halfway between the equator and the North Pole, but the surveyor missed the mark. Not only does the line weave back and forth, but it is located 1/4 mile too far north.

Today, the international border goes through gardens, backyards, and even through the Haskell Binational Free Library and Opera House. In 1979, it even separated two sides of a house belonging to the Bolducs (Blampied 1979). Citizenship in the family is split by virtue of which hospital was used for the birth of their children. Their son is Canadian while the daughter is an American. Duty must be paid on items purchased if they are brought to the other side of the house. The furnace is on the Canadian side and had to be purchased there. A factory even straddles the border. On the one side it is a Canadian corporation turning Canadian materials into Canadian products. Across a line on the factory floor is the American company turning American materials into American products. No materials or machinery can cross the borderline in the center of the building unless customs is notified and a duty is paid.

Like all classifications of the real world, the border between these towns is of human origin. Classification is used to make distinctions between things in our environment. Once defined, they represent convenient labels that can be mapped and discussed. Classification makes it possible for us to talk about the world.

Classification is only done for our convenience because there are no clearly defined classes in reality. We should be wary of how they are incorporated in maps.

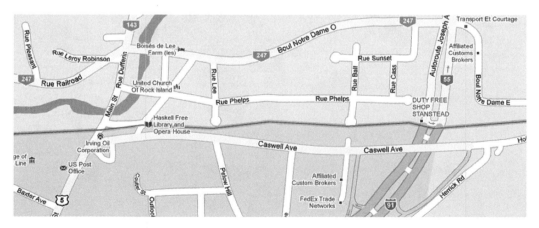

FIGURE 7.10. The U.S.–Canadian border divides the towns of Derby Line, Vermont, and Rock Island, Quebec. Misplaced by an early survey, the border runs through houses, a library, and a factory. Copyright 2013 Google.

In the end, classes represent a compromise between the notion that everything in reality is unique or that individual features are only specific examples of general groupings—like the 20 questions game "Animal, Vegetable, or Mineral." Classification lies in the middle between the two thoughts that everything in the world is different from everything else and that everything is the same.

Qualitative data is classified to form the nominal (name-based) categories that we see on maps. These categories are influenced by our language and vary between languages. What represents a road to one culture may only be a path to another. This difference is a particular problem when constructing geographical databases across international borders. The study of geographical ontology attempts to define an authoritative nomenclature in the geographical domain across all languages. Until a standard is defined and universally adopted, consistency in classification and representation cannot be assumed.

Quantitative data are often classified in mapping to facilitate the symbolization of features. Here, a continuous set of numeric data is divided into finite categories. As with most things in cartography, there is no single best solution to data classification.

A common method of data classification is the equal interval or equal step. If we are dividing percentages between 0 and 100 into five categories, the equal interval approach would divide the range by the number of desired classes (100/5). The result, 20, would be used to divide the data into the following categories: 0–20, 20–40, 40–60, 60–80, and 80–100. The problem with these categories is that a value of 20, for example, could be in the first or second category. A solution is to redefine the categories as: 0–19.9, 20–39.9, 40–59.9, 60–79.9, and 80–100. A major limitation of the equal-interval method of data classification is that while the first and last categories are assured of having a data value, this is not the case for the classes in between.

A major limitation of the equal-interval method of data classification is that some categories may have no observations. The quantile method ensures that every category has a relatively equal number of observations. With this method, the number of observations in the data set is divided by the number of desired categories. In mapping the United States by state, a quantile map with five classes would have 10 states per category (50/5). The exact class breaks would be established at the midpoint values between the data values at the class boundaries.

The natural breaks approach makes class divisions where large differences exist between adjacent data values in the rank-ordered list. The class break boundaries are also set at the midpoint between the surrounding data values. While making sure that there is at least one observation per class, the natural breaks approach can have many class breaks at one end of the data spectrum, leaving most of the observations in a single class.

Finally, a classification approach based on the standard deviation is commonly used because it can be justified based on statistical work. Here, the standard deviation for a data set is calculated. Tables on the chi-square distribution tell us how to divide a normal bell-shaped distribution into an equal number of observations. For example, if five classes are desired, the four class boundaries are made at 0.26 and 0.84 of a standard deviation from the mean (see Figure 7.11).

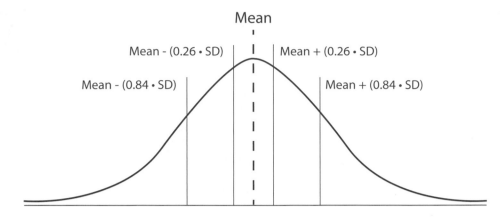

FIGURE 7.11. A method of data classification based on the standard deviation. The normal distribution (bell curve) is divided into categories based on chi-square values of 0.26 and 0.84 to create five equal divisions.

7.3.3 Simplification

Another major process involved in map generalization is simplification. While simplification can take several forms, it essentially involves the elimination of features. It may refer to the creation of one composite variable based on a series of underlying variables, such as a "quality of living" index. Most often, simplification refers to the various procedures used to generalize a series of points, lines, or areas so that they can be displayed at a smaller scale. Any type of simplification cannot occur in isolation because the simplification of one feature will influence the amount and type of simplification of another.

The easiest way to describe simplification is through the generalization of a line where the number of points in the line are systematically decreased to remove detail. Determining which points to remove is the challenge of this approach. The points chosen for elimination are those that don't significantly affect the appearance of the line. Figure 7.12 shows increasingly generalized maps of the world with country outlines. The beginning map has a total of 280,444 points. This is reduced to only about 4000 points in the second illustration and 2000 points in the third (Douglas and Peucker, 1973; Visvalingam and Whyatt, 1993).

7.3.4 Exaggeration

Most smaller scale maps would be essentially blank if features were not exaggerated in size. At a scale of 1:100,000, a 10-m-wide road would be depicted at a width of only 0.01 cm, or 0.1 mm. No drafting pen can make a line this thin. At 100 dpi, the thinnest line on the screen of the computer is 0.01 in or 0.254 mm thick, and a 1-pixel line is barely visible. To make the road appear on the map, it must be made at least three or four times its size. Exaggerated features consume more map space, further necessitating simplification and generalization of other neighboring features.

7.3.5 Symbolization

After all of the complicated selection and generalization processes comes the final step—the symbolization of the map. Decisions about symbolization have the strongest influence on the overall look and functionality of the map. Symbols can bear a strong resemblance to the features being mapped, or they can be totally arbitrary. Some symbols, like the star for a capital city, are based on convention.

The three basic types of symbols are point, line, and area (see Figure 7.13). Dot and graduated symbol maps are examples of a point map. Point symbols can be geometric (e.g., circles, triangles, and squares) or pictographic to look like the object being represented. A contour line is an example of a line symbol, as is a graduated line map that may show traffic along a road segment. Maps showing traffic commonly use lines of different colors to show the speed of traffic. The choropleth map uses shadings to show value by area. The cartogram, on the other hand, alters the size of the area being mapped to depict value.

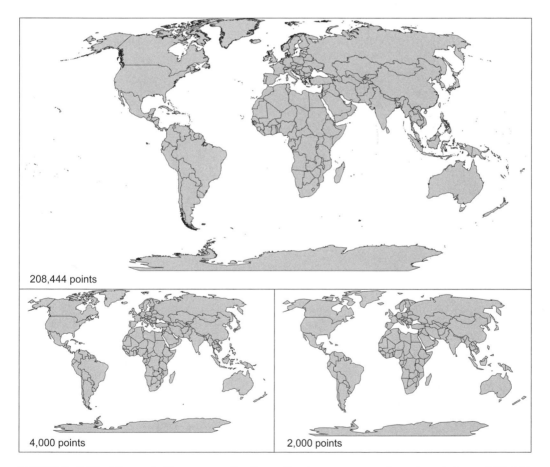

FIGURE 7.12. Through line generalization, the topmost map is reduced from 208,444 points to only 4,000 points in the map on the lower left and then 2,000 points in the map on the lower right. The operation was performed through MapShaper.org, a free online utility.

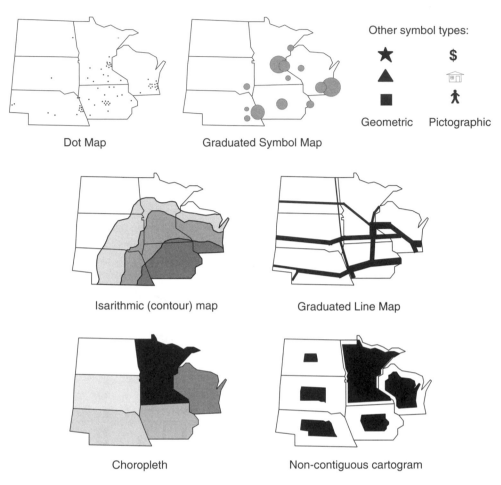

FIGURE 7.13. Examples of point, line, and area symbols.

7.4 Summary

The reduction of reality to create a useful map is an immensely complicated task. The smaller the scale, the more generalization is required. The small-scale map is only a caricature of reality. Cartographic abstraction, including the steps of selection, generalization, and symbolization, is at the heart of cartography. It includes all of the decision-making processes that go into creating a map. From a broader perspective, it is how humans have learned to communicate information about the world, with all of its various complexities.

7.5 Exercises

PART I. SCALE CONVERSIONS

1. Change the following to English and metric verbal scales:

   1:200,000,000   1 in: _____ mi   1 cm: _____ km
   1:600,000   1 in: _____ mi   1 cm: _____ km
   1:10,000,000   1 in: _____ mi   1 cm: _____ km
   1:23,678   1 in: _____ mi   1 cm: _____ km

2. Convert the following scales:

   1:63,360   1 in: _____ mi   1 cm: _____ km
   1: _____   1 in: 3.2 mi   1 cm: _____ km
   1: _____   1 in: _____ mi   1 cm: 10 km
   1:100,000,000   1 in: _____ mi   1 cm: _____ km
   1: _____   1 in: 150 mi   1 cm: _____ km
   1: _____   1 in: _____ mi   1 cm: 1,500 km

3. A map on which 1 in represents 5 mi has a metric scale of 1 cm: _____ km

4. Recall that in the metric system, 1 mm = 1/10 cm, 1 cm = 1/100 m,
   1 km = 1000 m.

   At a scale of 1:50,000, how many meters are represented by 1 cm? _____ m
   How many feet are represented by 1 in at that scale?   _____ ft
   Is this scale *larger* or smaller than 1 in:1 mi?   _____

5. Change the following to RFs and verbal scales.   Look carefully at the units.

	(RF)		(alternate verbal)	
6 in:1 mi;	_____	;	1 cm _____	m
1 in:20 mi;	_____	;	1 cm _____	km
1 cm:5 km;	_____	;	1 in _____	mi
1 cm:50 m;	_____	;	1 in _____	ft
5 cm:1 km;	_____	;	1 in _____	ft

6. How long should a graphic scale be, in inches, on a 1:50,000 map if it is to
   represent 2 miles? 2 km? _____ in _____ cm

PART II.   FINDING SCALES BASED ON AVAILABLE INFORMATION

1. One degree is about 69 mi. While degrees of latitude are relatively constant,
   the distance for a degree of longitude varies from about 69 mi at the equator to
   0 at the poles. At 60° of latitude, the distance between degrees of longitude is
   half the length of the equator, or about 34.5 mi. If the distance between 10° of
   longitude is 1.09 in on a map at the equator and at 60°N, what are the following two scales in RF?

The E-W scale at 60°N is _____

The N-S scale and the E-W scale at the equator _____

2. The rectangular land survey system, used in the Midwest and other parts of the United States, divides land area into townships of 6 × 6 mi (36 sq mi), which in turn are divided into sections of one square mile (640 acres), quarter-quarter sections (40 acres), and so on. If a section on a map is 4.2 in by 4.2 in, what is the scale of the map?

   RF _____   1 in: _____ mi

If a township is .5 × .5 in, what is the scale?

   RF _____   1 in: _____ mi

3. Two points are separated by 4.32 in on a 1:20,000 map. What is the distance between them in miles? _____ In km? _____

4. The distance between two cities on a map is 8.61 cm, and you know the actual distance between them is 84 mi. What is the scale of the map? _____

5. The length of the Nile River is 4149 mi. What would be its length in inches on a 1:16,000,000 map? _____

6. The maps presented by the different online map providers are not consistent in scale even between similar levels of detail. Use a ruler to calculate the RF for the following graphic scales for maps of similar scale from the following providers:

Google   2 mi   Bing   1 miles   1 km
              5 km

MapQuest   0   3200 m   Yahoo   2km
                 9600 ft   2mi

Google: 1:_____   Bing: 1:_____
MapQuest: 1:_____   Yahoo: 1:_____

PART III.   AREAL CONVERSIONS

1. On an air photo with a 1:10,000 scale, a cornfield measures 1.75 × 2.25 in. What is the area of the cornfield in square miles? In acres? (1 sq mi = 640 acres)

2. On a map with a scale of 1:24,000, a lake is measured with grid paper at 1/8 in on a side to be 87 full grid squares. Another 24 grid squares intersect with the edge of the lake. Typically, half of these grid squares are added to the total number of full squares. What is the area of the lake in square miles and square kilometers?

PART IV.   DATA CLASSIFICATION

Use a spreadsheet to classify a data set using the following four methods:

Equal Interval: This approach creates five equal steps in the data range. This equal data step is determined by dividing the difference between the maximum and minimum value by the number of classes. So, if the maximum is 100 and the minimum is 0, the equal step is 20. The number of observations in each category would depend on the distribution of the data. With this data classification approach, some categories may have no observations.

Shading	Range	no. of obs
	80.0–100.0	data dependent
	60.0–79.9	data dependent
	40.0–59.9	data dependent
	20.0–39.9	data dependent
	0.0–19.9	data dependent

Quantile: The quantile method divides the distribution into an equal number of observations. For example, if there are 100 observations (counties, census tracts) and we want five classes (quintile), then we would have 20 observations in each class.

Shading	Range	no. of obs
	midpoint between 80th and 81st obs—0.1 to maximum value	20
	midpoint between 60th and 61st obs—0.1 to midpoint between 80th and 81st obs	20
	midpoint between 40th and 41st obs—0.1 to midpoint between 60th and 61st obs	20
	midpoint between 20th and 21st obs—0.1 to midpoint between 40th and 41st obs	20
	minimum value—midpoint between 20th and 21st obs	20

Natural Breaks: Usually, selected arbitrarily based on largest perceived "breaks" between values; it should be done systematically by finding the midpoint between the four largest differences in the ranked data. A spreadsheet can easily calculate differences between ranked data values.

Shading	Range	no. of obs
	(midpoint between the two values that represent one of the four largest differences in the ranked values) + 0.1 to maximum value	data dependent
	(midpoint between the two values that represent one of the four largest differences in the ranked values) + 0.1 to (midpoint between the two values that represent one of the four largest differences in the ranked values)	data dependent
	(midpoint between the two values that represent one of the four largest differences in the ranked values) + 0.1 to (midpoint between the two values that represent one of the four largest differences in the ranked values)	data dependent
	(midpoint between the two values that represent one of the four largest differences in the ranked values) + 0.1 to (midpoint between the two values that represent one of the four largest differences in the ranked values)	data dependent
	minimum value to (midpoint between the two values that represent one of the four largest differences in the ranked values)	data dependent

Standard Deviation: Classification based on divisions of the standard deviation such that the area under the normal curve is divided into equal sections. These divisions are based on chi-square values. For a map with five classes, the class breaks are 0.84 and 0.26 from the mean in both positive and negative directions.

Shading	Range	no. of obs
	(Mean + 0.84 * stdev) + 0.1 to maximum value	about 20% of values
	(Mean + 0.26 * stdev) + 0.1 to (mean + 0.84 * stdev)	about 20% of values
	(Mean − 0.26 * stdev) + 0.1 to (mean + 0.26 * stdev)	about 20% of values
	(Mean − 0.84 * stdev) + 0.1 to (mean − 0.26 * stdev)	about 20% of values
	minimum value to (mean − 0.84 * stdev)	about 20% of values

If the number of observations in each class is significantly greater or less than 20% of the overall number of observations (for five classes), then the data is, by definition, skewed. To convert the skewed data to a normal distribution, each value can be changed to the Log10 equivalent and the standard deviation classification can be performed again. The Log10 values can be converted back to the unlogged values using the 10 to the power of x formula, where x is the Log10 value.

7.6   Questions

1. Fill in the blanks in the following expressions of scale:

1 in:80 mi	1: _____	1 cm: _____ km
1 in: _____	1:250,000	1 cm: _____ km
1 in: _____	1: _____	1 cm:1 km

2. Two points are 5.7 in apart on a 1:24000 map. What is the ground distance in miles and kilometers?

3. Two points are 5.7 in apart on a 1:24,000 map. On another map, they are 4.4 in apart. What is the scale of this map?

4. A field measures 5.7 in by 4.4 in on a 1:24,000 map. What is the area in:

    a) square miles   b) acres
    c) square kilometers   d) hectares

5. A lake measures 220 whole grid squares and 120 partial grid squares on a 1:50,000 map. Each grid square is 1/8 in on a side. Determine the following:

    a) square miles   b) square kilometers

6. Describe some aspects of cartographic selection.

7. What is meant by the statement: "There are no classes in reality."

8. What is the basic difference between equal interval and quantile classification?

9. Describe point, line, and area symbols.

7.7   References

Blampied, Phil (1979, Aug. 13) American Scene—Partly in Vermont: A Borderline Case. *Time*, p. 6.

Douglas, D. H., and Peucker, T. K. (1973) Algorithms for the Reduction of the Number of Points Required to Represent a Digitised Line or Its Caricature. *The Canadian Cartographer* 10(2): 112–122.

Visvalingam, M., and Whyatt, D. (1993) Line Generalization by Repeated Elimination of Points. *The Cartographic Journal* 30(1): 46–51.

CHAPTER 8
Programming the Web with JavaScript

The trouble with programmers is that you can never tell what a
programmer is doing until it's too late.

—Seymour Cray

8.1 Introduction

Programming is the language of computers. If we want the computers to do a specific task, then we must understand this language, at least on a basic level. The Internet has transformed programming into the art of using available online tools. Programs are now written by incorporating existing routines from servers. The function libraries, called application programming interfaces (APIs), do all the major work. Programming can now be viewed as simply understanding what these function libraries do and how to use them.

Formerly relegated to geeks, programming has again become a hot topic. Online programming sites like Codecademy have received increased use. Khan Academy has introduced a similar suite of free online programming exercises. Although the utility of these sites has been questioned, the new movement effectively argues that programming is a necessary skill that everyone should know, like reading and writing. Programming is now being viewed as a form of expression, as the "amplification of thinking," and as a necessary skill. This "coding as literacy" concept is promoting new ideas about how to teach programming. These ideas are particularly important in cloud computing, which is largely dependent on the use of existing resources. In many areas, like mapping, programming has become the ability to master the manipulation of many different online tools.

8.2 Programming Languages

Programming languages are either compiled or interpreted. A compiled programming language generates a separate executable file (e.g., .exe file) for a specific computer type and processor. An interpreted programming language is compiled and executed in the same step and does not create an executable version that is dependent on a specific computer. Compiled computer languages such as C++ and Java generate code that may run faster but interpreted languages are more transportable.

An example of an interpreted language that is part of many web pages is JavaScript. Used primarily to add interactively to web pages, JavaScript is a compact, object-oriented language for developing client-side applications (Haverbeke 2011). JavaScript statements are embedded in, or referenced from, a hypertext markup language (HTML) page and are executed when the browser page is opened. This feature was initially viewed as a problem because of slow execution speeds, but computers are now faster and so this is no longer considered a major disadvantage of JavaScript.

Originally developed by Netscape for its Navigator 2.0 browser, JavaScript was a cornerstone of the "browser wars" of the late 1990s. Microsoft developed its own version called Jscript, which was largely incompatible with JavaScript. Separate code within web pages had to be written for the Netscape Navigator and Microsoft Explorer browsers (Wilton and McPeak 2009). The third release of Netscape's JavaScript became sanctioned by ECMA, an international standards body. Even with the standard version, incompatibilities remained in the support for JavaScript between different browsers. For many years, the Internet Explorer browser executed JavaScript code much slower than other browsers.

The major disadvantage of JavaScript from a commercial standpoint is that the code is open and readable. Anyone can view the JavaScript code by simply choosing View Source (Page Source) from the browser. This, combined with the lack of a commercial debugging program and slower processing speeds, has limited the acceptance of the language by programmers, although most major websites incorporate a significant amount of JavaScript code. The language is ideal for educational purposes because it can be easily examined, although many JavaScript libraries have become encrypted, or compacted, so that the readability of the code is lost.

8.3 JavaScript Examples

By itself, HTML is simply a page formatting language. In combination with JavaScript, an HTML page can execute computer code (W3Schools.com 2011). The following examples demonstrate how JavaScript works and how it can be used to call other functions.

8.3.1 JavaScript within the Body of an HTML Document

HTML files are split between a head and a body. JavaScript code is usually in the form of functions within the head of an HTML document, or is placed in an external file referenced in the head. In the first two examples demonstrated here (Figures 8.1 and 8.2), the JavaScript code resides in the body of the HTML file. The code in Figure 8.1 uses the "document.write" function to write text.

Calculations can also be done within the script, as shown in Figure 8.2. Both of these initial examples do not use a separate JavaScript function but rather define a script within the body of the HTML code.

8.3.2 Defining JavaScript Functions in the Head of an HTML Document

Functions are the fundamental building blocks of JavaScript. A function is a procedure—a set of statements that performs a specific task. A function definition has four parts:

- The word "function."
- A unique function name.
- A comma-separated list of arguments in parentheses.
- The statements in the function in curly brackets ({}).

Defining the function specifies what will happen when the function is called. Calling the function actually performs the specified actions with the indicated arguments. Functions are generally defined in the head part of an HTML document. This ensures that all functions are defined before any content is displayed.

The example in Figure 8.3 defines a simple function in the head section of an HTML document. The function is then called in the body of the document. The function **square** takes one argument, called *number*, and the function consists of one statement:

```
return number * number
```

Code	Result

```<html>```
```<body>```
This is a normal HTML document.
``
``
```<script type="text/javascript">```
        document.write ("This is JavaScript!")
```</script>```
``
``
Back in HTML again.
```</body>```
```</html>```
	Mozilla Firefox
	File Edit View Go Bookmarks Too
	
	This is a normal HTML document.
	This is JavaScript!
	Back in HTML again.

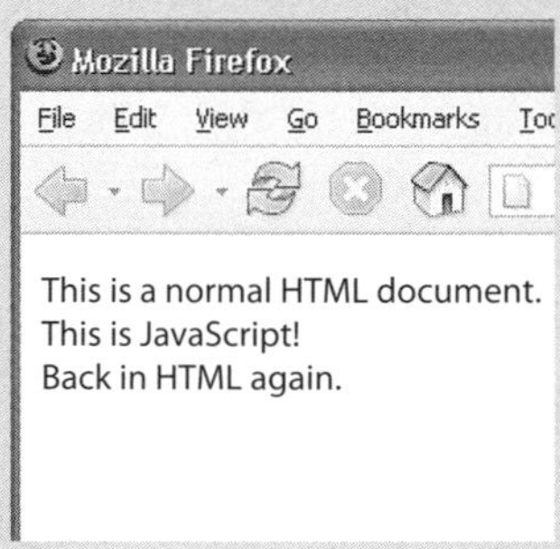

FIGURE 8.1. JavaScript code within HTML. The middle line of text is written by JavaScript, while the other two lines are written by HTML.

Code	Result

``html
<html>
<body>
<script type="text/javascript">
        var x = 2 * 2
        document.write("x = ", x)
</script>
</body>
</html>
``
	x = 4

FIGURE 8.2. A calculation is performed with JavaScript within the body of an HTML file.

that indicates it should return the argument of the function multiplied by itself. The return statement specifies the value that is returned by the function.

8.3.3   Specifying a File of JavaScript Code

Rather than embedding the JavaScript code directly in the HTML file, either in the body or the head, it is possible to place the JavaScript functions within a separate file. The SRC attribute of the <script> tag specifies the external file where the JavaScript code can be found. Figure 8.4 shows the external file called common. js and how it is referenced in the head part of an HTML document. The external JavaScript file may contain multiple functions but no HTML code.

   The external file that contains JavaScript functions can be on the same computer as the HTML file, as shown in Figure 8.4, or it can be on another computer or server. This is how API code is distributed. One reference to a library of API code makes it possible for a web page designer to access thousands of functions. Figure 8.5 shows how the Google Maps API code is referenced. To aid in debugging, the

Code	Result

``html
<head>
<script LANGUAGE="JavaScript">
    function square(number) {
    return number * number
 }
</script>
</head>
<body>
<script>
        document.write("The square of 5 is ", square(5), ".")
</script>
<P>All done.</P>
</body>
``
	The function returned 25.
	All done.

FIGURE 8.3. This function squares the number passed to it by a call to the function.

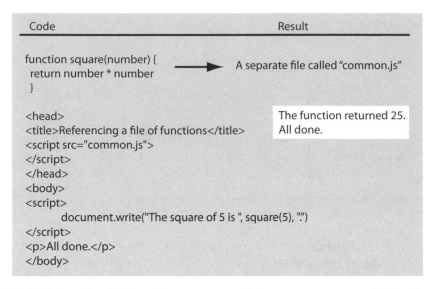

FIGURE 8.4. A function is placed into an external document, common.js. The function is then referenced from the HTML file with <script src="common.js">.

Google Maps API will work locally on a computer without the need to transfer the code to a server. In order for others to see the map, the code must reside on a server.

There are numerous APIs and JavaScript libraries to support almost any type of web page development. The basic variety support complex calculations, interaction, animated displays and other graphic effects. jQuery, a free and open source software library, is the most popular of these JavaScript libraries. Like most implementations, it is a single JavaScript file that can be included by linking a local or public server copy. Another library is the Dojo Toolkit that is specifically designed for JavaScript/Ajax-based applications and promotes internationalization, localization, and accessibility of web pages. A third option is MooTools (My Object-Oriented Tools) that is characterized by its small size and modular approach.

More specialized JavaScript libraries are also available. For example, Lightbox is a JavaScript library that is used to overlay images on the current web page. Lightbox

```
Code

<head>
 <title>Google Maps JavaScript API Example</title>
 <script type="text/javascript"
        src="http://maps.google.com/maps/api/js?sensor=false">
 </script>
</head>
```

FIGURE 8.5. Implementing a call to the Google Maps API. Here, the sensor is set to "false." The value of a sensor would be set to "true" if a mobile device is used that can provide the current position.

darkens the background web page while the foreground image is displayed. D3 (data-driven documents) is designed for various types of dynamic and interactive graphic displays. Written specifically as a tool for data visualization, it is used in many different areas—including mapping. Developed in 2011, it has been widely adopted although it has a complex interface.

8.3.4 Variables

In contrast to some other programming languages, JavaScript variables do not need to be defined at the beginning of the file but can be defined at any point. Variables may begin with upper- or lower-case A through Z and the "_" (underscore) character. The remaining characters can consist of these same characters and the numbers 0 through 9. Variable types include:

- Integers—whole numbers.
- Floating point—such as 3.14, –3.14, 314e-2.
- Boolean values—true or false.
- String values or text—May be enclosed by single or double quotes. The \' or \" sequence of characters will insert a quote character into a string.

The variable is defined when it is first used. Therefore, the example:

```
var example_var=3
```

creates an integer variable. The variable can subsequently be changed to a floating point, replacing the previous value:

```
example_var = 3.14
```

Or it can also be changed to a string:

```
example_var = "Text associated with the example var variable"
```

Note that that keyword "var" is used to initially define the variable. When referencing the variable after it is defined, or when changing its value, the keyword "var" is no longer needed. JavaScript also includes conversion functions that change variables from text to numbers, numbers to text, and floating point to integer.

8.3.5 Arrays

Arrays are a very important part of any programming language. Just think of an example in which you want to store 100 different names. You could define 100 separate variables and assign the different names to each, but this would be very time consuming. Arrays can be seen as a bundle of variables that can be accessed through one variable name and a number.

Arrays of multiple values may be defined for any of the variable types, but they must be declared before they are used. For example:

```
Car = new Array (4)
```

would define an array Car with four elements. Arrays in JavaScript are zero indexed, so the elements of this array would be referenced with the number 0 to 3. An array would be defined in this way:

```
Car[0] = "Toyota"
Car[1] = "Honda"
Car[2] = "Volkswagen"
Car[3] = "Volvo"
```

The array may also be defined and populated in a single step:

```
Car = new Array('Toyota','Honda','Volkswagen','Volvo')
```

In addition, the array size may be expanded by simply adding additional values. The following statement would increase the array to six values:

```
Car[5] = "BMW"
```

Arrays may also contain different variable types. The following statement would place a number into the fifth array element:

```
Car[4] = 3.14
```

Finally, arrays may also be nested or two-dimensional, so that each array element may have its own array. The following statement would define subtypes for the first array element:

```
Car = new Array('Toyota',new Array('Camry','Corolla','Prius'))
```

The following line would write "Prius."

```
Document.write(Car[0][2])
```

Figure 8.6 shows how the minimum and maximum value is determined from an array of 93 values. First, the popdata array is defined. Then, the min and max variables are set to values that are outside of the range for population values. Finally, a loop cycles through the 93 values and replaces the value of min and max when values lower than the previous min and max are found.

The code in Figure 8.7 ranks the population values from low to high. In this case, two functions are defined. The first, initialize, is called in the opening of the body section. This function defines the data values and writes out the unsorted values. Then, it calls the function called sort; this function sorts the values using the bubble sort algorithm. This algorithm progressively examines the array to find new low values. The sorted array is returned to initialize, which then writes out the sorted array.

Code	Result

```
<html>
<title>Calculate the min and max of a dataset</title>
```
The minimum value is 372.
The maximum value is 492003.
```
<h3>Calculating the min and max of a dataset</h3>
<script type="text/javascript">
   var popdata = new Array (33185,6931,372,783,492,5668,11132,2185,3354,
43954,7341,8595,25963,8819,3811,5934,9865,6564,10113,9660,11242,20587,8466,
25018,1958,6170,36171,492003,2109,6259,3348,2729,5003,23365,1995,1790,1978,
660,2454,55555,9490,3446,1029,2926,10610,756,6736,7874,4683,6701,8250,892,
3710,8812,267135,35865,749,656,497,35279,7954,5171,3705,7247,4650,15747
2804,2992,9442,7564,31962,5349,10865,8656,1544,14155,142637,20344,36546
16835,5571,3083,1403,6570,5317,629,7273,4373,20044,9196,3701,823,14502);

//initialize min and max to values beyond the range
        var min=100000000;
        var max=-100000000;

//set n equal to the number of values in the array
        var n=93;

//loop through all values and reset min and max
        for (var i = 0; i < n; i++) {
                        if (popdata[i] < min) { min=popdata[i] }
                        if (popdata[i] > max) { max=popdata[i] }}

//write dataset and min and max
        document.write('<b>Nebraska populations by county</b>: '+ popdata)
        document.write('<p>The minimum value is ' + min + '.')
        document.write('<p>The maximum value is ' + max+ '.')

</script>
</html>
```

FIGURE 8.6. The calculation of the minimum and maximum values for an array of population values.

8.3.6   Objects

Objects are combinations of functions and data all packaged under a single name. Each object has properties, methods, and event handlers. An object named car, for example, would possess properties such as make, model, year, and color. Methods might be attributes like 'go' and 'stop.' An event handler includes pressing a button or moving the mouse over a link.

JavaScript incorporates several built-in objects. Table 8.1 describes the Date object that includes the getDate, getDay, getHours, etc., methods. These are used in the script in Figure 8.6 to determine the current time. Note how the function Time calls the AddZero function to add a zero in front of numbers that are less than 10. This demonstrates how functions can themselves call other functions.

Code

```
function initialize() {
        var popdata = new Array (33185,6931,372,783,492,5668,11132,2185,3354,
43954,7341,8595,25963,8819,3811,5934,9865,6564,10113,9660,11242,20587,8466,
25018,1958,6170,36171,492003,2109,6259,3348,2729,5003,23365,1995,1790,1978,
660,2454,55555,9490,3446,1029,2926,10610,756,6736,7874,4683,6701,8250,892,
3710,8812,267135,35865,749,656,497,35279,7954,5171,3705,7247,4650,15747
2804,2992,9442,7564,31962,5349,10865,8656,1544,14155,142637,20344,36546
16835,5571,3083,1403,6570,5317,629,7273,4373,20044,9196,3701,823,14502);
        var n=93;

//write unsorted data
  document.write('<h3>Unsorted Nebraska populations by county:</h3> '+ popdata);

// call sort function
        sortdata(n,popdata);

//write sorted data
  document.write('<h3>Sorted Nebraska populations by county:</h3> '+ popdata);}
</script>

<script type="text/javascript">
function sortdata(n,data) {
  var i, j, swapped
  for(i = 0; i < n; i++) {
   var swapped = false;
   for(j = 0; j < (n-1); j++) {
    if(data[j] > data[j+1]) {          // Chk if variables need swapping
      swap = data[j+1];                // The following 3 lines swap the variables
      data[j+1] = data[j];
      data[j] = swap;
      swapped = true;
    }
   }
   if (swapped==false) break;}         // if the swapped flag isn't changed then there is
return data}                           //point continuing as all values are in order
</script>

</head>
<body onLoad="initialize()">
</body>
```

FIGURE 8.7. Sorting the array of population values. The sorting algorithm used is called a bubble sort.

8.3.7 Conditional if Statement

The if statement is used to evaluate a certain condition, and, if the condition is true, specific steps are performed. It is used in Figure 8.8 to add a "0" in front of numbers that are less than 10. The example in Figure 8.9 uses the if statement to write "Good Morning" if the current time is before noon and "Good Afternoon" if the time is 12 PM or after.

TABLE 8.1. A Description of the Built-in JavaScript Date Object

Object	Properties	Methods	Event handlers
Date	none	getDate getDay getHours getMinutes getMonth getSeconds getTime getTimeZoneoffset getYear parse prototype setDate setHours setMinutes setMonth setSeconds setTime setYear toGMTString toLocaleString UTC	none

The example in Figure 8.10 uses an if statement to convert the time to an AM or PM. "PM" is assigned if the time is greater than or equal to 12; otherwise an AM is printed after the time. In addition, since hours are defined between 0 and 24, 12 is subtracted from the current time if the time is greater than 12.

In the example in Figure 8.11, the JavaScript function, CalculateDays, computes the number of days before or since a certain date. The current date is subtracted from a date defined as theevent. This number is divided by 1000 to determine the number of seconds to/from the theevent. The number of minutes, hours, and days are computed from the number of seconds. Finally, the function CalculateDays is called in the body definition statement.

8.3.8 Recursion

Recursion occurs when a function calls itself. The code in Figure 8.12 also computes the days, hours, minutes, and seconds before/since a certain day and time. In addition, the function updates constantly by calling itself at the end of the function. A recursive function is a function that repeats constantly. In the function code, you can see the call to itself at the end of the function. The example also demonstrates the use of forms, a way of displaying output from a calculation. Forms can also be used to accept input from the user.

8.3.9 Complex Calculations

JavaScript includes a number of internal functions called Math Objects (see Table 8.2) and Math Object properties (see Table 8.3). Most of these are trigonometric

Code	Result

``html
<script type="text/javascript">
function Time()
{
var currentTime = new Date()
var hours = currentTime.getHours()
var minutes = currentTime.getMinutes()
var seconds = currentTime.getSeconds()
 	
// call the AddZero function to add a
// zero in front of numbers that are less
// than 10
hours = AddZero(hours)
minutes = AddZero(minutes)
seconds = AddZero(seconds)
 	
// write out the time
document.write("<b>" + hours + " hours " + minutes +
        " minutes " + seconds + " seconds </b>")
}
 	
function AddZero(i)
{
if (i<10)
        {i="0" + i}
  return i
}
</script>
</head>
 	
// call the Time function
<body onload="Time()">
</body>
``
	19 hours 23 minutes 06 seconds

FIGURE 8.8. Calling the built-in date object to determine the current time in hours, minutes, and seconds.

functions that might be used to determine the distance between two points on the surface of the Earth, as in Figure 8.13. Other functions compute numbers like the absolute value or square root.

Another function included in JavaScript is a random number generator, denoted by random(). The random number function returns a value between 0 and 1. Thus, on average, half the values returned by the random number generator should be below 0.5 and half should be above. The JavaScript in Figure 8.14 links to two different websites based on the result of the random number.

8.3.10   *User Input*

Input may also be requested from the user based on a prompt. In Figure 8.15, the user is requested to enter a name. This name is then put into a question. In Figure 8.16, two numbers are requested that are subsequently multiplied.

146

Code	Result
```html<script type="text/javascript">var d = new Date()var time = d.getHours()if (time < 12){ document.write("<b>Good morning</b>")}else{ document.write("<b>Good Afternoon</b>")}</script>```	*If it is before 12:* **Good Morning** *If it is after 12:* **Good Afternoon**

FIGURE 8.9. The use of an "if" statement to determine if the current time is morning or afternoon.

Code	Result
```html<script type="text/javascript">function Time2(){ var currentTime = new Date() var hours = currentTime.getHours() var minutes = currentTime.getMinutes() var suffix = "AM"; if (hours >= 12) { suffix = "PM"; hours = hours - 12; } if (hours == 0) { hours = 12; } if (minutes < 10) minutes = "0" + minutes document.write("<b>" + hours + ":" + minutes + " " + suffix + "</b>")}</script></head>// call the Time2 function<body onload="Time2()"></body>```	*7:54 PM*

FIGURE 8.10. The use of an "if" statement to add AM or PM to the current time.

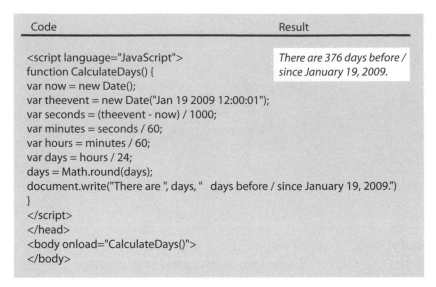

Code	Result

``<script language="JavaScript">
function CalculateDays() {
var now = new Date();
var theevent = new Date("Jan 19 2009 12:00:01");
var seconds = (theevent - now) / 1000;
var minutes = seconds / 60;
var hours = minutes / 60;
var days = hours / 24;
days = Math.round(days);
document.write("There are ", days, " days before / since January 19, 2009.")
}
</script>
</head>
<body onload="CalculateDays()">
</body>```
	There are 376 days before / since January 19, 2009.

FIGURE 8.11. Calculating the number of days before or since a certain date.

8.4 Mouseover Maps

The mouseover changes the display based on whether the mouse is located over a specific feature. A click is not required. Mouseovers are implemented with Java-Script, and like the clickable image, the mouseover is also based on hotspots. In this example, when a mouse is over a hotspot as defined with "<area = shape" in the body of the HTML, the initial image is substituted with another image. This can be demonstrated with the four identically sized images in Figure 8.17. The JavaScript function hiliter, shown in Figure 8.18, loads the images into memory and then swaps them out when the mouse is over a defined hotspot. As with the clickable map, the map name section defines the four hotspots and the usemap option in the <img> tag makes the association.

A more useful mouseover map can be created using different-sized images that can be placed anywhere against a background. Rather than simply overlaying images, it should be possible to place text on top of a background image. In the case of a map, you could have a whole series of smaller images that would be displayed as a result of moving the mouse. A large number of such hotspots could be defined. This mouseover method is based on the div layer or block—a way of positioning elements on a page using absolute positioning. These elements can be graphic files or text strings. The code for a div layer is shown in Figure 8.19 along with an explanation of the elements.

In the example in Figure 8.20, three div layers are defined: newdehli, mumbai, and indiamap. The last-named is a map of India that is positioned at 1,1 (the upper left corner) with a z-index or layer value of 1. The other two layers display text and are positioned to label the cities of New Delhi and Mumbai. They have a z-index value of 2. The text for New Delhi is positioned at 58, 100. The text for Mumbai, which is located further south, is placed at 72, 200. As with all previous examples,

Code	Result

``html
<html>
<head>
<script LANGUAGE="JavaScript">
// The function update is called by
// itself to constantly update the values
// for days, hours, minutes, seconds
var now = new Date();
var theevent = new Date("Jan 19 2009
    12:00:01");
var seconds = (theevent - now) / 1000;
var minutes = seconds / 60;
var hours = minutes / 60;
var days = hours / 24;
ID=window.setTimeout("update();", 1000);
function update() {
        now = new Date();
        seconds = (theevent - now) / 1000;
        seconds = Math.round(seconds);
        minutes = seconds / 60;
        minutes = Math.round(minutes);
        hours = minutes / 60;
        hours = Math.round(hours);
        days = hours / 24;
        days = Math.round(days);
        document.form1.days.value = days;
        document.form1.hours.value = hours;
        document.form1.minutes.value = minutes;
        document.form1.seconds.value = seconds;
        //this function calls itself to update the time
        ID=window.setTimeout("update();",1000);
}
</script>
</head>
<body>
<p>Countdown To January 19, 2009, at 12:00:</p>
// forms are used to display the values
<form name="form1"><p>Days <input type="text" name="days"
        value="0" size="3"> Hours
 	
<input type="text" name="hours" value="0" size="4"> Minutes
 	
<input type="text" name="minutes" value="0" size="7"> Seconds
 	
<input type="text" name="seconds" value="0" size="7"></p></form>
</body>
</html>
``
	Countdown To January 19, 2009, at 12:00:
	Top of Form
	__ Days
	__ Hours
	__ Minutes
	__ Seconds
	
	(These numbers are continuously updated on the webpage.)

FIGURE 8.12. Calculating the number of days, minutes, and seconds before or since a certain date and time. The numbers are continuously updated on the web page by placing a call to the function "update" within the function. This programming method is called recursion.

TABLE 8.2. JavaScript Internal Math Objects

Method	Description
abs(x)	Returns the absolute value of a number
acos(x)	Returns the arccosine of a number
asin(x)	Returns the arcsine of a number
atan(x)	Returns the arctangent of x as a numeric value between –PI/2 and PI/2 radians
atan2(y, x)	Returns the angle theta of an (x, y) point as a numeric value between –PI and PI radians
ceil(x)	Returns the value of a number rounded upwards to the nearest integer
cos(x)	Returns the cosine of a number
exp(x)	Returns the value of Ex
floor(x)	Returns the value of a number rounded downwards to the nearest integer
log(x)	Returns the natural logarithm (base E) of a number
max(x, y)	Returns the number with the highest value of x and y
min(x, y)	Returns the number with the lowest value of x and y
pow(x, y)	Returns the value of x to the power of y
random()	Returns a random number between 0 and 1
round(x)	Rounds a number to the nearest integer
sin(x)	Returns the sine of a number
sqrt(x)	Returns the square root of a number
tan(x)	Returns the tangent of an angle
toSource()	Represents the source code of an object
valueOf()	Returns the primitive value of a Math object

TABLE 8.3. JavaScript Internal Math Object Properties

E	Returns Euler's constant (approx. 2.718)
LN2	Returns the natural logarithm of 2 (approx. 0.693)
LN10	Returns the natural logarithm of 10 (approx. 2.302)
LOG2E	Returns the base-2 logarithm of E (approx. 1.414)
LOG10E	Returns the base-10 logarithm of E (approx. 0.434)
PI	Returns PI (approx. 3.14159)
SQRT1_2	Returns the square root of 1/2 (approx. 0.707)
SQRT2	Returns the square root of 2 (approx. 1.414)

Code

```
function haversine (lat1, lon1, lat2, lon2) {
    var R = 6371;                   // earth's mean radius in km
    var dLat = toRad(lat2-lat1);    // difference in latitude
    var dLon = toRad(lon2-lon1);    // difference in longitude
    lat1 = toRad(lat1);             // convert to radians
    lat2 = toRad(lat2);             // convert to radians
    var a = Math.sin(dLat/2) * Math.sin(dLat/2) + Math.cos(lat1) * Math.cos(lat2) *
        Math.sin(dLon/2) * Math.sin(dLon/2);
    var c = 2 * Math.atan2(Math.sqrt(a), Math.sqrt(1-a));
    var d = R * c;                  // distance is the earth's mean radius times c
    return d;
```

FIGURE 8.13. This JavaScript example uses a series of internal JavaScript functions in the calculation, including sin, cos, atan2, and sqrt.

Code	Result

`<p>The following is an example of random link in javascript </p>`
`<script type="text/javascript">`
`var r=Math.random()`
`if (r>0.5)`
`  {`
`   document.write("<a href='http://`
`   www.unomaha.edu'>Visit Our School's`
`   Website!</a>")`
`  }`
`  else`
`  {`
`   document.write("http://maps.unomaha.`
`   edu/Mikep/biography.htm'>Visit Dr.`
`   Peterson's Homepage!</a>")`
`  }`
`</script>`
	Either:
	
	Visit our school's Website!
	
	or
	
	Visit Dr. Peterson's Homepage!

FIGURE 8.14. The random number function is called to return a number between 0 and 1. If the number is greater than 0.5, a link is made to a certain website. The function directs the user to another website if the number is less than or equal to 0.5.

Code	Result

`<script type="text/javascript">`
`function disp_prompt()`
`  {`
`   var name=prompt("Please enter your`
`   name")`
`   if (name!=null && name!="")`
`   {`
`    document.write("Hello " + name + "!`
`    How are you today?")`
`   }`
`  }`
`</script>`
	Please enter your name
	
	Hello Michael! How are you today?

FIGURE 8.15. A JavaScript function that requests the user's name.

the `map  name` command is used to define the hotspots. In this case, a circle is defined for each of the cities with a radius of 12 pixels. If the mouse passes over this circle, the appropriate text is displayed at the predefined *x*, *y* location. The result is displayed in Figure 8.21.

Code

```html
<html>
  <title>Multiplication w/dialog</title>
  <h3> Mulitiplication with JavaScript</h3>
  <body>
    <button onClick="myFunction()">
      START
    </button>
    <p id="demo">
    </p>
    <script type="text/javascript">

      function myFunction(){
        var Result;
        var Number1 = prompt("Please enter first number");
        var Number2 = prompt("Please enter second number");

        if (Number1 != null) {
          Result = Number1 * Number2;
          document.getElementById("demo").innerHTML = Number1 + " x " +
              Number2 + " = " + Result;
        }
      }

    </script>
  </body>
</html>
```

FIGURE 8.16. A JavaScript function that requests two numbers for multiplication.

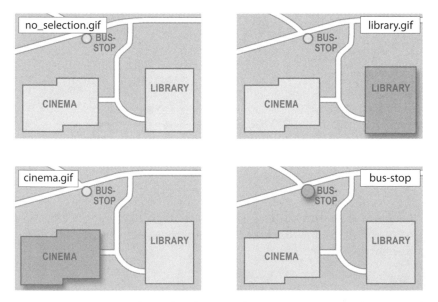

FIGURE 8.17. Four individual graphic files used in a mouseover application. Placing the mouse over an object displays an alternate graphic file. The images are preloaded, resulting in their immediate display.

```
<html>
<head>
<title>Sample JavaScript</title>
<script type="text/javascript">
// TASK: PRELOAD IMAGES
{
no_selection = new Image(330,220);
no_selection.src = "no_selection.gif";
selection_lib = new Image(330,220);
selection_lib.src = "library.gif";
selection_cin = new Image(330,220);
selection_cin.src = "cinema.gif";
selection_bus = new Image(330,220);
selection_bus.src = "busstop.gif";
}
function hiLiter(imgBase,imgSwap)
// TASK: SWAP IMAGES
// imgBase - name of the document image to be replaced
// imgSwap - name of the image object to be swapped in
{
document.images[imgBase].src = eval(imgSwap + ".src")
}
</script>
</head>
<body>
<a name="top"></a>
<img border="0" src="no_selection.gif" name="map"
        usemap="#mousemap" width="330" height="220" alt="Selector" />
<map name="mousemap">
<area shape="rect" coords="229,97,316,209" href="#top"
        onMouseOver="hiLiter('map','selection_lib');
        window.status='selection_lib'; return true"
        onMouseOut="hiLiter('map','no_selection');
        window.status='selection'; return true">
<area shape="poly" coords="15,208,15,127,75,127,75,115,147,115, 147,196,85,196,85,208" href="#top"
        onMouseOver="hiLiter('map','selection_cin');
        window.status='selection_cin'; return true"
        onMouseOut="hiLiter('map','no_selection');
        window.status='selection'; return true">
<area shape="circle" coords="127,45,10" href="#top"
        onMouseOver="hiLiter('map','selection_bus');
        window.status='selection_bus'; return true"
        onMouseOut="hiLiter('map','no_selection');
        window.status='selection'; return true">
</map>
</body>
</html>
```

FIGURE 8.18. HTML and JavaScript code that alternates the display of four identically sized graphic files based on the current position of the mouse.

```
<div id="layer1" style="position:absolute; top:25; left:35; width:200;
height:200; z-index:1; visibility:hidden; padding:8px; border: #000000
2px solid; background-color:#000000;">
   The specific content, text or graphic, would be referenced here.
</div>
```

id	Many div elements can be defined and the id value is used to identify each. "layer1" is the id for this layer.
style	Defines all of the attributes of the layer, including all of the following:
position	How the layer will be positioned. In this case, an absolute position is being defined relative to an origin.
top	The number in pixels that the div element is positioned from the top of the page. In this case, 25 pixels down from the top of the page.
left	The number in pixels that the div element is positioned vertically from the left side of the page. In this case, 35 pixels from the left.
width	The width of the div element in pixels.
height	The height of the div element in pixels.
z-index	The order in which the div elements are stacked on top of each other. The number 1 would be the first layer, the layer closest to the background. The number 2 would be stacked on top of number 1, etc.
padding	Similar to the cellpadding tag for tables, this is a margin around the div element in pixels.
border	An optional border line may be drawn the div element. The width of the line is specified in pixels. You can have a border that is solid, dotted, dashed, double, groove, ridge, inset, or outset.

FIGURE 8.19. Definition of a div layer in JavaScript and explanation of parameters.

8.5 Summary

Programming is what happens behind the curtains with computers, and only a basic understanding of programming is needed to program with the Internet. Many tools are available to assist in writing programs, including function libraries and APIs. Sufficient knowledge is needed only to understand how operations are done so that they can be emulated. Much as learning a few words in a second language is useful, learning some programming will help you to not only survive in a different environment but also to better understand how computers work and how programmers think.

```
     <html>
     <head>
     <title>MouseOver with JavaScript and DIV-Layers</title>
❶   <script type="text/javascript">
       function Show(id) {
       if (document.getElementById)
         document.getElementById(id).style.visibility = "visible";
       }
       function Hide(id) {
       if (document.getElementById)
         document.getElementById(id).style.visibility = "hidden";
       }
     </script>

     </head>

     <body>
❷   <div id="newdelhi" style="position: absolute; left: 58;
       top: 100; z-index: 2; visibility: hidden">New Delhi</div>

     <div id="mumbai" style="position: absolute; left: 72;
       top: 200; z-index: 2; visibility: hidden">Mumbai</div>

     <div id="indiamap" style="position: absolute; z-index: 1;
       left: 1; top: 1; visibility: visible"><img src="india.gif"
           width="280" height="355" border="0"
       usemap="#Map"></div>

     <map name="Map">
❸     <area shape="circle" coords="114,97,12"
       onMouseOver="Show('newdelhi')" onMouseOut="Hide('newdelhi')">
❹     <area shape="circle" coords="62,211,12"
       onMouseOver="Show('mumbai')" onMouseOut="Hide('mumbai')">
     </map>
     </body>
     </html>
```

❶ JavaScript code is inserted in a <script> tag in the head part of the html file.
The function Show() searches in the document for an element with a certain
    id and sets the visibility of that element to "visible".
The function Hide() works in the same way to hide elements.

❷ A <div> element is a layer. By giving that element an id-attribute, it can be
    addressed by scripting. The style-attributes are stylesheets and written in css-syntax.
  ▸ position: absolute; Sets the position of the layer. In this case, 58 px from the left and 100
    px from the top.
  ▸ z-index: 2; Defines how layers are positioned over each other. The layer with the highest
    number is on the top.
  ▸ visibility: hidden; Hides this layer when the page is first loaded.

❸ OnMouseOver is one of the most common event handlers. It invokes JavaScript when the
    mouse is positioned over a hotspot. In this case, the JavaScript function Show is called
    with the id 'newdelhi'. Then, JavaScript searches for the element 'newdelhi' and
    makes it visible. The OnMouseOut event hides the layer in the same way.
❹ Mumbai text appears when the mouse is positioned over this hotspot.

FIGURE 8.20. The code for the mouseover map uses JavaScript event handlers to set effects for mouse events. In this example, JavaScript is used to show and hide layers, based on the position of the mouseover predefined hotspots.

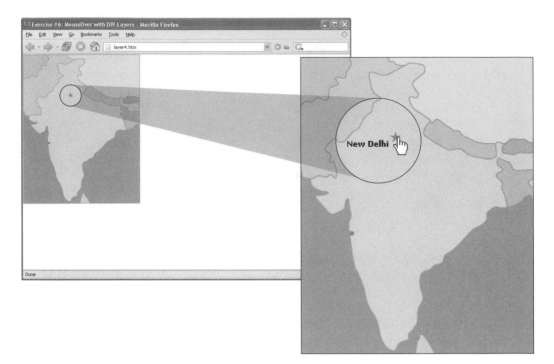

FIGURE 8.21. Using layers defined in JavaScript. Moving the mouse over the symbol for the capital of India displays its name.

8.6   Exercises

1. Enter the complete code examples presented in the web page associated with this chapter. In some cases, only the function is presented and the body will need to be added to call the function. You will likely make typing mistakes, and you will need to debug them to make them work properly. Make sure they work before you go on to step 2.

2. Change some aspect of these code examples, either in the HTML or JavaScript. A change might be the use of a different formula, different output text, or different function call.

3. Put these files in a folder called JavaScript and create an index.htm file for your new JavaScript folder. This index.htm file would look like the following:

```
<html>
<h1> My Exercise #1 JavaScript  programs </h1>
<h3> _____ Name _____ </h3>
<hr>
<a href=8-1.html> 8-1 </a>
<br>
<a href=8-2.html> 8-2 </a>
<br>

</html>
```

4. Upload the JavaScript folder to your web hosting site.

8.7 Questions

1. Do you believe that learning how to code is the "amplification of thinking"?

2. Why was JavaScript developed?

3. What is the difference between an internal and external JavaScript function?

4. Describe an example of recursion.

5. What internal JavaScript functions are used to calculate the distance between two points on a spherical surface?

6. An average of one thousand calls to the JavaScript random function should result in a number close to what?

7. Define a two-dimensional array that would store a series of *x* and *y* coordinates.

8. What is an object in JavaScript? Provide an example.

9. What is the difference between the two mouseover examples presented in this chapter?

10. What would be some possible applications of the mouseover in mapping?

8.8 References

W3Schools.com (2011) JavaScript Tutorial. [http://www.w3schools.com/js/default.asp]. (search: Learning JavaScript)

Haverbeke, Marijn (2011) *Eloquent JavaScript: A Modern Introduction to Programming.* San Francisco, CA: No Starch Press.

Wilton, Paul, and Jeremy McPeak (2009) *Beginning JavaScript.* Indianapolis, IN: Wiley Publishing.

CHAPTER 9
Map Digitizing and GPS

Any sufficiently advanced technology is
indistinguishable from magic.

—Arthur C. Clarke

9.1 Introduction

As part of the digital revolution, everything that was formerly stored in an analog
fashion—sound, pictures, film—is being converted to digital form. A good example
is music. On a phonograph record, sound is stored as a "carved sound wave" within
the record groove. The shape of the groove is related in an analog fashion to the
sound. Analog describes values as a continuous variable. In contrast, with a digital
recording, sound is sampled 44,056 times a second and each sample is converted
to a number. The number is then decoded through a digital music player into a
specific sound.

Photographic film is also an analog representation. Film contains silver halide
crystals that react to light. The number of silver halide crystals that are converted
per unit area is directly related to the amount of light incident on the crystals. The
result is an analog representation of light. The digital sampling done by modern
cameras is accomplished in a gridlike pattern with thousands of pixels in each of
the two dimensions. The value of each pixel represents the amount of light incident over that area. The pixel values for a portion of a grayscale picture are shown
in Figure 9.1.

Although maps can be stored and transmitted in the same raster format as
pictures, the potential for manipulating and analyzing the map in this way is limited. In contrast to sound and pictures, maps consist of information that cannot
be simply sampled. This is because geographic information is composed of both

158

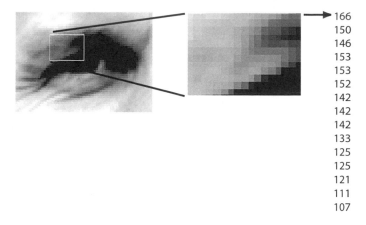

166
150
146
153
153
152
142
142
142
133
125
125
121
111
107

FIGURE 9.1. A picture of an eye represented in the form of a grid with individual "pixels." This grayscale picture is coded with values between 255 (white) and 0 (black). A small section of the area above the eye has been enlarged and individual grayscale values are shown on the right. The original picture has a width of 440 pixels and a height of 334. This equals 146,960 total pixels. In terms of computer storage, each pixel requires 1 byte (equivalent to 8 bits that can be used to represent numbers between 0 and 255). There are 1024 bytes in a kilobyte; therefore, the entire picture requires about 140 kb of storage.

graphic and nongraphic elements. The points, lines, and shapes on a map constitute the graphic aspect, while the labels and numbers represent the nongraphic or "attribute" component. These two sides of a map are coded separately by the computer. Cartographers speak of a "surface structure"—the map that we see—and the "deep structure"—the information that underlies the map (Nyerges 1980). More sophisticated methods than the simple grid approach are needed to input and process maps. In this chapter, we take a closer look at methods of map input—called digitizing—and the global positioning system (GPS).

9.2   The Vector Map

The vector method of encoding represents the primary approach for capturing and storing maps by computer. With vectors, vertices are converted to x, y values with two-dimensional Cartesian coordinates (see Figure 9.2). The Cartesian coordinate system, devised by the French mathematician Rene Descartes (1596–1650), is based on perpendicular x- and y-axes. A vector is an individual line segment that is defined with a pair of x and y coordinates. Ultimately, these planar coordinates are converted and stored in geographic coordinates of latitude and longitude.

Although not as commonly used, other methods of capturing a graphic include polar coordinates and the Freeman chain code (see Figure 9.3). Points in a polar coordinate system are defined with an angle from an origin and distance from the center, using a concentric grid of circles and intersecting lines as shown on the left. In its simplest form, the Freeman chain coding uses numbers to code standard movements from a current position. These numbers range from 0 to 7 and indicate

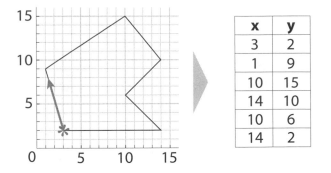

FIGURE 9.2. An example of vector encoding. The vertices of an area are converted to a series of x and y coordinates for representation by the computer.

the eight possible movements. The shape on the right in Figure 9.3 is encoded as 6, 7, 0, 2, 0, 3, 4, 4, beginning with an origin at the lower left of the figure. Notice that the diagonal movements in the 1, 3, 5, 7 directions are longer than the horizontal and vertical 0, 2, 4, 6 lengths.

9.3   Digital Map Input

9.3.1   *Manual Map Input*

Maps can be digitized (literally, converted to numbers) by hand but the process is extremely time-consuming. The only tools required are a map and a piece of graph paper. Placed on top of each other on a light table, the lines in the graph paper will be visible through the map. An origin is established at the lower-left corner, and vector coordinates are derived relative to this origin. It is not necessary to record many points; rather, points need only be taken where lines change direction. Some

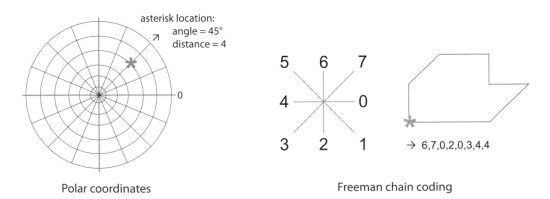

FIGURE 9.3. Less commonly used methods of graphic input. The polar coordinate system at left defines points at an angle and distance from the center. The Freeman chain code uses short line segments in one of eight directions.

lines, such as contour lines, change direction almost constantly, and their encoding would result in a large number of points.

Two approaches have been used for manually recording raster data, and both also involve the use of graph paper. In the first, grid squares are assigned values that match the ID of the area on the map. These IDs might be the land-use designation, the political unit, or a physical characteristic like soil type. This approach would produce a long list of numbers divided into rows. There would be a great deal of repetition in this list of numbers. The second and far easier approach for manually coding a map in raster form involves counting the number of identical pixels, called runs, and then specifying the length of the run. A list of pixels such as 0, 0, 0, 0, 0, 1, 1, 0, 0, 0, 0, 0 would be encoded as 5, 0, 2, 1, 5, 0. This system saves space and time if there is considerable duplication between adjacent pixels. However, if there is no duplication, the run-length method of encoding essentially doubles the number of numbers by adding an extra "1" in front of each pixel.

9.3.2 Automated Map Input

Two basic types of hardware devices are used to input maps: the coordinate digitizer and the scanner (see Figures 9.4 and 9.5). The manually operated coordinate digitizer captures vector data while the scanner inputs in raster form. Although both types of map input are automated, a considerable amount of effort is still required to make a usable digital map.

Map input with the coordinate digitizer is a time-consuming and tedious process. The operator moves a stylus, called a puck, to a point on the map and presses a button. After a feature is digitized, the attribute for the feature is entered by typing a name or a number on a keyboard. Input errors are common and require additional time to correct. A 45-minute time limit is often applied to operators between

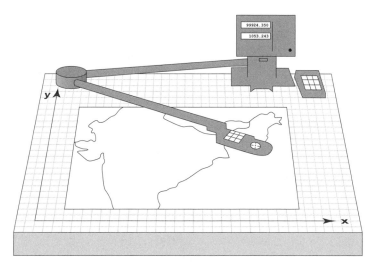

FIGURE 9.4. Example of a coordinate digitizer. The table-sized device incorporates a magnetic grid. A pointing device or "puck" is used to position a cursor over a point. Clicking on the puck outputs an *x, y* coordinate pair to a computer, where it is scaled and stored in a file.

FIGURE 9.5. Large-format scanner used for scanning maps. Page-size scanners are less than US $100. Large-format scanners, like the scanner pictured, can cost over US $10,000. Image courtesy of Contex A/S.

required breaks. The number of errors produced after 45 minutes of continuous operation increases to the point where further work is counterproductive.

Digitizing maps requires so much time that labor costs become very expensive. To save on costs, digitizing is often done in developing countries. In offshore digitizing, maps are often digitized twice by different operators and then compared for errors. The cost of digitizing the map twice is less than digitizing the map once in a country where wages are higher.

In contrast to the coordinate digitizer, scanners capture the graphic of a map in a matter of seconds. The values within the grid may simply be 0 or 1 for a map with only black lines, or may range up to millions of colors for a color map or a photograph. Inexpensive desktop scanners have resolutions of 300–1200 dots per inch (dpi; 118–472 dots per cm). The larger and considerably more expensive scanners that are more commonly used for scanning maps (see Figure 9.5) can resolve elements as small as 1/10,000 of an inch (3937 dots per cm).

The amount of data collected by high-resolution scanners is extremely large. At 2400 dots per inch, a page-size scanner would capture an 8 in × 10 in (20.3 cm × 25.4 cm) document at 19,200 × 24,000 pixels. At 24-bits per pixel for a color image (three bytes per pixel), the size of the file in bytes would be 1.4 gb (19,200 × 24,000 × 3). Even scanning at 100 dots per inch creates a file of 2.4 mb before compression. For distribution through the Internet, graphic file sizes should generally be less than 500 kb. Large-format scanners with a resolution of 10,000 dots per inch create documents that are well over 100 gb. A 20 in × 30 in map scanned at 10,000 dots per inch would result in a file size of 180 gb. Such a file cannot be reasonably distributed through the Internet. The compression of files is usually required for distributing scanned maps.

Some large online map collections, particularly of historical maps, have used a proprietary compression method called Mr. Sid™ that substantially reduces file sizes for scanned material without loss of detail. The online distribution of these documents has created a new interest in old maps that very few people had the opportunity to see or examine before the age of the Internet.

The scanning process also records everything that is on the map, including extraneous information. In addition, features on the map may be obscured by other features and therefore would not be encoded. For this reason, maps may need to be re-drafted, making several maps out of one, before scanning. The drafting process may take as much time as would be required to input the map with the coordinate digitizer.

Scanned maps are normally converted to vectors through a process called raster-to-vector conversion. This process requires considerable computing resources, including a large amount of computer memory and a fast processor. It also requires subsequent postprocessing in the form of manual editing. However, the total labor costs can be far less than the cost of the coordinate digitizer, making this form of map input less expensive than manual digitizing, at least in the developed world. Figure 9.6 describes two alternative methods.

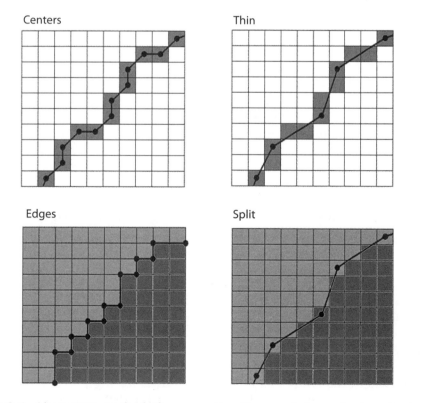

FIGURE 9.6. Alternative methods for converting lines and shapes from raster to vector. The methods depicted on the right result in far fewer points.

An important consideration in scanning a map for the Internet is copyright. Any contemporary map produced by a company and by most governments is protected by copyright. Maps produced by the U.S. government are not copyrighted, including maps from organizations such as the Central Intelligence Agency (CIA). Not surprisingly, then, most scanned maps available through the Internet are from U.S. government agencies. The Perry–Castañeda online map collection at the University of Texas consists almost entirely of maps from the U.S. government.

Finally, it should be remembered that maps are a generalized version of reality that have been expressly designed for a certain scale. When digitizing a map, the points, lines, and areas and the accompanying scale-dependent generalization of these features are being input. Once digitized, the size of a map can be easily changed without any indication that the underlying map was designed for a particular scale. Most digital maps used today are based on paper maps that incorporate some inherent level of generalization from the original map. This level of generalization is often unknown to the user of the digital version. In retrospect, digitizing maps was not a good idea. It will be many years before these generalization "artifacts" are corrected or even identified.

9.4   Maps from GPS

Global positioning systems are being used for a wide variety of applications. Auto GPS devices help drivers navigate by displaying the current position of their vehicle on an electronic map. Trucking companies use GPS to monitor the location of their entire fleet. Small, hand-held receivers are used by hikers to find their way through the wilderness. GPS has produced a revolution in how position is determined and how maps are made. Rather than digitizing a map, GPS makes it possible to directly digitize the Earth. This new ability is particularly useful in the developing world where accurate maps do not yet exist. The significance of this system can only be understood by examining how location is defined both today and before the advent of this technology.

9.4.1   Location

A location can be given in relative or absolute terms. An example of a relative description of place is as follows: "The house next to mine." This type of description is used in normal conversation because it can be easily understood. If the description is not understood, we choose a different landmark until the person makes a connection with a particular feature.

Absolute location is always defined based on the origin of a coordinate system. An example would be the numbered streets of a city arranged in a grid format. For example, "42nd Street and 5th Avenue" would refer to a specific intersection that is a certain number of blocks away from 1st Street and 1st Avenue. Cartesian coordinates are used in a similar way to define location on a flat surface.

In the spherical coordinate system that is used for the Earth, latitude can be related to the y coordinates and longitude corresponds to the x (see Figure 9.7). Lines of latitude and longitude are also referred to as parallels and meridians. The

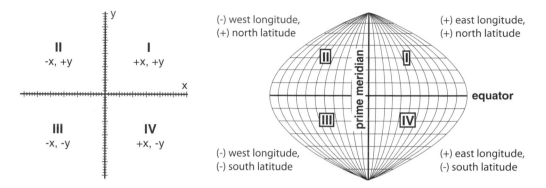

FIGURE 9.7. The Cartesian coordinate system uses *x* and *y* coordinates to locate points on a plane. The four quadrants specify the position of points in plus and minus directions from an origin. The spherical coordinate system of latitude and longitude has its origin at the intersection of the equator and the prime meridian. It is also divided into four sectors, with each being either north, south, east, or west of the origin. The *x* coordinates can be compared to longitude and y coordinates to latitude.

origin of this spherical coordinate system is located at the intersection of the prime meridian and the equator.

The equator is a natural origin for lines of latitude, but no similar line of origin exists for longitude. In the 2nd century AD, Ptolemy used a zero meridian through the Canary Islands that marked the western boundary of the known world. Until the 1870s, each country defined its own zero meridian. Portugal placed its zero meridian at Lisbon, for Spain it was the coastal southern city of Cadiz, for Norway it was Christiana, for Sweden Stockholm, for Russia Pulkowa, for Germany Ferro, for France Paris, for the United States Washington, DC, and for Great Britain it was Greenwich.

In 1871, at the first International Geographical Congress (IGC) in Antwerp, Belgium, it was decided that whenever ships exchanged longitudes at sea, they should be based on a prime meridian that passed through an observatory in Greenwich, England. Another meeting in Rome in 1875 failed to resolve the standardization of the prime meridian at Greenwich. France tied its acceptance of Greenwich to the British acceptance of the metric system.

Finally, at the International Meridian Conference in 1884 in Washington, DC, 25 countries voted to adopt the meridian at Greenwich as the prime meridian for the world. The precise location of the prime meridian was defined by the crosshairs in the eyepiece of the large "Transit Circle" telescope at Greenwich. It was agreed that longitude would be measured in two directions from this line: "east longitude being plus and west longitude being minus."

The system of latitude and longitude is only one of many coordinate systems that is used to define location on the Earth. The other coordinate systems are used for smaller areas and consider the Earth to be flat. Although this practice causes some error, especially further away from the origin, the error is so slight that these coordinate systems are still useful for many applications such as public works departments in urban areas. These flat coordinate systems, such as the Universal

Transverse Mercator (UTM) and state plane coordinates, were used because latitude and longitude were difficult to determine. With the advent of GPS, almost all locations are now being defined with latitude and longitude.

9.4.2 Radio-Based Navigation

Before GPS there was radio-based navigation. First introduced in the early part of the 20th century, the LOng-RANge accurate radio navigation (LORAN) system was used extensively by ships and aircraft by the middle part of the 1900s. LORAN measures the time-of-arrival difference between two signals transmitted by two different ground stations. The pulse from the first station, called the master, triggers the second station, called the slave, into transmitting a similar pulse after a set time delay. A chart is then consulted that contains a series of hyperbolic curves that define constant time differences between particular station pairs. The position of the ship or airplane will be somewhere along the curve that corresponds to the measured time difference. By taking another reading from a second pair of stations whose curves intersect those of the first pair, a definitive geographic location can be established.

An advanced system called Loran-C was developed to provide radio navigation service for U.S. coastal waters. It was later expanded to include complete coverage of the continental United States as well as most of Alaska. U.S. Loran-C stations worked in international partnership with Canadian and Russian stations. Loran-C provided better than 0.25 nautical mile absolute accuracy. It was possible to navigate to a previously determined position with an accuracy of 50 meters or better using Loran-C in the so-called time difference repeatable mode. Advances in technology allowed greater automation of LORAN-C operations. Even with improvements, however, the system could not compete with the accuracy of GPS. LORAN ceased to operate in 2010 (search: Coast Guard terminates LORAN-C broadcast).

9.4.3 Development of the Global Positioning System

Similar to the Internet, GPS is also a product of the Cold War. The U.S. Department of Defense wanted a system that could target missile silos in the Soviet Union. U.S. missiles were placed in submarines to avoid Soviet detection. Being underwater, it was difficult for the submarine to determine exactly where it is located, and accurate targeting by the submarines required that its position be precisely determined. A system was needed to quickly and accurately determine the position of the submarine once it surfaced.

The U.S. Department of Defense established the defense navigation satellite system (DNSS) in 1969 (Kaplan and Hegarty 2005, p. 2). By 1977, ground stations had been installed, although no satellites had yet been launched. Eleven satellites were put into orbit between 1978 and 1985. The initial system, called NAVSTAR (navigation system with timing and ranging) was strictly for military use. An airline tragedy in 1983 prompted the U.S. government to make the system available for civilian use. A Korean Airlines flight from New York to Seoul, South Korea, with a stop in Anchorage, Alaska, mistakenly strayed over Kamchatka Island, a Soviet possession. The Boeing 747 was shot down by Soviet interceptors, and all 269 people aboard were killed.

Civilian access to NAVSTAR was discontinued in 1990 owing to the first Gulf War. In 1993, public use was restored, and GPS became fully operational with a full complement of satellites in 1995. In 2000, Selective Availability, a system that had degraded the civilian signal, was turned off by an order of President Clinton. Overnight, the accuracy of GPS for civilian use went from no better than 100 m to 10–15 m. Per U.S. policy and law, the GPS standard positioning service is now available free of charge to civilian users worldwide for their peaceful transportation, scientific, and other uses. GPS is widely regarded as one of the major human accomplishments of the latter part of the 20th century.

9.4.4   *GPS Constellation*

GPS consists of a constellation of satellites that orbit the Earth at an altitude of 20,200 km (12,550 mi) between 60°N and 60°S latitude at a speed of 1.9 mi per second (El-Rabbany 2006, p. 2). Each satellite is 17 ft (5.8 m) in size with solar panels extended and weighs almost a ton (see Figure 9.8). The satellites orbit the Earth twice a day, and this, along with the number of satellites, guarantees that signals from at least six of the satellites can be received from any point on Earth at almost any time. A GPS receiver collects signals from satellites in view and then displays the user's position, velocity, and time (see Figure 9.8). Some receivers display additional data, such as distance and bearing to selected waypoints or digital charts.

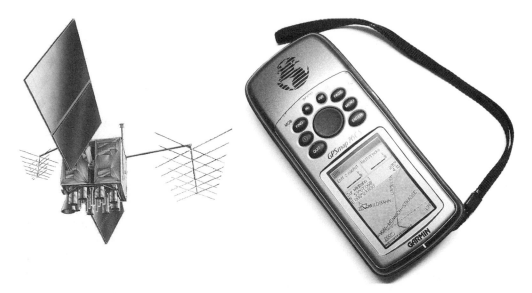

FIGURE 9.8. Each GPS satellite (left image) weighs 1900 lb and is 5.8 m (17 ft) wide with solar panels extended. They orbit the Earth at an altitude of 20,200 km (12,550 mi) at a speed of 3.06 km per second (1.9 mi per second). A hand-held GPS receiver with integrated map display is shown on the right. It receives low-energy signals from at least four satellites to determine its position. Left from www.gps.gov/multimedia/images/; right from en.wikipedia.org.

The satellites are designed for a life span of eight years, but many last longer. Of the 49 GPS satellites put into orbit, 32 were still active in 2012. A total of 20 of these satellites are older than 2005—with the oldest dating to 1992 (search: U.S. Coast Guard Navigation Center). A total of 24 satellites are needed to keep the system operational. The remaining satellites serve as orbiting spares.

The satellites are continuously tracked by unmanned monitoring stations spread around the globe at Ascension Island, Diego Garcia, in the Indian Ocean, Kwajalein, Hawaii, Cape Canaveral, Florida, and Colorado Springs, Colorado. The orbit puts each satellite over a ground control station at least once a day when their position is precisely measured. The station continuously determines the distance to each satellite in view. These measurements are used to update the master control station's precise estimate of each satellite's position in orbit through the master ground control station at Schriever Air Force Base in Colorado Springs, Colorado. Any discrepancy found in a satellite's position is transmitted to the satellite, and this data is transmitted to the GPS receivers. These corrections in the orbits of the satellites are small because their high altitude ensures that their orbits are very stable and precise. Any change in orbit is usually caused by the gravitational pull from the Moon or the Sun and by the pressure of solar radiation on the satellite.

9.4.5 *Positioning with GPS*

GPS is based on satellite ranging. The position of the receiver is determined by measuring the distance to a group of satellites in space. Because their positions are known, the satellites act as precise reference points.

Each GPS satellite transmits a signal of 1500 bits at millisecond intervals. Called a pseudorandom code, the message contains the satellite ID, orbit, and current time. The complicated code appears to be random but is carefully chosen. The codes are purposely made complicated so that they can be easily and unambiguously compared.

The GPS receiver generates a similar internal code at exactly the same time. The time difference is measured between the code sent by the satellite and the code generated by the receiver (see Figure 9.9). The speed of the radio signal is equal to the speed of light and is precisely known (186,282 mi per second or 299,792.458 km per second). The receivers multiply the speed of light by the travel time to find the distance. To determine latitude, longitude, and elevation, the distance between a GPS receiver and at least four orbiting satellites is calculated.

Use of the pseudorandom code has a number of advantages. The first is that the code can be transmitted with a very weak signal that uses only a very small amount of power from the satellite. In fact, the GPS satellite uses less electricity than a 50-watt light bulb. A second benefit is that satellites can share the same frequency without interfering with each other, and each satellite has its own code so that it can be easily identified. GPS is controlled by the United States Air Force. The Air Force can interrupt civilian access for any region of the world, rendering the system useless for anyone but the U.S. military.

The signal sent by the satellite is so faint that it is just barely above the Earth's inherent background noise. To distinguish between the two, the signal is divided into time slices and then moved back and forth until a match is made between the

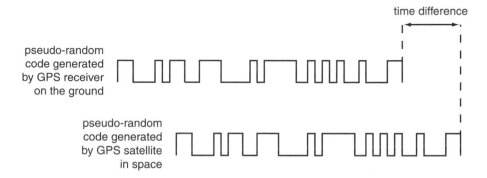

FIGURE 9.9. The GPS receiver and satellite generate the same pseudorandom code at exactly the same time. When the code arrives from the satellite it is compared to the code generated by the receiver, and a time difference is calculated. This difference is multiplied by the speed of light (186,000 mi per second) to determine the distance to the satellite.

identical signal produced by the satellite and receiver (see Figure 9.10). The receiver tracks multiple satellites at once to determine its position.

The sending of the pseudorandom code is controlled by very precise atomic clocks. Although the name implies that these clocks use atomic energy, they actually just measure time through the oscillations of a particular atom. Each clock costs $100,000, and every satellite has four clocks to be sure that one is always operational. The clocks are accurate to one-billionth of a second. It would therefore take 30 million years for it to gain or lose a single second. The precision is important because the signal travels so fast and so far that slight errors in timing can cause huge errors in the calculation of distance. For example, the time it takes

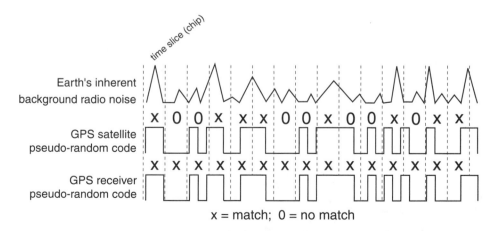

FIGURE 9.10. The GPS satellites transmit a very weak signal, about the same as the Earth's background radio noise. Both the GPS signal and the background noise are random so that when we divide the signal up into time slices or chips, the number of signal matches (X's) will equal the number of nonmatches (0's). If we slide the receiver's pseudorandom code back and forth until it lines up with the signal from the satellite, we will be able to distinguish the signal from the Earth's background noise.

the signal to travel from a satellite that is directly overhead is 6/100th of a second (12,550/186,000).

Calculating the location from a series of distance measurements is essentially a form of triangulation involving trigonometric calculations. Imagine that each satellite is in the middle of a sphere (see Figure 9.11). The distance to a satellite defines the periphery of a sphere. If the receiver is 13,000 miles from satellite "A," it is located somewhere on the surface of a sphere with a 13,000-mile radius. At the same time, the receiver might be 15,000 miles from a second satellite. The imaginary sphere that is formed by this satellite intersects with the first 13,000-mile sphere. The distance measurement from the third satellite narrows down the position even more because the only place that three spheres can come together is in two points. The distance to a fourth satellite is used to distinguish between these two points and resolve clock accuracy problems in the receivers.

To calculate the position on the ground, we need to know not only the distance to the satellites, but also the position of the satellites in space. Fortunately, the orbits are very predictable and change so little that GPS receivers can calculate their positions from an internal "almanac." This almanac tells the receiver where each satellite is located at any given moment. The ground-based monitoring

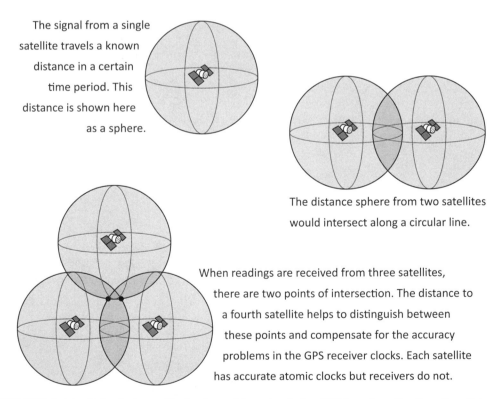

The signal from a single satellite travels a known distance in a certain time period. This distance is shown here as a sphere.

The distance sphere from two satellites would intersect along a circular line.

When readings are received from three satellites, there are two points of intersection. The distance to a fourth satellite helps to distinguish between these points and compensate for the accuracy problems in the GPS receiver clocks. Each satellite has accurate atomic clocks but receivers do not.

FIGURE 9.11. Calculation of latitude and longitude by GPS involves finding the distance to at least three different satellites. Each distance constitutes a sphere and the three spheres converge on two points. The distance to a fourth satellite is used to distinguish between these points and to compensate for the lack of accuracy of the internal clocks in the GPS receivers.

stations provide information about any changes in the orbits of the satellites, and this is transmitted back to each receiver through the satellites.

9.4.6 Sources of GPS Error

Clock accuracy leads to the most serious potential source of GPS error. A variation of only a second will cause a distance calculation error of 186,000 miles. Although each satellite uses four atomic clocks at $100,000 each, the clock associated with the receiver only has the accuracy of a quartz watch. The two clocks are rarely in sync. With measurements from only two satellites, such an error would not be noticeable, but three erroneous distances from three separate satellites does not lead to a viable solution as the distances will not intersect at a point. Use of the fourth satellite is critical in compensating for the slight time variations in the much less expensive clock in the receiver.

GPS receivers are so programmed that when they get a series of measurements from satellites that cannot intersect at a single point, they realize that something is wrong and suspect the internal clock is not correct. The computer in the GPS receiver then begins adding and subtracting a constant amount of time from all the measurements until it determines an answer that intersects at a point. Rather than iteratively adding and subtracting, this compensation is done through an algebraic expression. The fourth satellite is also needed to determine the elevation of the point that is also provided by the GPS receiver.

A second source of error in calculating distance is the particular constellation of satellites. If the GPS satellites are all in a similar position, then the satellite geometry is poor and the receiver is unable to provide an accurate position reading. Because the satellites are in a similar position, the common area where these distance measurements intersect is fairly large (see Figure 9.12). This is called geometric dilution of precision (GDOP) and results in errors of as much as 300 to 500 ft (~90–152 m). Use of a GPS receiver that can track a larger number of satellites simultaneously reduces this form of error. Most modern GPS receivers can track up to 12 satellites at once.

Tall buildings and mountains lead to a third source of error for GPS and may result in a total loss of signal. In a city, where buildings block the signal from some or all of the satellites, this phenomenon is referred to as an *urban canyon*. If buildings are on both sides, signals can only be received from satellites that are directly above. Even if enough signals can be obtained to establish a position, the position will be inaccurate because of GDOP. If less than three or four satellites are in view, the current position cannot be determined.

A fourth source of error is called multipath and results when signals are reflected by objects in the environment. Multipath errors occur with GPS when the signal bounces off a building or terrain before reaching the GPS receiver's antenna. The signal takes longer to reach the receiver than if it had traveled a direct path. As a result, the GPS receiver calculates a longer distance to the satellite. Multipath errors are typically less than 15 ft (~5 m).

Further sources of error are propagation delay due to atmospheric effects. Propagation delay is the "slowing down" of the GPS signal as it passes through Earth's ionosphere and troposphere. Radio signals travel at significantly slower speeds

FIGURE 9.12. Geometric dilution of precision (GDOP). When the satellites are close together, the overlap between the estimated distances is larger than when the satellites are close together. More accurate GPS measurements are possible when the available satellites are further apart.

once they enter our atmosphere. GPS receivers are designed to compensate for these effects, but very small positional errors still occur.

9.4.7 Differential GPS

All of the possible sources of GPS error can add up to 60 to 500 ft (18 to 152 m). GPS measurements can be made much more accurate using two receivers. One receiver is placed in a known location and is used to calculate the current error in the satellite range data. It acts as a static reference point because its position is precisely known. The correction data that it calculates can then be applied to all other receivers in the same area, thereby eliminating most measurement error (see Figure 9.13). The concept works because the satellites are so far away that any errors measured by one receiver will be the same for all other receivers in the area. Differential correction simultaneously resolves all errors in the GPS signal, whether the error is associated with the receiver clocks, satellite geometry, or delays to the signal caused by atmospheric conditions.

Various methods have been devised to provide differential GPS data, including the construction of beacons that transmit on a long-wave frequency over a several-hundred-mile area. Special receivers are needed to capture the differential information. The larger and more expensive GPS receivers all integrate beacon radio receivers so that they can simultaneously receive data from the GPS satellites and ground-based beacons. These receivers and associated antenna are larger than the actual GPS unit.

In 1999, the Federal Aviation Administration implemented a similar system to improve the accuracy of GPS. The wide area augmentation system (WAAS) utilizes a network of ground-based reference stations that transmit corrections to geosynchronous communications satellites, which then transmit the corrections to the user. The position of these reference stations has been accurately surveyed.

The two WAAS satellites are in geostationary orbits over the United States, restricting coverage to this part of the world. These satellites transmit the correction information to the GPS user on the GPS frequency. The GPS receiver then decodes this information and applies it to its calculated position to significantly improve accuracy. WAAS was designed to allow aircraft to use GPS for "instrument only" landings. In practice, WAAS has a horizontal accuracy of 1 m and a vertical accuracy of 1.5 m. This compares to a 15-m horizontal accuracy and a 23-m vertical accuracy for non-WAAS GPS. WAAS reception is incorporated in many GPS receivers, even smaller hand-held units, although it requires more power to operate.

9.4.8 Applications of GPS

A variety of applications of GPS developed, spurring a multibillion dollar business (Spencer et al. 2003, p. 12). GPS is a prime example of how a government program can create an entire business sector employing tens of thousands of people. GPS is used in airplane, car, and marine navigation, land survey, mapping and GIS, marine survey, mining, public safety, and timing and synchronization applications. In 1998 the cost of a car navigation that included GPS was over $2000. The devices now cost less than $200 and have become nearly ubiquitous in cars, especially in some parts of the world where the street layout makes navigation difficult. The small device provides visual and voice prompts to guide the driver turn by turn to the destination. Smart phones have implemented similar functionality.

One of the most intriguing uses of GPS is in agriculture. The method, called

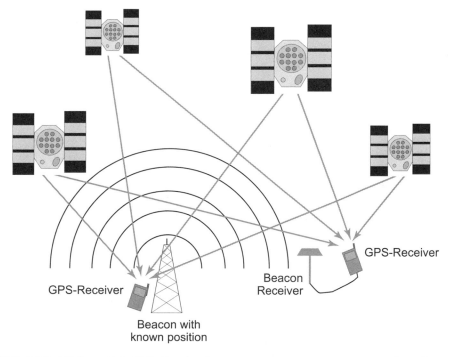

FIGURE 9.13. Differential GPS. To obtain more accurate measurements than is possible from a single GPS unit, a fixed GPS receiver broadcasts the current error in the GPS signal. The GPS unit in the field simultaneously receives data from the GPS satellites and the GPS receiver on the ground. The GPS error from the known position is used to correct that of the GPS receiver in the field.

precision agriculture, involves gathering data on yields as the field is being harvested. This data gathering is done by installing a GPS on the harvester and measuring yield in real time. The following spring, fertilizer is applied at a different rate across the field according to the yields of the previous year. GPS also controls the differential application of fertilizer. The purpose of the procedure is to apply fertilizer to only those parts of the field where it would actually increase the yield. In many cases, the savings in fertilizer in a single year compensate for the cost of the GPS receiver and associated computers.

9.5 Summary

Digitizing the Earth, first through maps and then directly through GPS, has been a major effort since the 1970s. The resulting digital representation varies in detail and accuracy. Digital maps that have been captured from paper maps incorporate the generalization of the map. Removing this generalization and constantly updating the underlying base map will require considerable resources in the future.

GPS accuracy has increased enormously. Initially, the civilian signal was intentionally degraded. In 2000, this Selective Availability was turned off, and accuracies for civilian uses increased from only 100 m to less than 15. WAAS, implemented by the U.S. Federal Aviation Administration (FAA), further improved system accuracy, especially for elevations. Reception of the WAAS signal is incorporated on many GPS receivers. A combination of a variety of error correction methods, particularly the use of differential GPS, can lead to accuracies in the range of centimeters.

Other countries are developing their own GPS systems. At 21 satellites, the Russian GLONASS (Globalnaya navigasionnaya sputnikovaya sistema) system is just short of the 24 satellites needed for global coverage. High-end GPS receivers incorporate the reception for both US and Russian satellites. The first of 30 satellites in the Galileo system that is being established by the European Union was launched in October 2011. The system is expected to be fully operational by 2019. The Chinese Compass Navigation System was put into operation for China in December 2011 and is expected to have global coverage by 2020.

9.6 Exercise

Use a handheld receiver or another mobile device with GPS to capture a series of 10 points of interest. These points will be mapped in Chapter 10.

9.7 Questions

1. Describe the difference between an analog and digital picture.

2. What is the "deep structure" of a map that makes it different from a picture?

3. What is the relationship between the Cartesian (x, y) and spherical coordinates (lat/long)?

4. Describe two methods of manual map input.

5. Compare the input of maps through a coordinate digitizer and scanner.

6. Describe the raster-to-vector conversion process.

7. What defines the current prime meridian, and when was it finally accepted as the starting point of the longitude system?

8. What was the primary method of navigation by ship during the 20th century?

9. What was the impetus for the development of GPS?

10. Describe some milestones in the development of GPS. When did the system become fully operational?

11. Describe the GPS satellite constellation. How many satellites are used, and how far are they from the Earth?

12. What is a GPS almanac?

13. How is the location of the GPS receiver determined?

14. What is multipath error with GPS?

15. What are a GPS beacon and WAAS?

16. Describe some possible applications of precision agriculture.

9.8 References

El-Rabbany, Ahmed (2006) *Introduction to GPS: The Global Positioning System*. Boston: Artech House.

Kaplan, Elliott, and Christopher Hegarty (2005) *Understanding GPS: Principles and Applications*, 2nd ed. Norwood, MA: Arctech House.

Nyerges, T. (1980) *Modelling the Structure of Cartographic Information for Query Processing*. Unpublished Ph.D. dissertation, Ohio State University, Columbus, Ohio. 203 pp.

Spencer, John, Brian Frizzelle, Philip Page, and John Vogler (2003) *Global Positioning System: A Field Guide for the Social Sciences*. London: Blackwell Publishing.

CHAPTER 10
Map Mashups

When people use the term Web 2.0, I always feel a little bit stupider for the rest of the day.

—JOEL SPOLSKY

10.1 Introduction

The term *mashup* was first used for a movement in pop music that involved the digital mixing of songs from different artists and genres. In technology, the term refers to the melding of web resources and information. A mashup combines tools and data from multiple online sources. The most common type of mashup is the mapping of data.

Mashups are an integral part of what is commonly referred to as Web 2.0. Web 2.0 represents a variety of innovative resources, and ways of interacting with, or combining web content that began in about 2004. In addition to mashups, Web 2.0 also includes wikis, such as Wikipedia, blog pages, podcasts, RSS feeds, and Ajax. Social networking sites like Facebook and Google+ are also seen as Web 2.0 applications.

Central to mashups are Application Programming Interfaces (APIs). These are function libraries that are made freely available. Many different APIs have been written for the user-driven web. APIs are the tools that facilitate the melding of data and resources from multiple web resources by providing the means to acquire, manipulate, and display information from a variety of sources.

In a strict sense, a *map mashup* combines data from one website and displays it with a mapping API. The term has come to be used for any mapping of data using an API, even data supplied by the user. The popularity of mapping APIs should be no surprise considering the amount of data that has a location component. The ease of mapping this information has resulted in all kinds of different maps, many showing information that has never been mapped before.

One particular advantage of using an API from a major online mapping site is that the maps represent a standard and immediately recognizable representation of the world. Overlaying features on top of these maps provides a familiar and comfortable frame of reference for the map user. This has created a different way of making thematic maps. In the past, thematic maps limited the display of reference information such as cities and transportation networks—partly for simplicity and to emphasize the distribution being mapped. In doing so, the cartographer may have left out critical information to help the user understand the locational aspects of the spatial pattern. Adding locational information may also detract from the spatial pattern that is being communicated.

Map mashups have had a major influence on how spatial information is presented. This chapter provides an introduction to map mashups and associated APIs. Specific examples are presented using Google Maps and a variety of other mapping APIs from Microsoft, MapQuest, Nokia, OpenStreetMap, Baidu, AutoNavi, and Leaflet. All of these implementations will be seen to be very similar, using almost identical functions and objects. Finally, an API called Mapstraction is introduced that attempts to bridge the differences between the major mapping APIs.

10.2   Google Maps API

Introduced soon after Google Maps in 2005, the Google Maps API is by far the most commonly used. The API consists of a series of functions that control the appearance of the map, including its scale and location, and any added information in the form of points, lines, or areas and associated descriptions. There are multiple versions of the Google Maps API, the most widely used being the Google Maps JavaScript API.

The use of Google Maps API is essentially free, provided the site does not charge for access. Google places a limitation on the number of maps that can be served. A site cannot generate more than 25,000 map loads a day for 90 consecutive days. A map load is one map displayed with the Google Maps API. Once loaded, the degree to which a user interacts with a map has no impact on the map load number.

It would be extremely difficult for the average user of the Google Maps API to reach 25,000 map loads. Even if a site were to go "viral" with a topic that generates considerable interest, it would need to sustain 25,000 map loads per day for 90 consecutive days before the limit would be reached. Usage limits can be placed on a site so that it does not exceed that number. If the site consistently exceeds 25,000 maps a day, Google charges US $0.50 for each 1000 map views beyond this limit. Serving 100,000 Google maps a day would cost $37.50 (75,000/1000 * 0.5) a month.

Specialized Google Maps API web services have additional usage limits, including:

- Directions—provides directions in text form—limited to 2500 a day.
- Distance Matrix—returns travel distance and time—limited to 100 elements per query and 2500 a day.

- Elevation—elevation at points—limited to 2500 requests per day where each request returns up to 512 elevations for a total of 1,280,000.
- Geocoding—converts a street address to latitude and longitude—limited to 2500 a day.
- Places—returns business establishments and other points of interest around a point—requires an API key and limited to 1000 requests a day.

A Google Maps API key is a numeric code that registers your site with Google. It is not needed for normal applications and will be required only if the usage limits are exceeded, or the Places web service is used. A $10,000 a year pay service, Google Maps API for Business, provides an unlimited number of map downloads and no limits on these specialized services. Over 99% of Google Map API users pay nothing for use of the service.

10.2.1 Basic Google Map

Figure 10.1 shows the JavaScript code and API calls for displaying a simple map that is centered at a specific location. The zoom level, which can range from 0 to 21, is set to 15 under myOptions. The center is defined with a specific latitude and longitude value, and the ROADMAP option is selected to define the map style. All of the API calls are made in the initialize function. This function is called using onload within the body of the HTML file.

A simple change to this code can be made by substituting new latitude and longitude values. The latitude/longitude for a specific point can be determined in a number of different ways:

1. In Google Maps, clicking on the map will show the latitude and longitude of the point will appear in the top line of the Google Maps window.
2. A right-mouse click with MapQuest displays the values in a pop-up window.
3. In Bing Maps, the mouse location is displayed with a right-click.
4. To display the coordinates in the decimal degrees format with Google Earth, select Tools/Options and click on the decimal degrees option.
5. Finally, there are a number of online utilities. Searching for "Finding latitude and longitude" will lead you to a site. Most use Google Maps. One of these sites is:

http://www.findlatitudeandlongitude.com/

Another change to the basic Google Map is the type or style of map that is displayed. Google offers four views:

- **MapTypeId.ROADMAP** displays the default road map view.
- **MapTypeId.SATELLITE** displays Google Earth satellite images.
- **MapTypeId.HYBRID** displays a mixture of normal and satellite views.
- **MapTypeId.TERRAIN** displays a physical map based on terrain information.

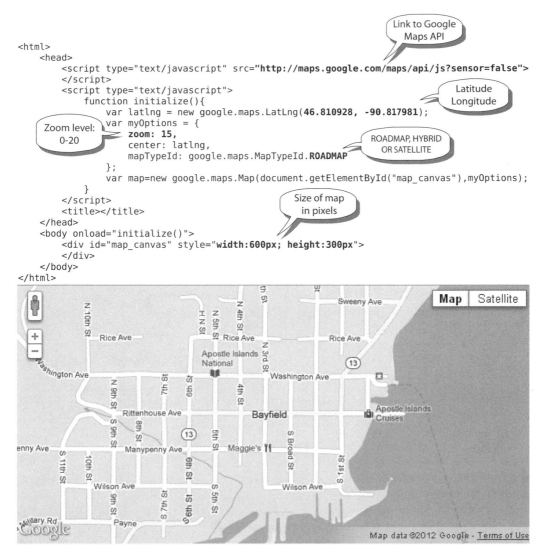

```
<html>
    <head>
        <script type="text/javascript" src="http://maps.google.com/maps/api/js?sensor=false">
        </script>
        <script type="text/javascript">
            function initialize(){
                var latlng = new google.maps.LatLng(46.810928, -90.817981);
                var myOptions = {
                    zoom: 15,
                    center: latlng,
                    mapTypeId: google.maps.MapTypeId.ROADMAP
                };
                var map=new google.maps.Map(document.getElementById("map_canvas"),myOptions);
            }
        </script>
        <title></title>
    </head>
    <body onload="initialize()">
        <div id="map_canvas" style="width:600px; height:300px">
        </div>
    </body>
</html>
```

Callout labels: Link to Google Maps API · Latitude Longitude · Zoom level: 0-20 · ROADMAP, HYBRID OR SATELLITE · Size of map in pixels

FIGURE 10.1. A basic Google map. The center of the map can be changed along with the zoom level and map type. Copyright 2013 Google.

The initial zoom level can also be changed. A value of "0" would draw a zoomed-out, small-scale map of the world showing the extreme distortion of the Mercator projection. As the zoom level number increases, so too does the scale of the map. The upper value varies for different parts of the world. Generally, 20 levels of details are always available. Some parts of the world have more than 20 zoom levels.

10.2.2 Google Map with Marker

The default Google marker is an upside-down raindrop symbol, but a large number of alternative symbols are available. It is even possible to design symbols using a program like Adobe Photoshop™. The example in Figure 10.2 places the basic Google upside-down teardrop marker at the center of the map. The `event.addlistener` option sets the zoom level to 17 when the marker is clicked. The initial zoom level is 15. The title text, "Hello World," is displayed when you hover the mouse over the marker.

10.2.3 Google Map with Clickable Marker

In the example in Figure 10.3, the variable `contentString` is defined with text formatted in HTML. This is associated with an `infoWindow` variable that is

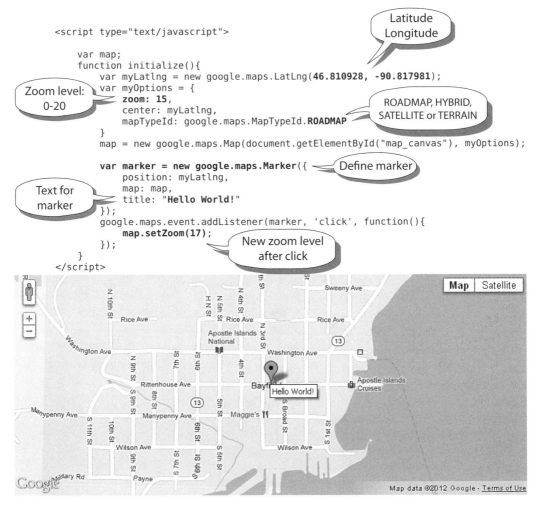

FIGURE 10.2. A single marker with a mouseover function that displays "Hello World!" (Copyright 2013 Google.

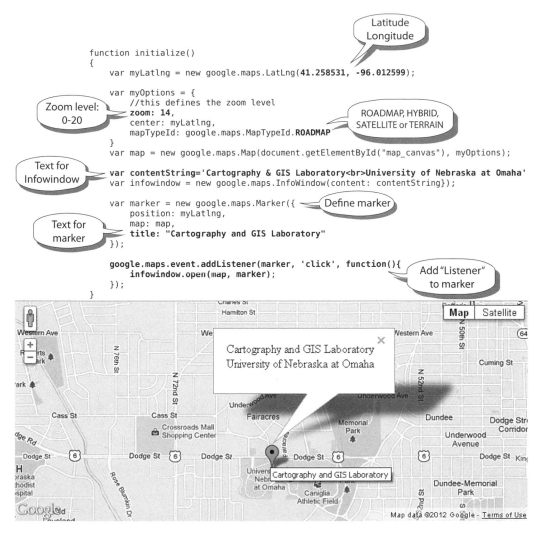

FIGURE 10.3. Example of a clickable marker. The `contentString` text variable is defined in HTML. Copyright 2013 Google.

subsequently associated with `google.maps.event.addListener`. When the user clicks on the marker, the text is displayed in a pop-up bubble. The HTML for this bubble could reference a picture or even a video.

10.3   Other Mapping APIs

Within a few years after Google Maps was introduced in 2005, Microsoft, Yahoo!, and MapQuest changed their online mapping service to a tile-based system. Eventually, they also released their own mapping APIs. Like Google's initial API, each of these required the use of an electronic key, an alphanumeric string within the initial reference to the API. While the key is made freely available, it limits its use to the server that is specified when the key is requested. The primary benefit of the key is that it controls access to the API. A user who exceeds the limits on use, or uses the API for illegal purposes, can easily be denied access. The key can be obtained for free.

10.3.1 Bing Maps API

Microsoft's Bing Maps (formerly LiveLocal) followed Google's lead and now includes a street map, an aerial view, a Bird's-Eye view, a StreetSide view, and 3D Maps, although 3D maps services are not as developed. The oblique Bird's-Eye view provided through Bing Maps generally has more detail than Google's satellite view. In the next examples, the key is defined following "credentials." Although a key is required, it is no longer associated with a specific server.

Samples of code using Bing Maps' API are provided in Figures 10.4 through 10.7. Figure 10.4 is Bing's API code for a simple map centered at a specific location.

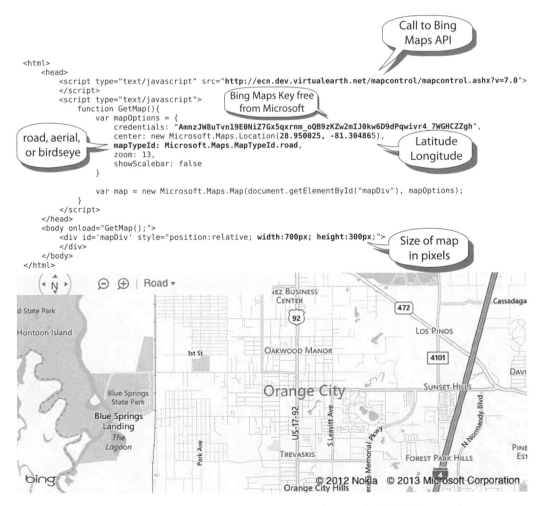

FIGURE 10.4. A basic map with the Bing Maps API. Copyright 2013 Microsoft.

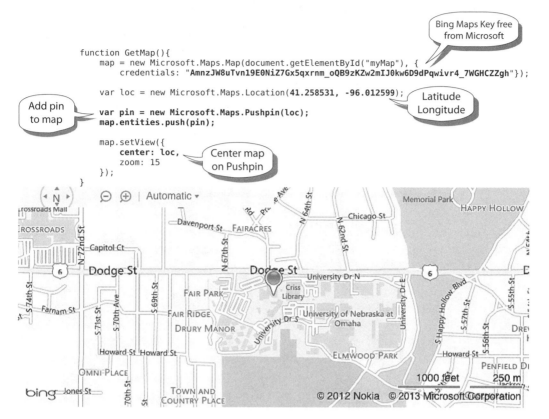

FIGURE 10.5. Placing a marker with the Bing Maps API. Copyright 2013 Microsoft.

Figure 10.5 uses the Bing API to add a marker called a pushpin. Figure 10.6 adds an info window to the pushpin. In Figure 10.7, the contents of this window are visible by simply hovering the mouse over the marker.

10.3.2 Nokia HERE API

Initially, mobile phone maker Nokia concentrated on supplying a stand-alone mapping application for its phones. Using a large map file stored in the memory of the phone, the user could make a map without being connected to the Internet. With the purchase of the mapping giant NavTeq in 2007, Nokia turned its attention to the Internet with the introduction of OviMaps in 2007, renamed Nokia Maps in 2011 and eventually HERE in 2012. HERE is also available as an "app" for mobile devices.

The Nokia mapping API requires use of a key, as shown in Figure 10.8, that makes a basic map with a zoom level of 10. Like many APIs, Nokia uses a document object model (DOM) to define the interface between objects. Figure 10.9 adds a marker to the Nokia's map. This marker is styled in Figure 10.10, and converted to text in Figure 10.11. The text option for markers is not easily implemented with other mapping APIs.

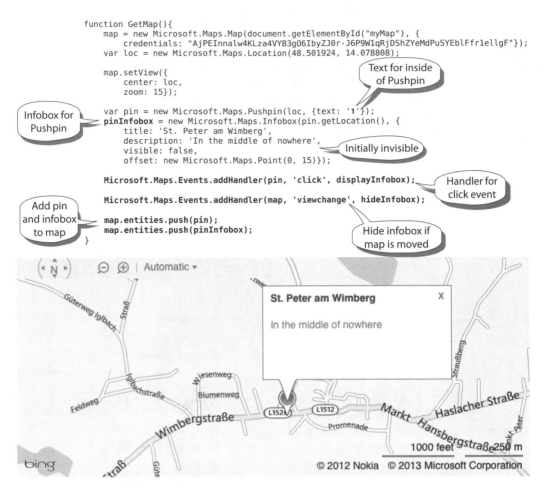

FIGURE 10.6. Placing a marker with the Bing Maps API. Copyright 2013 Microsoft.

10.3.3 MapQuest API

MapQuest first introduced its mapping website in 1996 and enjoyed many years as the primary online mapping website. Caught off-guard by the introduction of Google Maps in 2005, it adapted slowly to the new tile-based, API-based method. MapQuest lost its number one position to Google in 2009. As with the Nokia API, use of the MapQuest API also requires a numeric key. The examples in Figures 10.12–10.15 demonstrate how to make a basic map and add controls to the map, a marker, and a line.

10.3.4 OpenStreetMap API

OpenStreetMap is the major volunteered geographic information (VGI) website, also referred to as crowd-sourcing. The site has thousands of contributors who have uploaded everything from GPS traces to new points of interest. One major advantage of OpenStreetMap is its ability to access the underlying vector data, although

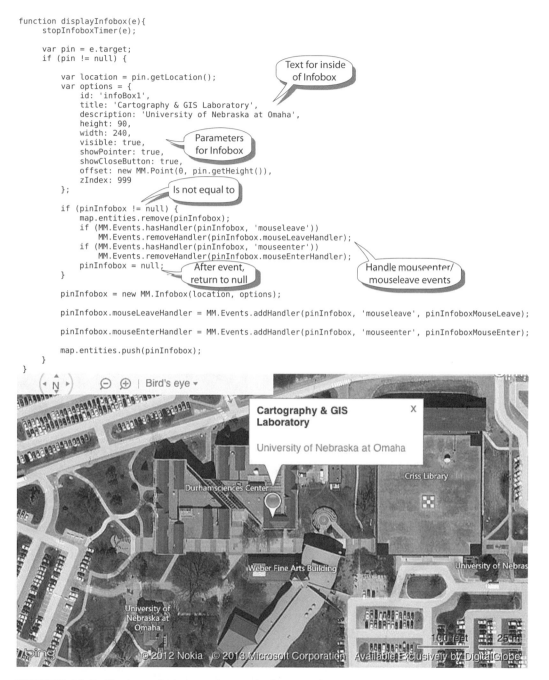

FIGURE 10.7. Placing a hover marker with the Bing Maps API. Copyright 2013 Microsoft.

Provide key to
use Nokia HERE

```
<script type="text/javascript" id="exampleJsSource">
    nokia.Settings.set("appId", "_peU-uCkp-j8ovkzFGNU");
    nokia.Settings.set("authenticationToken", "gBoUkAMoxoqIWfxWA5DuMQ");

    var mapContainer = document.getElementById("mapContainer");
    var map = new nokia.maps.map.Display(mapContainer, {

        center: [52.51, 13.4],
        zoomLevel: 10,
        components: [new nokia.maps.map.component.Behavior()]
    });
</script>
```

Get DOM
for the map

Add pan/zoom
behavior

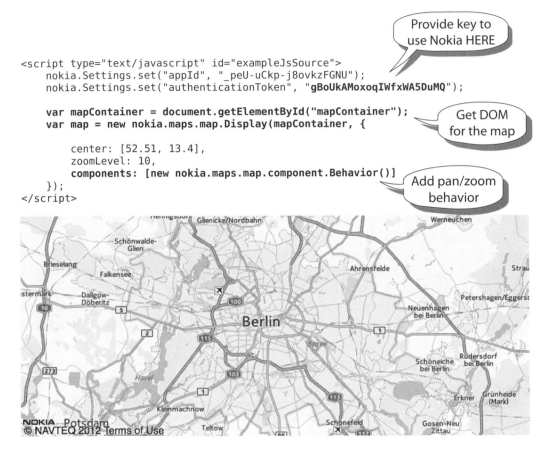

FIGURE 10.8. A basic map using the Nokia API. Copyright 2013 Nokia. Nokia map content is used with permission.

Provide key to
use Nokia API

```
<script type="text/javascript" id="exampleJsSource">

    nokia.Settings.set("appId", "_peU-uCkp-j8ovkzFGNU");
    nokia.Settings.set("authenticationToken", "gBoUkAMoxoqIWfxWA5DuMQ");

    var mapContainer = document.getElementById("mapContainer");
    var map = new nokia.maps.map.Display(mapContainer, {

        center: [52.51, 13.4],
        zoomLevel: 10,
        components: [new nokia.maps.map.component.Behavior()]
    });

    var standardMarker = new nokia.maps.map.StandardMarker(map.center);
    map.objects.add(standardMarker);

</script>
```

Get DOM
for the map

Add pan/zoom
behavior

Add marker at
map center

Put marker
in map object

FIGURE 10.9a. A Nokia map with marker. Copyright 2013 Nokia. Nokia map content is used with permission.

FIGURE 10.9b.

```
var standardMarkerProps = [null,          1st marker with
{                                          null properties
    text: "42",
      brush: {color: "#F80"}}, {           2nd marker
    text: "@",
      brush: {color: "#F80"},
      textPen: {strokeColor: "#333"}}, {   3rd marker
    text: "Hi",
      textPen: {strokeColor: "#333"},
      brush: {color: "#FFF"},              4th marker
      pen: {strokeColor: "#333"}}],
                                           Number of markers
Counter goes    i = 4;
backwards       while (i--)
    map.objects.add(new nokia.maps.map.StandardMarker(
    map.center.walk(360 / standardMarkerProps.length * i, 6000),
standardMarkerProps[i]));
                                           Places markers
    </script>                              around middle
```

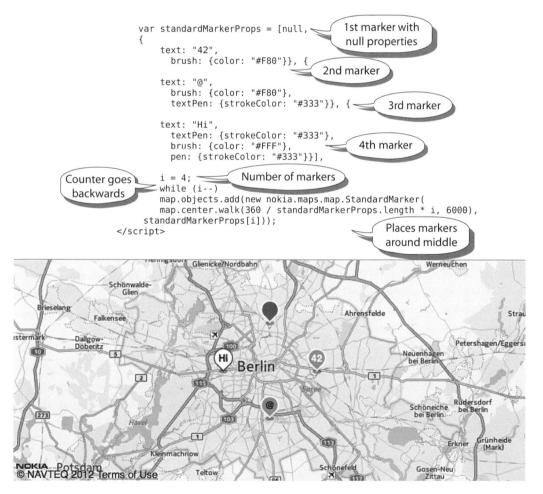

FIGURE 10.10. Nokia API coding to place a styled marker at a specified location. Copyright 2013 Nokia. Nokia map content is used with permission.

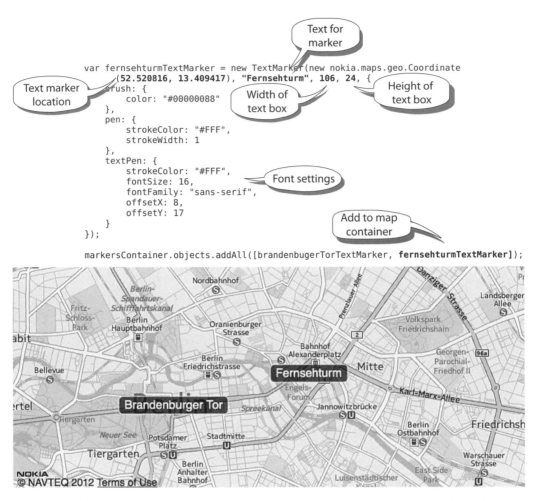

```
var fernsehturmTextMarker = new TextMarker(new nokia.maps.geo.Coordinate
    (52.520816, 13.409417), "Fernsehturm", 106, 24, {
brush: {
    color: "#00000088"
},
pen: {
    strokeColor: "#FFF",
    strokeWidth: 1
},
textPen: {
    strokeColor: "#FFF",
    fontSize: 16,
    fontFamily: "sans-serif",
    offsetX: 8,
    offsetY: 17
}
});

markersContainer.objects.addAll([brandenbugerTorTextMarker, fernsehturmTextMarker]);
```

Text for marker

Text marker location

Width of text box

Height of text box

Font settings

Add to map container

FIGURE 10.11. Nokia API coding to place a text marker at a specified location. The map labels the Brandenburg Gate and television tower. Copyright 2013 Nokia. Nokia map content is used with permission.

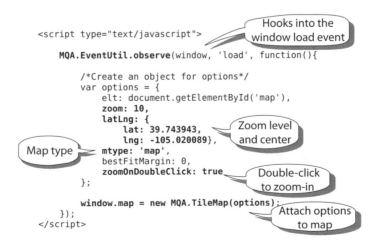

```
<script type="text/javascript">

    MQA.EventUtil.observe(window, 'load', function(){

        /*Create an object for options*/
        var options = {
            elt: document.getElementById('map'),
            zoom: 10,
            latLng: {
                lat: 39.743943,
                lng: -105.020089},
            mtype: 'map',
            bestFitMargin: 0,
            zoomOnDoubleClick: true
        };

        window.map = new MQA.TileMap(options);
    });
</script>
```

Hooks into the window load event

Zoom level and center

Map type

Double-click to zoom-in

Attach options to map

FIGURE 10.12a. A basic map with the MapQuest API. Copyright 2013 MapQuest—Portions; Navteq, Intermap.

FIGURE 10.12b.

Create an
options object

```
window.map = new MQA.TileMap(options);

MQA.withModule('largezoom', function() {
  map.addControl(
  new MQA.LargeZoom(),
  new MQA.MapCornerPlacement(MQA.MapCorner.TOP_LEFT, new MQA.Size(5,5))
  );
});
```

Add a large
zoom control

Distance in pixels
from top left

Place zoom
control at top left

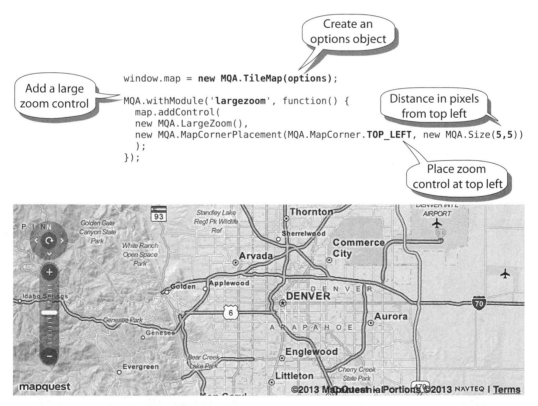

FIGURE 10.13. Adding a zoom control with the MapQuest API. Copyright 2013 MapQuest—Portions; Navteq, Intermap.

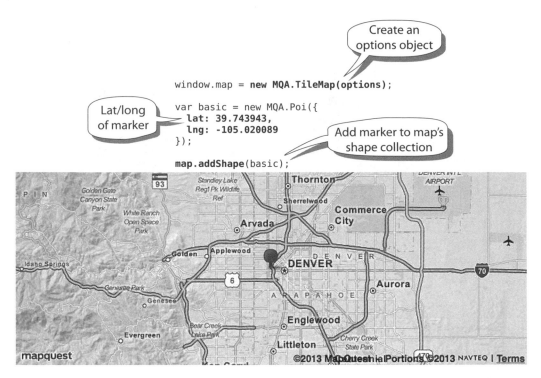

FIGURE 10.14. Adding a marker with the MapQuest API. Copyright 2013 MapQuest—Portions; Navteq, Intermap.

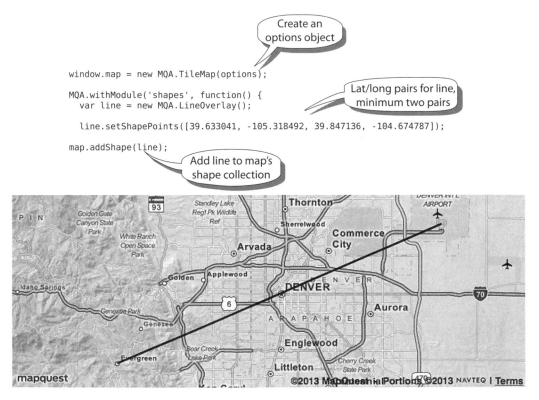

FIGURE 10.15. Adding a line using the MapQuest API. Copyright 2013 MapQuest—Portions; Navteq, Intermap.

the rendered, tile-based map is used here. The examples in Figures 10.16 through 10.21 show the basic OpenStreetMap, the adding of controls and a scale bar, a single marker, multiple markers, and a hover marker. The OpenStreetMap code is longer and more complicated than other mapping APIs.

10.3.5 *Leaflet API*

The Leaflet API is specifically designed for mobile devices. Developed by Vladimir Agafonkin (2013) and other contributors, it is just 28 kb of JavaScript code—much smaller than other APIs. The resulting code is simpler and much more readable. The code also makes use of CloudMade tiles that have been rendered from OpenStreetMap vector data. Figures 10.21 to 10.24 demonstrate use of the Leaflet API.

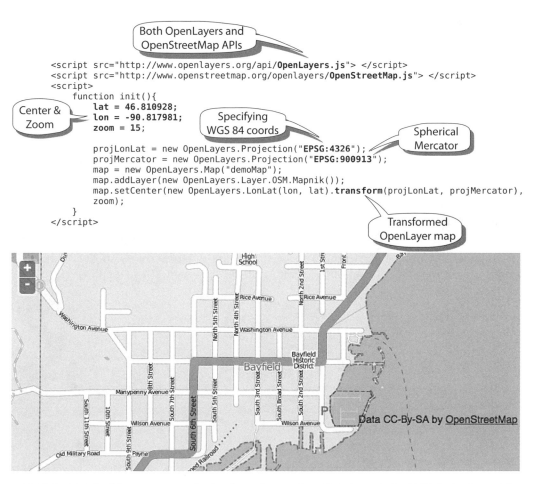

Both OpenLayers and OpenStreetMap APIs

```
<script src="http://www.openlayers.org/api/OpenLayers.js"> </script>
<script src="http://www.openstreetmap.org/openlayers/OpenStreetMap.js"> </script>
<script>
    function init(){
        lat = 46.810928;
        lon = -90.817981;
        zoom = 15;

        projLonLat = new OpenLayers.Projection("EPSG:4326");
        projMercator = new OpenLayers.Projection("EPSG:900913");
        map = new OpenLayers.Map("demoMap");
        map.addLayer(new OpenLayers.Layer.OSM.Mapnik());
        map.setCenter(new OpenLayers.LonLat(lon, lat).transform(projLonLat, projMercator),
        zoom);
    }
</script>
```

Center & Zoom

Specifying WGS 84 coords

Spherical Mercator

Transformed OpenLayer map

FIGURE 10.16. A basic map with the OpenStreetMap API. Copyright 2013 OpenStreetMap contributors CC-BY-SA.

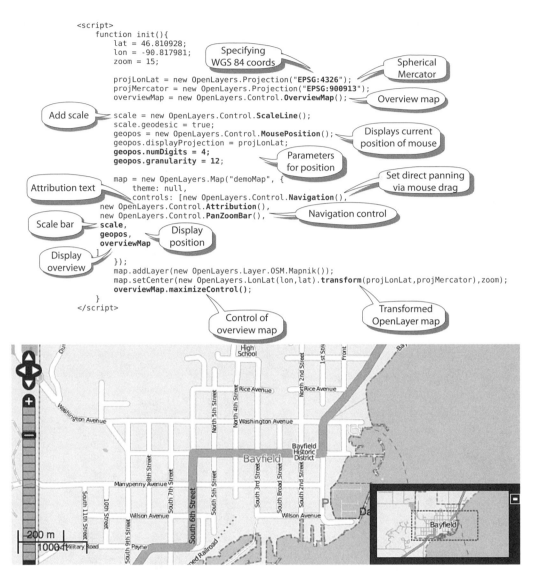

```
<script>
    function init(){
        lat = 46.810928;
        lon = -90.817981;
        zoom = 15;

        projLonLat = new OpenLayers.Projection("EPSG:4326");
        projMercator = new OpenLayers.Projection("EPSG:900913");
        overviewMap = new OpenLayers.Control.OverviewMap();
        scale = new OpenLayers.Control.ScaleLine();
        scale.geodesic = true;
        geopos = new OpenLayers.Control.MousePosition();
        geopos.displayProjection = projLonLat;
        geopos.numDigits = 4;
        geopos.granularity = 12;

        map = new OpenLayers.Map("demoMap", {
            theme: null,
            controls: [new OpenLayers.Control.Navigation(),
        new OpenLayers.Control.Attribution(),
        new OpenLayers.Control.PanZoomBar(),
        scale,
        geopos,
        overviewMap
        ]
        });
        map.addLayer(new OpenLayers.Layer.OSM.Mapnik());
        map.setCenter(new OpenLayers.LonLat(lon,lat).transform(projLonLat,projMercator),zoom);
        overviewMap.maximizeControl();
    }
</script>
```

Callouts: Specifying WGS 84 coords · Spherical Mercator · Overview map · Add scale · Displays current position of mouse · Parameters for position · Attribution text · Set direct panning via mouse drag · Scale bar · Navigation control · Display position · Display overview · Control of overview map · Transformed OpenLayer map

FIGURE 10.17. Adding controls and a scale bar with the OpenStreetMap API. Notice the positioning of the scale on top of the zoom control. Copyright 2013 OpenStreetMap contributors CC-BY-SA.

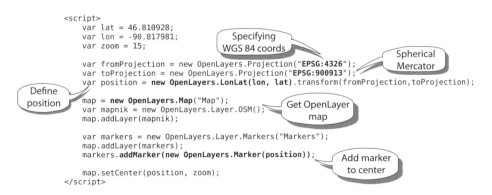

```
<script>
    var lat = 46.810928;
    var lon = -90.817981;
    var zoom = 15;

    var fromProjection = new OpenLayers.Projection("EPSG:4326");
    var toProjection = new OpenLayers.Projection("EPSG:900913");
    var position = new OpenLayers.LonLat(lon, lat).transform(fromProjection,toProjection);

    map = new OpenLayers.Map("Map");
    var mapnik = new OpenLayers.Layer.OSM();
    map.addLayer(mapnik);

    var markers = new OpenLayers.Layer.Markers("Markers");
    map.addLayer(markers);
    markers.addMarker(new OpenLayers.Marker(position));

    map.setCenter(position, zoom);
</script>
```

Callouts: Specifying WGS 84 coords · Spherical Mercator · Define position · Get OpenLayer map · Add marker to center

FIGURE 10.18a. Adding a single marker with the OpenStreetMap API. Copyright 2013 OpenStreetMap contributors CC-BY-SA.

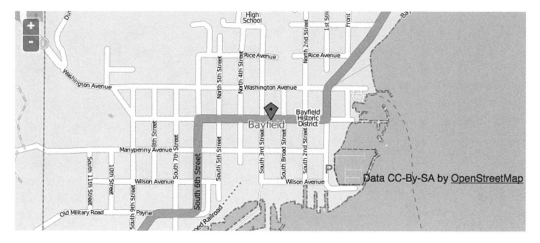

FIGURE 10.18b.

FIGURE 10.19a. Adding multiple markers with the OpenStreetMap API. Copyright 2013 OpenStreetMap contributors CC-BY-SA.

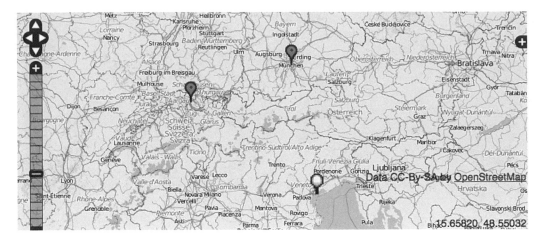

FIGURE 10.19b.

```
var markerClick = function(evt) {
    if (this.popup == null) {
        this.popup = this.createPopup(this.closeBox);
        map.addPopup(this.popup);
        this.popup.show();
    } else {
        this.popup.toggle();
    }
    OpenLayers.Event.stop(evt);
};
marker.events.register("mousedown", feature, markerClick);

marker.events.register("mouseover", feature, markerClick);

marker.events.register("mouseout", feature, markerClick);

layer.addMarker(marker);
```

MouseDown event

MouseOver event

MouseOut event

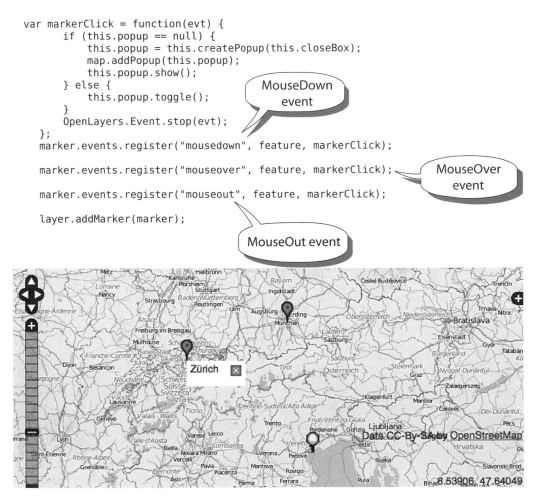

FIGURE 10.20. Adding a hover marker with the OpenStreetMap API. Copyright 2013 OpenStreetMap contributors CC-BY-SA.

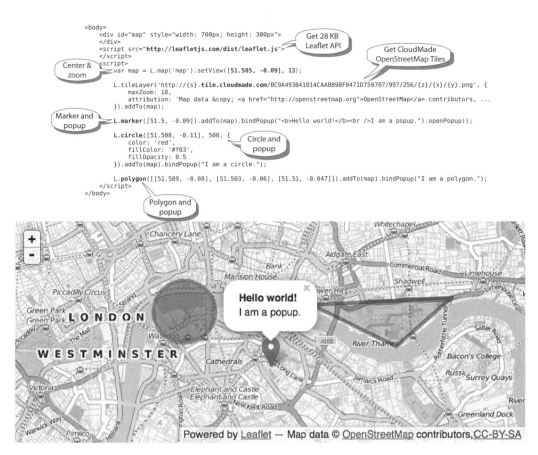

```
<body>
    <div id="map" style="width: 700px; height: 300px">
    </div>
    <script src="http://leafletjs.com/dist/leaflet.js">        Get 28 KB
    </script>                                                    Leaflet API
    <script>
        var map = L.map('map').setView([51.505, -0.09], 13);    Get CloudMade
                                                                 OpenStreetMap Tiles
Center &
zoom        L.tileLayer('http://{s}.tile.cloudmade.com/BC9A493B41014CAABB98F0471D759707/997/256/{z}/{x}/{y}.png', {
                maxZoom: 18,
                attribution: 'Map data &copy; <a href="http://openstreetmap.org">OpenStreetMap</a> contributors, ...
            }).addTo(map);

Marker and  L.marker([51.5, -0.09]).addTo(map).bindPopup("<b>Hello world!</b><br />I am a popup.").openPopup();
popup
            L.circle([51.508, -0.11], 500, {
                color: 'red',                   Circle and
                fillColor: '#f03',              popup
                fillOpacity: 0.5
            }).addTo(map).bindPopup("I am a circle.");

            L.polygon([[51.509, -0.08], [51.503, -0.06], [51.51, -0.047]]).addTo(map).bindPopup("I am a polygon.");
    </script>
</body>
                    Polygon and
                    popup
```

FIGURE 10.21. A basic Leaflet API map. Copyright 2013 OpenStreetMap contributors CC-BY-SA.

```
<body>
    <div id="map" style="width: 700px; height: 300px"></div>
                                                                    Get 28 KB
    <script src="http://leafletjs.com/dist/leaflet.js"></script>   Leaflet API
    <script>
        var map = L.map('map').setView([51.5, -0.09], 13);

        L.tileLayer('http://{s}.tile.cloudmade.com/{key}/22677/256/{z}/{x}/{y}.png', {
            attribution: 'Map data &copy; 2013 OpenStreetMap contributors',
            key: 'BC9A493B41014CAABB98F0471D759707'
        }).addTo(map);                              Get one shadow
                                                    for all 3 icons
        var LeafIcon = L.Icon.extend({
            options: {
                shadowUrl: 'http://leafletjs.com/docs/images/leaf-shadow.png',
                iconSize:     [38, 95],
                shadowSize:   [50, 64],
                iconAnchor:   [22, 94],
                shadowAnchor: [4, 62],          Link to
                popupAnchor:  [-3, -76]         three icons
            }
        });

        var greenIcon = new LeafIcon({iconUrl: 'http://leafletjs.com/docs/images/leaf-green.png'}),
        redIcon = new LeafIcon({iconUrl: 'http://leafletjs.com/docs/images/leaf-red.png'}),
        orangeIcon = new LeafIcon({iconUrl: 'http://leafletjs.com/docs/images/leaf-orange.png'});

        L.marker([51.5, -0.09], {icon: greenIcon}).bindPopup("I am a green leaf.").addTo(map);
        L.marker([51.495, -0.083], {icon: redIcon}).bindPopup("I am a red leaf.").addTo(map);
        L.marker([51.49, -0.1], {icon: orangeIcon}).bindPopup("I am an orange leaf.").addTo(map);

    </script>
</body>                Put icons
                       on map
```

FIGURE 10.22a. Adding custom icons with the Leaflet API. Copyright 2013 OpenStreetMap contributors CC-BY-SA.

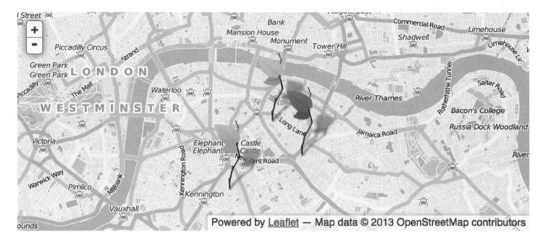

FIGURE 10.22b.

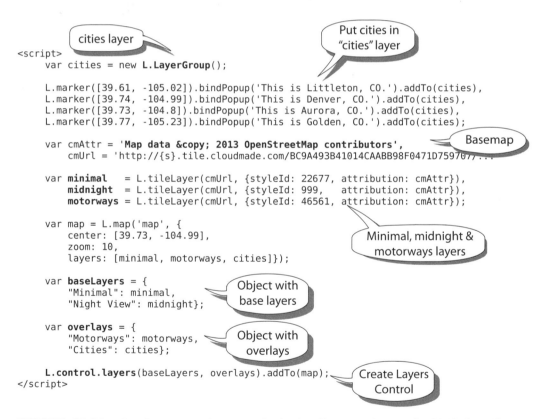

FIGURE 10.23a. Implementing layers with the Leaflet API. Copyright 2013 OpenStreet-Map contributors CC-BY-SA.

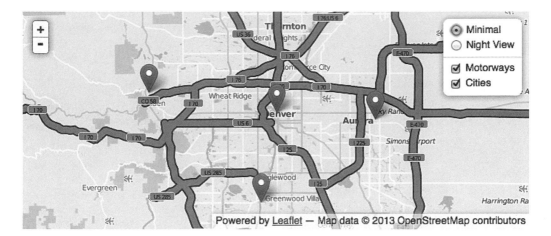

FIGURE 10.23b.

FIGURE 10.24a. Choropleth map with the Leaflet API. The code (overleaf, top) shows part of the us-states.js file that contains the state outlines. Copyright 2013 OpenStreetMap contributors CC-BY-SA.

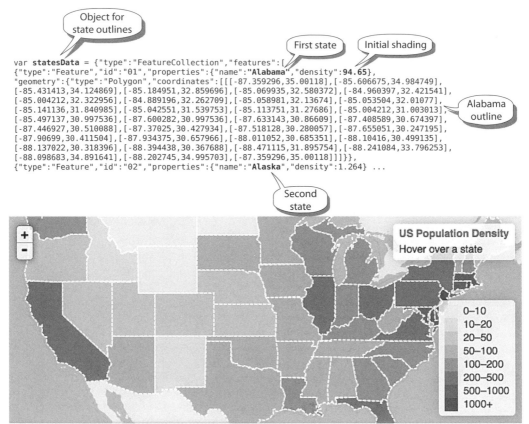

FIGURE 10.24b & c.

10.4   Mapping APIs from China

The government of China has exercised a level of control over the Internet that is unmatched by other countries. Access to Google in China is intermittent, making it essentially useless. Although Google Maps works better than the Google search engine and gmail, the blocking of Google has fostered the development of online mapping entities within China. The two major companies are Baidu and AutoNavi. Both companies provide detailed coverage for China and Taiwan but not other parts of the world.

Baidu is the major search engine in China, analogous to Google for the rest of the world. With headquarters in Beijing, the company offers a Chinese-language search engine for websites, audio files, and images, in addition to Baidu Map. Baidu became the first Chinese company to be listed on the NASDAQ-100 index. The Baidu mapping API is very similar to other APIs we have examined. Figure 10.25a shows a basic Baidu map code that activates the various controls for the map. Figure 10.25b adds a marker to the map.

```html
<html>
  <head>
    <meta http-equiv="Content-Type" content="text/html; charset=utf-8" />
    <script type="text/javascript"
            src="http://api.map.baidu.com/api?v=1.5&ak=927cb23887926d2b345b0c762045feb3"></script>
    <title>Baidu Map</title>
    <script type="text/javascript">
      function initialize () {
        var map = new BMap.Map("allmap");                        // Create a new map instance
        map.centerAndZoom(new BMap.Point(116.404, 39.915), 11); // Initialize map, set center and zoom level
        map.addControl(new BMap.NavigationControl());           // Add navigation control for zoom and pan
        map.addControl(new BMap.ScaleControl());                // Add a control to show the scale of the map
        map.addControl(new BMap.OverviewMapControl());          // Add an overview control
        map.enableScrollWheelZoom();                            // Enable mouse wheel for zooming
        map.addControl(new BMap.MapTypeControl());              // Add a control for choosing map types
        map.setCurrentCity("北京");                             // Set current city for the map, this step is necessary (北京 = Beijing)
      }
    </script>
  </head>
  <body onLoad="initialize()">
      <div id="allmap" style="width:700px; height:300px"></div>
  </body>
</html>
```

Baidu Map API Key

```html
<html>
  <head>
    <meta http-equiv="Content-Type" content="text/html; charset=utf-8" />
    <script type="text/javascript"
            src="http://api.map.baidu.com/api?v=1.5&ak=927cb23887926d2b345b0c762045feb3"></script>
    <script type="text/javascript">
      function initialize() {
        var map = new BMap.Map("allmap");                       // Create a map
        var point = new BMap.Point(116.404, 39.915);           // Make a point
        map.centerAndZoom(point, 15);                          // Set map center and zoom level
        var marker = new BMap.Marker(point);                   // Create a marker
        map.addControl(new BMap.NavigationControl());          // Add Navigation Control for zoom and pan
        map.addControl(new BMap.ScaleControl());               // Show the scale of the map 添加比例尺控件
        map.addControl(new BMap.OverviewMapControl());         // Add an overview control
        map.enableScrollWheelZoom();                           // Enalbe wheel of mouse for zooming
        map.addControl(new BMap.MapTypeControl());             // Add a control for choosing map types
        map.addOverlay(marker);                                // Add the marker to the map
        marker.setAnimation(BMAP_ANIMATION_BOUNCE);            // set animation type of the marker

        var infoWindow = new BMap.InfoWindow("Marker Infomation");  // Create an info window
        marker.addEventListener("click", function(){this.openInfoWindow(infoWindow);});  //Add a listener
      }
    </script>
    <title>Animated Marker</title>
  </head>
  <body onLoad="initialize()">
    <div id="allmap" style="width:700px; height:300px"> </div>
  </body>
</html>
```

Baidu Map API Key

FIGURE 10.25. The top code is a basic implementation of the Baidu Map API. The bottom adds a bouncing marker. Both require a key from Baidu, a major Chinese website and map provider.

AutoNavi is another online mapping company from China. The company would be comparable to NavTeq or TeleAtlas in that it concentrates on updating its underlying database. Like Google Maps, it can also present buildings in 3D for larger cities. Like Baidu, it also requires a key. Figure 10.26 shows the fairly extensive AutoNavi API code for adding a marker to the map.

```html
<html>
  <head>
    <meta http-equiv="Content-Type" content="text/html; charset=utf-8">
    <title>Add Marker</title>
    <script language="javascript"
            src="http://webapi.amap.com/maps?v=1.2&key=eac2bfeea107a0189110a195d552a807"></script>      AutoNavi Map API Key
    <script language="javascript">
      var mapObj,tool,view,scale;
      function mapInit(){
      mapObj = new AMap.Map("map",{
        center:new AMap.LngLat(116.392936,39.919479)
        });
      mapObj.plugin(["AMap.ToolBar","AMap.OverView,AMap.Scale"],function(){

        tool = new AMap.ToolBar({               //Add toolbar
          direction:false,
          ruler:false,
          autoPosition:false                    //Auto positioning disabled
        });
        mapObj.addControl(tool);                //Add overview
        view = new AMap.OverView({visible:false});
        mapObj.addControl(view);                //Add scale
        scale = new AMap.Scale();
        mapObj.addControl(scale);
      });

      var marker = new AMap.Marker({            //Create a marker
        map:mapObj,
        position:new AMap.LngLat(116.373881,39.907409),  //position 基点位置
        icon:"http://webapi.amap.com/images/marker_sprite.png", //Icon of the marker url
        offset:{x:-8,y:-34}                     //Marker Offset
      });
      var info = [];                            //Information window: to show the address of AutoNavi
      info.push("<b>  AutoNavi</b>");
      info.push("  TEL: 010-84107000   Zip Code: 100102");
      info.push("  Address: 北京市望京阜通⊠大街方恒国⊠中心A座16⊠");

      var inforWindow = new AMap.InfoWindow({
        offset:new AMap.Pixel(0,-30),
        content:info.join("<br />")
      });

      AMap.event.addListener(marker,'click',function(e){      //Add a listener
        inforWindow.open(mapObj,marker.getPosition());
      })
      mapObj.setCenter(marker.getPosition());       //Set the marker to the center
    }
  </script>
</head>
  <body onLoad="mapInit()">
    <div id="map" style="width:700px;height:400px;border:#F6F6F6 solid 1px;"> </div>
  </body>
</html>
```

FIGURE 10.26. Mapping API code from the Chinese company, AutoNavi. The service presents detailed maps for China but not the rest of the world.

10.5   Mapstraction

Mapstraction is an open-source library that provides a common API for various online mapping services (Duvander 2010). Its purpose is to allow the user to easily switch between APIs without having to worry about the unique implementation of each. The example in Figure 10.27 shows Mapstraction code that allows using either the OpenLayers or Google API. To access the Google Maps API, "openlayers" is simply changed to "googlev3" (see Figure 10.28). Note that a link to the mapping API is also changed on the third line of the code.

10.6   Summary

APIs are the building blocks for mapping in the cloud. They provide the tools for adding information to the underlying base map. The particular coding of each API

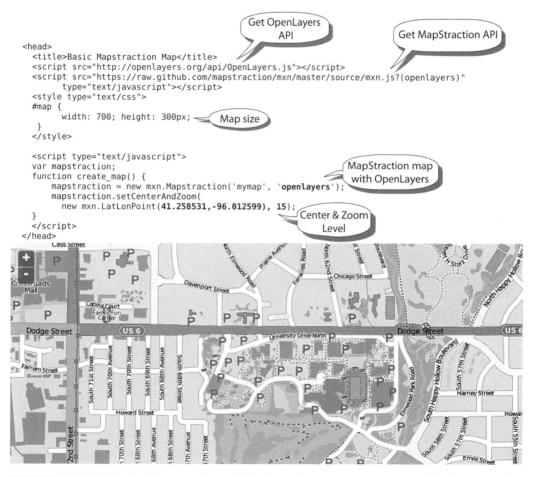

FIGURE 10.27. An implementation of Mapstraction, an open-source API that provides a common interface to multiple online mapping sites. This example creates a map using OpenLayers/OpenStreetMap. Copyright 2013 OpenStreetMap contributors CC-BY-SA.

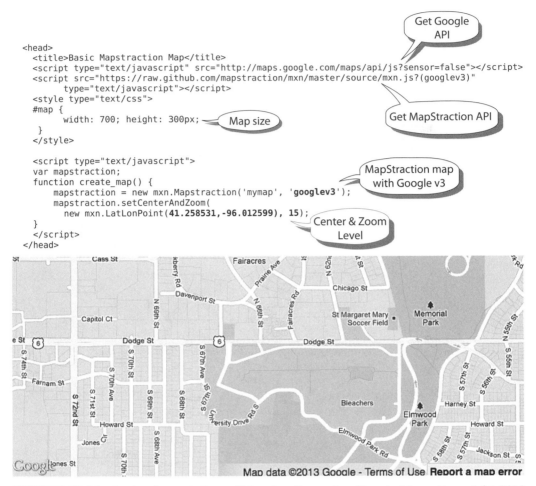

```
<head>
  <title>Basic Mapstraction Map</title>
  <script type="text/javascript" src="http://maps.google.com/maps/api/js?sensor=false"></script>
  <script src="https://raw.github.com/mapstraction/mxn/master/source/mxn.js?(googlev3)"
        type="text/javascript"></script>
  <style type="text/css">
  #map {
        width: 700; height: 300px;
  }
  </style>

  <script type="text/javascript">
  var mapstraction;
  function create_map() {
      mapstraction = new mxn.Mapstraction('mymap', 'googlev3');
      mapstraction.setCenterAndZoom(
        new mxn.LatLonPoint(41.258531,-96.012599), 15);
  }
  </script>
</head>
```

Get Google API

Get MapStraction API

Map size

MapStraction map with Google v3

Center & Zoom Level

FIGURE 10.28. An implementation of Mapstraction using Google Maps. Copyright 2013 Google.

varies slightly and the open-source Mapstraction API makes it possible to easily switch between many of them. The development of mapping APIs is in an early stage, and it may be too early to try to define a standard set of calls. For example, certain functions are implemented in the Google Maps API that are not possible with other APIs, including Mapstraction.

While there are slight variations in the coding, the major difference between the mapping APIs may be the rendering of the underlying base map, and the speed of map delivery. The delivery of the tiles to a large number of simultaneous users at an acceptable speed may be the single most important factor in judging the viability of any particular API.

10.7   Exercise

Download the code for this chapter and implement the examples. Change the center of the map, information associated with the markers, and/or other aspects of the map. Obtain the appropriate keys, if necessary, and upload the examples to your web server.

10.8   Questions

1. What is the relationship between mashups and APIs?

2. What are some advantages of using a common mapping API?

3. Describe current limitations placed on the usage of the Google Maps API and related services.

4. What are some different views (not overlays) offered by Google Maps and other online map services?

5. What are some limitations of the API method for mapping?

6. Describe an API key, and what are the advantages?

7. What are some advantages of the Leaflet API?

8. What are the advantages and disadvantages of an API like Mapstraction?

9. What is the current status of the major mapping APIs, including those from Google, Bing, Yahoo, OpenStreetmap, and Nokia?

10. Explain the lines of code following code segments labeled A through E:

```
<head>
<meta name="viewport" content="initial-scale=1.0, user-scalable=no" />
```

A)

```
<style type="text/css">
  html { height: 100% }
  body { height: 100%; margin: 0px; padding: 0px }
  #map_canvas { height: 100% }
</style>

<script type="text/javascript"
```

B)

```
src="http://maps.google.com/maps/api/js?sensor=false">
</script>

<script type="text/javascript">
  function initialize() {
```

C)

```
    var latlng = new google.maps.LatLng(41.258531,-96.012599);
    var myOptions = {
      zoom: 15,
      center: latlng,
```

D)

```
        mapTypeId: google.maps.MapTypeId.HYBRID
    };
    var map = new google.maps.Map(document.getElementById("map_canvas"),
        myOptions);
}
</script>
</head>
<body onload="initialize()">
```

E)

```
    <div id="map_canvas" style="width:100%; height:100%"></div>

</body>
</html>
```

10.9 References

Agafonkin, Vladimir (2013) Leaflet: An Open-Source JavaScript Library for Mobile-Friendly Interactive Maps. [http://leafletjs.com/]

Duvander, Adam (2010) *Map Scripting 101: An Example-Driven Guide to Building Interactive Maps with Bing, Yahoo!, and Google Maps*. San Francisco, CA: No Starch Press.

Google Maps JavaScript API Basics (2011) (search: Google Maps Javascript API Basics).

CHAPTER 11
Points and Point Data

I once was lost, but now am found.

—*Amazing Grace*

11.1 Introduction

The global positioning system (GPS) has made it almost effortless to determine latitude and longitude. The problem is that the GPS device needs to be at or near the specific location in order to acquire a position. In many cases, a location needs to be determined for other places. Determining these positions is based on using the available information.

Embedded within most maps are coordinate systems. Lines of latitude and longitude are almost always shown. Other coordinate systems on maps vary by country or region. In the United States, a combination of lines or tick marks on USGS topographic maps shows coordinates for the Universal Transverse Mercator (UTM), the state plane coordinate (SPC) system, and the United States Public Land Survey (USPLS). Locations provided in all of these systems can be converted to latitude and longitude.

Another method for deriving latitude and longitude is through a street address. This *geocoding* has been implemented in a select few countries where locations of street intersections have been determined. Once the locations of street intersections are known, the geographic coordinates of addresses between them can be approximated. In some countries, the exact latitude and longitude of each residence have been determined.

Latitude and longitude are also defined differently depending on the datum, a reference point for the size and shape of the Earth. In this chapter, we examine the datum, coordinate systems, and corresponding conversion methods.

11.2   Geodesy and the Datum

As early as the 1600s, it was known that ground distances were not consistent between degrees of latitude at different locations on the Earth. These inconsistent measurements raised the suspicion that the Earth was not a perfect sphere. In the 1700s, the French Academy of Sciences, established in 1699 by King Louis XIV, sent an expedition to Lapland and Peru. They found that a degree of latitude in Lapland, near the Arctic Circle, was 9/16th of a mile shorter than at the equator (Burritt 1833, p. 201). Modern measurements show that the Earth's diameter at the equator is nearly 43 km (27 mi) more than the diameter that runs through the north and south poles (12,756.28 km or 7927 mi vs. 12,713.56 km or 7900 mi). The shape of the Earth is termed an oblate spheroid, exhibiting so-called polar flattening or equatorial bulge.

In addition to being nonspherical, the Earth's landmass also has mountains and valleys, the deepest of which are at the bottom of the oceans. The difference between the highest mountain and the lowest trench is nearly 20 km, from Mount Everest at +8948 m to the Mariana Trench at –10,916 m. The true shape of the Earth is described as a geoid—by definition, Earth-shaped. The three different ways for describing the shape of the Earth are shown in Figure 11.1.

The irregular shape of the geoid prevents it from being used for horizontal positioning on the Earth's surface. Instead, a representative spheroid is used that approximates the geoid. Geodesy is the science that deals with measuring the shape of the Earth and defining a spheroid that best approximates the geoid. This results in a datum based on a pole-flattened oblate spheroid. Three example datums include the following.

1. The North American Datum of 1927 (NAD27) was defined based on the Clarke spheroid of 1866, with an origin in the middle of the continent at Meades Ranch, Kansas.
2. NAD83, defined in 1983 and an improvement on NAD27, provides a datum for worldwide use. NAD83 is based on the adjustment of 250,000 points

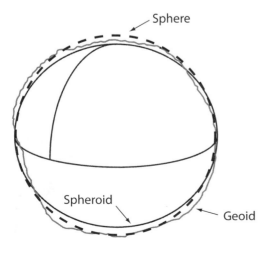

FIGURE 11.1. The Earth is not a perfect sphere. Polar flattening leads to a spheroid shape. Mountains and valleys further alter the shape of the Earth from a perfect sphere, making what is defined as a geoid.

that constrain the solution to a geocentric origin (*Geodetic Glossary*, p. 57). Coordinates in NAD83 are 70 to 100 meters different from those of NAD27.

3. The World Geodetic System of 1984 (WGS84) is the reference coordinate system used by the global positioning system. The coordinate origin of WGS 84 is meant to be located at the Earth's center of mass. NAD83 and WGS84 are nearly identical over North America, with an average difference of only one meter. Originally, both NAD83 and WGS84 were to use the Geodetic Reference System of 1980 (GRS 80) as a reference spheroid. As it happened, the WGS84 spheroid differs by 0.0001 m in the semi-minor axis from NAD83 (Wilson 1995).

No direct mathematical method exists that can accurately transform latitude and longitude coordinates from one system to the other. All conversion programs are only approximations that have a general accuracy of 0.15 m. The online NADCON utility converts between NAD27 and NAD83/WGS84 (search: NADCON).

In modern practice, all latitude and longitude coordinates are expected to be in WGS84 and are normally defined in decimal degrees. Thus, the seconds are always given in decimal format, but the minutes can be provided in either degrees or decimals. Valid formats for defining latitude and longitude are given in Table 11.1.

Converting minutes and seconds to decimal degrees involves dividing by 60, the number of minutes in a degree and the number of seconds in a minute. To convert 40° 05′ 18″ N to degrees and decimal minutes, we convert the seconds to a decimal part of a minute by dividing the seconds (18) by the number of seconds in a minute (60). This gives us the decimal part of a minute that 18 seconds represents (18/60 = 0.3). We now add the decimal minutes to the minutes and "discard" the seconds (05 + 0.3 = 5.3). The answer is 40° 05.3′ N.

To convert the 5.3 minutes to degrees, we divide the number of minutes by 60 and add that value to the number of degrees. In this example, we divide the minutes (5.3) by the number of minutes in a degree (60) to give the decimal part of a degree that 5.3 minutes represents (5.3/60 = 0.08833). Then, we add the decimal degrees to the degrees and discard the minutes (40 + 0.08833 = 40.08833). The final answer is 40.08833°.

If seconds are not provided, we can simply convert the minutes. For example, for 96° 0.926′, we would divide 0.926 by 60 to get .01543, resulting in a decimal longitude of 96.01543.

TABLE 11.1. Valid Formats for Specifying Decimal Degrees

The seconds are always expressed in decimal format. The minutes may also be expressed in decimal, as shown in the last example.

Latitude	Longitude
N43°38′19.39″	W116°14′28.86″
43°38′19.39″N	116°14′28.86″W
43 38 19.39	–116 14 28.86
43.63871944444445	–116.2413513485235

11.3   Coordinate Systems

The geographical coordinates of latitude and longitude are based on a spherical coordinate system. Most other coordinates on maps are flat and are only usable over small parts of the Earth. They were developed to help locate features before GPS was available. We begin by examining latitude and longitude on maps.

11.3.1   *Geographic Coordinates*

Topographic maps are themselves bounded by lines of latitude and longitude. Maps in the United States represent 7 minutes and 30 seconds of latitude and longitude. The width of the map will vary from almost square in Hawaii to rectangular in the mainland. Topographic maps of Alaska are at a smaller scale and depict 15 minutes in latitude and from 20 to 36 minutes of longitude. The corners of the map are labeled with latitude and longitude, and these can be used to approximate any point on the map (see Figure 11.2). Nearly all topographic maps were made before 1984 and use the older NAD27 datum. Any latitude and longitude coordinates derived from these maps would therefore need to be converted to WGS84. The online NADCON utility can perform this conversion.

11.3.2   *The Universal Transverse Mercator*

The Universal Transverse Mercator (UTM) was developed in 1940 as a more practical way of determining location. It divides the Earth into 60 sections of 6° longitude each. Each section is projected using the Mercator projection but positioned

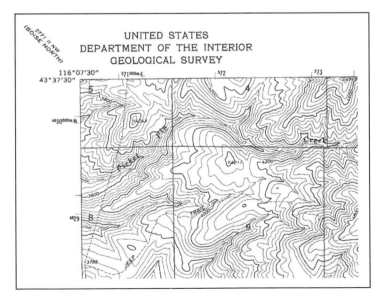

FIGURE 11.2. Latitude and longitude on topographic maps are indicated at each corner. The map also shows tick marks and lines for the UTM, SPC, and UPLSS. Courtesy of U.S. Geological Survey, Department of the Interior/USGS.

transverse to the equator along the central meridian for each 6° zone (Mercator's area distortion is minimal in the 6° section). A false origin is established 500,000 m to the west of the zone-specific central meridian. For the Northern Hemisphere, the origin in the *y*-axis is the equator. The South Pole is usually used for the Southern Hemisphere, but it may be defined 10,000,000 m south of the equator. This system ensures that all coordinates are in the positive quadrant (see Figure 11.3).

Figure 11.4 shows the zones over the contiguous United States. The UTM zone is indicated on the bottom left of the topographic map (see Figure 11.5) along with the information on the projection and the datum. UTM tick marks are at 1000 m increments, and are abbreviated by leaving off the trailing zeros (see Figure 11.6). Points can be defined with UTM by measuring between the 1000-m tick marks. UTM is closely related to the military grid reference system (MGRS). Online conversion utilities can be used to convert UTM measurements to WGS84 latitude and longitude.

11.3.3 State Plane Coordinates

In the 1930s, the state plane coordinate (SPC) system was created in the United States to help define the location and features and avoid the necessity of determining

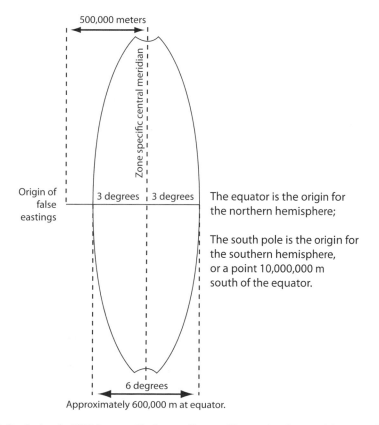

FIGURE 11.3. A single UTM zone. To keep all coordinates in the positive quadrant, a false origin is established that is 500,000 m west of the zone-specific central meridian.

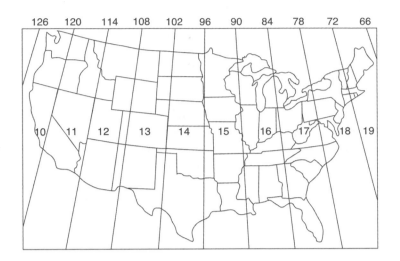

FIGURE 11.4. UTM zones over the 48 contiguous United States. Each zone constitutes 6° of longitude. Courtesy of U.S. Geological Survey, Department of the Interior/USGS.

latitude and longitude. Approximately 120 zones were formed for the 50 states, with all states having at least one zone and some larger states having as many as six (see Figure 11.7). The projection used for zones longer than wide was the transverse Mercator. Zones wider than long were projected with the conic Lambert conformal. Originally defined in NAD27, most zones have been updated to NAD83. The

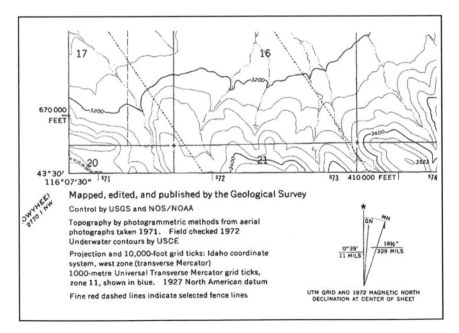

FIGURE 11.5. Map sheet information for a United States Geological Survey topographic map. Courtesy of U.S. Geological Survey, Department of the Interior/USGS.

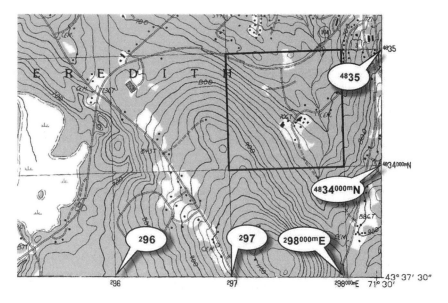

FIGURE 11.6. UTM designations on a topographic map. Labels for most tick marks are abbreviated by removing the trailing zeros. Courtesy of U.S. Geological Survey, Department of the Interior/USGS.

origin is defined at a certain distance to the west and south of the zone so that all coordinates are positive. NAD27 measurements are all defined in feet, and some NAD83 systems have been converted to meters. Online utilities can convert from either system to WGS84.

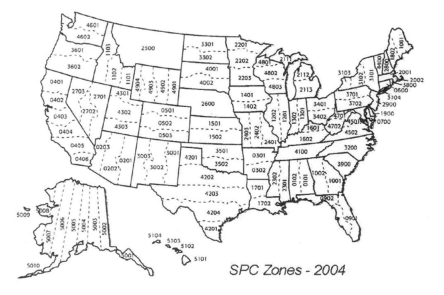

FIGURE 11.7. U.S. state plane coordinate zones. Courtesy of U.S. Geological Survey, Department of the Interior/USGS.

11.3.4   United States Public Land Survey System

Used to define land ownership boundaries for much of the United States, the USPLS consists of a series of separate surveys. Each survey begins at an origin, and townships are surveyed north, south, east, and west from that point. The north–south line of the origin is a line of longitude and is called the principal meridian. There are 37 principal meridians, each having a name that is used to distinguish the various surveys. The east–west line that runs through the initial point is called a base line, and it is perpendicular to the principal meridian (see Figure 11.8). Townships are defined relative to each origin (see Figure 11.9). Each section is further subdivided and designated as indicated in Figure 11.10. Online utilities are available to convert section centers to WGS84. With the advent of GPS, many states have converted all USPLS designations to geographic coordinates.

11.4   Geocoding

Geocoding is the general process of finding geographic coordinates of latitude and longitude from the available information. It is most commonly based on street addresses, but city names or zip codes may also be used.

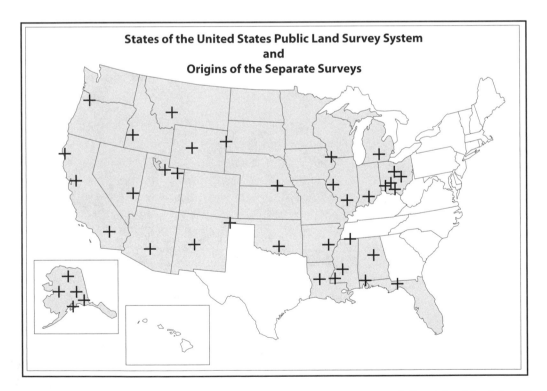

FIGURE 11.8. The United States Public Land Survey system was defined for 30 of the 50 states. It consists of 37 different surveys, each based on a separate origin. Townships are defined to the northeast, southeast, northwest, and southwest of each origin with designations like T5N, R3E.

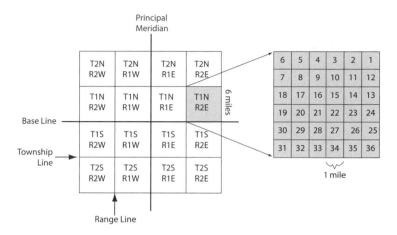

FIGURE 11.9. Six-by-six-mile townships are defined relative to the principal meridian and base line. Each township is divided into 36 sections, each a mile on a side.

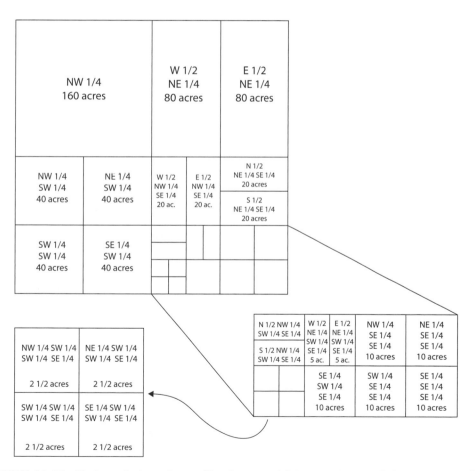

FIGURE 11.10. Various designations of land areas within a section, and the corresponding number of acres.

Related to the development of the U.S. Census Bureau's TIGER file in 1990, the first national street database for the United States, was the coding of latitude and longitude values for all of the nation's street intersections. This, along with the address range for each street segment, made it possible to approximate the latitude and longitude of every address in the country. Other countries used similar methods to convert street addresses, but most countries have yet to implement such a system. Thus, addresses cannot be shown on a map.

The best way to understand geocoding from street addresses is to view the street network as a series of individual segments. A street segment is the link between street intersections. In formal terms, we call the link an arc, and the intersections are referred to as nodes. Each arc would have a certain range of addresses along either side. In many cities in the United States, the even-numbered street addresses are along the north and west sides of the street and the odd-numbered street addresses are along the south and east sides. A specific street segment would also likely represent a certain block with numbers in the same range.

As depicted in Figure 11.11, the 500 block has addresses that range from 502 to 548 on the west, even-numbered side of the street. In finding the location of 516 S. 10th Street, you would first determine that 516 is about one-third of the way from 502 to 548. A latitude value is then calculated that is one-third of the way between the latitudes of the two nodes that define the end-points of the street segment. The street runs in a north–south direction, so the longitude could remain unchanged, but a small factor is usually added or subtracted to place the marker on the correct side of the street. In this case, a small amount is subtracted because the address is on the west side of the street.

Different formats for defining an address, or missing street segments in the database, may make geocoding impossible. In the beginning of geocoding, it would

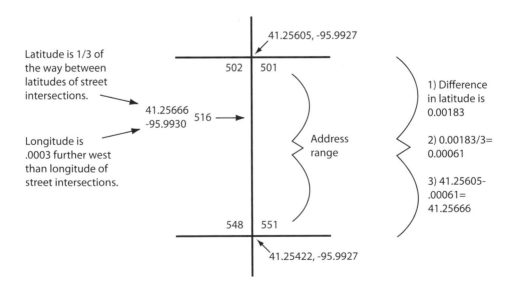

FIGURE 11.11. Geocoding of street addresses involves interpolating the latitude and longitude based on the coordinates of the street intersections.

be considered fortunate if 80% of the addresses provided could be properly geocoded. With improvements in methods for parsing the address—the way the computer reads the address—and more up-to-date databases, geocoding "hits" are now generally above 90%.

These levels of accuracy are not sufficient for many applications, especially emergency services. Many cities have improved the accuracy by determining the center of parcels and have provided this information to the online mapping services. Parcel-based locating is evident when parcels are indicated on the map and the marker is placed in the center. Figure 11.12 compares a detailed street map between Google and MapQuest. Parcel land-ownership lines are shown on the Google map, and the location is clearly indicated at the middle of the parcel.

All of the major mapping services—Google, Bing, Yahoo, OpenStreetMap,

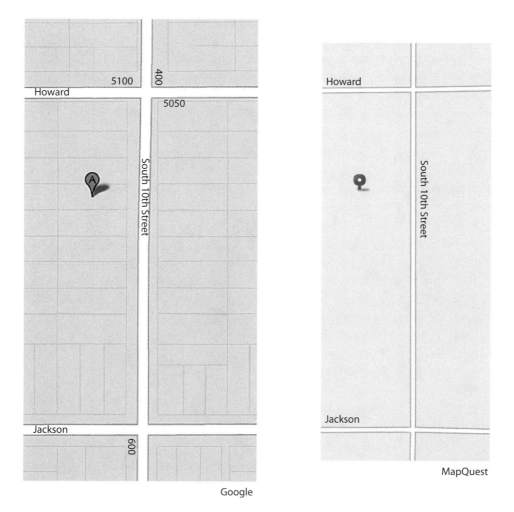

FIGURE 11.12. Google finds locations of addresses based on land-ownership boundaries. Parcels are not shown on the corresponding map from MapQuest. Copyright 2013 Google; copyright 2013 MapQuest—Portions Copyright 2013 NAVTEQ, Intermap or Copyright 2013 MapQuest—Map data Copyright OpenStreetMap and contributors, CC-BY-SA.

MapQuest, and Nokia—perform geocoding. Because the process requires considerable computing resources, they all place limits on the number of addresses that can be geocoded in a particular time period—generally on the order of 2500 geocoding requests a day. Websites have been created to use the geocoding facility of the online mapping services to geocode a set of addresses (see Figure 11.13). These sites may also place a limit on the number of geocoding requests, but once the addresses have been converted to latitude/longitude and saved, the locations can be mapped without need for the geocoding service.

11.5   Summary

Accurately defining positions on the Earth's surface has been a major human accomplishment. A system of latitude and longitude developed fairly early—dating to the work of Ptolemy nearly 2000 years ago. It was determined by at least the

FIGURE 11.13. Online geocoding site for converting addresses to latitude and longitude. The addresses for murders in Omaha in 2001 are copied and pasted into a window on the website. Clicking on "Map Now" converts the addresses to latitude and longitude and places them on a map. Copyright 2013 http://batchgeo.com/; copyright 2013 Google.

1600s that the Earth was not completely spherical and that measurements were needed to establish the exact shape. Since then this determination of shape has been further refined, resulting in a variety of datums, most recently WGS84.

The difficulty of determining location before the development of GPS led to establishing different coordinate systems that could be used for local areas to define both points and areas. UTM is one such coordinate system that can be applied to the entire world. Country-specific coordinate systems for the United States include the state plane coordinate system and the United States Public Land Survey. Coordinates for all these systems can be converted back to latitude and longitude.

Location can also be determined for street addresses with geocoding. The process involves using the known coordinates for the end-points, or nodes, of street segments and the corresponding range of addresses. The location is then estimated based on the street address number. Reverse geocoding involves finding a street address based on the latitude/longitude provided by a spatially enabled mobile device.

11.6 Exercises

A. Convert the following latitude, longitude coordinates into decimal degrees and determine the location:

Degrees, minutes, and seconds	**Decimal degrees**
Latitude: N 41° 15′ 31.4963″	_____
Longitude: W 96° 0′ 46.019″	_____
Latitude: N 44° 51′ 13.8582″	_____
Longitude: W 92° 37′ 18.3459″	_____
Latitude: N 34° 1′ 7.5628″	_____
Longitude: W 118° 24′ 20.8624″	_____

B. Convert the following UTM coordinates to latitude and longitude.

UTM	**Decimal degrees**
Zone: 14	
Easting: 750259	_____
Northing: 4571786	_____
Zone: 11	
Easting: 370202	_____
Northing: 3765128	_____
Zone: 15	
Easting: 529887	_____
Northing: 4966785	_____

C. Convert the following state plane coordinates to latitude and longitude.

State Plane coordinates   **Degree decimal**

2600-Nebraska State Plane
Easting: 833920.520m Northing: 165958.572m   _____
Easting: 2735954.240ft Northing: 544482.416ft   _____

4802-Wisconsin State Plane Central Zone
Easting: 392805.708m Northing: 116743.289m   _____
Easting: 1288730.061ft Northing: 1911525.142ft   _____

0405-California State Plane Zone 5
Easting: 1962518.944m Northing: 557620.539m   _____
Easting: 6438697.570ft Northing: 1829460.053ft   _____

D. Provide the location of the following land divisions within the United States Public Land Survey:

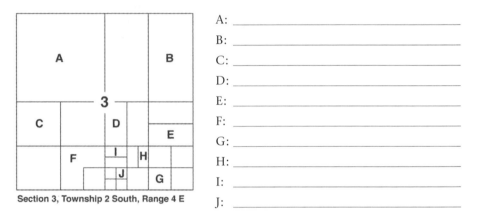

Section 3, Township 2 South, Range 4 E

A: _____

B: _____

C: _____

D: _____

E: _____

F: _____

G: _____

H: _____

I: _____

J: _____

11.7 Questions

1. List all coordinate systems on a typical USGS topographic map.

2. Explain the distinctions among the geoid, the sphere, and the spheroid.

3. Describe NAD27, NAD83, and WGS84.

4. Convert 18° 23' 40" to decimal format.

5. What is the UTM zone of your current location? How far are you from a UTM boundary?

6. How are east measurements made in the UTM coordinate system?

7. What are similarities and differences between UTM and SPC?

8. How many acres is a quarter section in the USPLS?

9. How is the geocoding of addresses performed?

10. What would be an application of reverse geocoding?

11.8 References

Burritt, Elijah H. (1833) *Geography of the Heavens, and Class Book on Astronomy*. 5th ed., F. J. Huntington & Co.

Geodetic Glossary (1986) Washington, DC: National Geodetic Survey, U.S. Department of Commerce.

Wilson, Stanley W. (1995) *Use of the "NAD/GWS 84" Datum Tag on Mapping Products*. Federal Register, Vol. 60, No. 157.

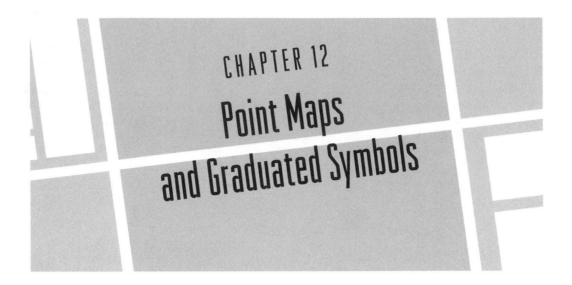

CHAPTER 12
Point Maps
and Graduated Symbols

I have an existential map; it has "you are here" written all over it.

—STEVEN WRIGHT

12.1 Introduction

Point symbols are normally used to show the location of features. They may also be used to depict qualitative attributes or quantitative values. Qualitative differences are represented through the use of different symbols, or the symbols may be graduated in size to show quantitative value. We have already examined how to define the location of a marker using a mapping application programming interface (API). This chapter shows how this symbol can be changed and how a series of locations can be input through different types of external files. Finally, two examples of graduated point symbols are presented.

12.2 Markers

Markers are small graphic files defined in the PNG format that are normally 32×32 pixels but can be of any size (Svennerberg 2010, p. 101). A large variety of different icons are made available to denote the location of almost any type of feature, including icons for restaurants, airports, movie theaters, and so on (see Figure 12.1). A shadow was associated with the icon to provide a 3D appearance, as shown with the restaurant symbol at the bottom of Figure 12.1, but this practice is now less common. Custom icons can be designed using a pixel-based graphic editing

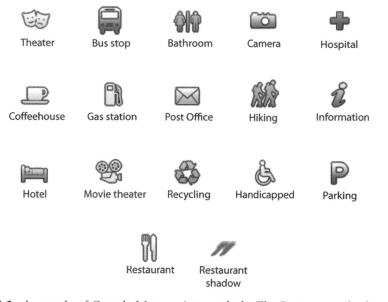

FIGURE 12.1. A sample of Google Maps point symbols. The Restaurant shadow symbol at the bottom is intentionally blurred to appear as a shadow. It would normally be combined with the restaurant symbol to create an impression of 3D. Copyright 2013 Google.

program such as Adobe Photoshop™ as well. Thousands of Google Map icons can also be found by searching for sites like Nicolas Mollet icons, Benjamin Keen icons, and Mapito icons.

In Google Maps API, the google.map.Marker object specifies the marker attributes. Many properties can be defined using this object, including the size of the icon, its position relative to the point, the icon shadow file, and the shape and size of the clickable region of the icon (see Figure 12.2). A majority of the code in this example is used to define the image and shadow attributes.

The next example demonstrates how to change an icon with mouse events like *hover* and *click*. We begin with a large 98 × 75 pixel icon file that contains a total of six icons (see Figure 12.3). We want to display the top row of icons based on three mouse events—mouseout, mouseover, or mouseclick. A different set of pixels are accessed with wifi, wifiHover, and wifiClick, with wifiHover and wifiClick offset from the first by 33 and 66 pixels, respectively. The following Hover and Click events specify which of the three icons to display based on the position and state of the mouse. The example also demonstrates that a series of icons can be placed in a single file and accessed through an "offset technique." It also shows how to make the icon change by either moving the mouse over, or clicking on, the icon.

12.3 Mapping Multiple Points

12.3.1 Random Points

The example in Figure 12.4 computes five random points using the Math.random JavaScript function. This function generates a random number between 0

FIGURE 12.2. Map with user-defined marker and shadow. The marker icon is a 32 × 32 pixel PNG file. Copyright 2013 Google.

and 1, which is then multiplied by the difference between the longitude (lngspan) and latitude (latspan). Adding this to the corner latitude and longitude values, southwest.lat and southwest.lng, results in a random point within the specified area (Google 2011). For example, if the random number happens to be 0.5, then the point will be halfway between the two sides of the bounding box.

This example also uses numbered icons to indicate the points. These icons are in a "NumberedMarkers" folder. Subfolders within this folder contain both square (SQ) and teardrop (TD) numbered markers in blue, green, red, and yellow. Markers between 0 and 99 are provided for all styles, resulting in a total of 800 distinct graphic files.

Because the filename of the icon changes for every icon drawn, the name of

```
function initialize() {
  var myLatlng = new google.maps.LatLng(41.258531,-96.012599);
  var myOptions = {
    zoom: 15,                                          Map
    center: myLatlng,                                  center
    mapTypeId: google.maps.MapTypeId.ROADMAP
  }
  map=new google.maps.Map(document.getElementById("map_canvas"),myOptions);

      var wifi = new google.maps.MarkerImage('img/markers.png',
Icon    new google.maps.Size(32, 37),        Defining
position  new google.maps.Point(0, 0),       initial icon
        new google.maps.Point(16, 35));

      var wifiHover = new google.maps.MarkerImage('img/markers.png',
Icon    new google.maps.Size(32, 37),        Defining
position  new google.maps.Point(33, 0),      hover icon
        new google.maps.Point(16, 35));

      var wifiClick = new google.maps.MarkerImage('img/markers.png',
Icon    new google.maps.Size(32, 37),        Defining
position  new google.maps.Point(66, 0),      click icon
        new google.maps.Point(16, 35));

      var shadow = new google.maps.MarkerImage('img/shadow.png',
        new google.maps.Size(51, 37),                Defining
        new google.maps.Point(0, 0),                 shadow icon
        new google.maps.Point(16, 35));

      var marker = new google.maps.Marker({
        position: new google.maps.LatLng(41.258531,-96.012599),
        map: map,
        icon: wifi,              Draw the
        shadow: shadow           icon
      });

google.maps.event.addListener(marker, 'mouseover', function() {
Hover    this.setIcon(wifiHover);
event   });
        google.maps.event.addListener(marker, 'mouseout', function() {
          this.setIcon(wifi);
        });

google.maps.event.addListener(marker, 'mousedown', function() {
Click    this.setIcon(wifiClick);
event   });
        google.maps.event.addListener(marker, 'mouseup', function() {
          this.setIcon(wifiHover);
        });

  google.maps.event.addListener(marker, 'click', function() {
    map.setZoom(17); });
                     Reset
                     zoom
```

marker.png

FIGURE 12.3a. A different icon is displayed based on hover and click mouse events. The three icons reside in the same file and are referenced according to their position. If the map is clicked, the map zooms from level 15 to 17. Copyright 2013 Google.

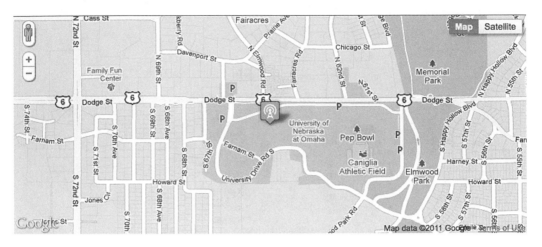

FIGURE 12.3b.

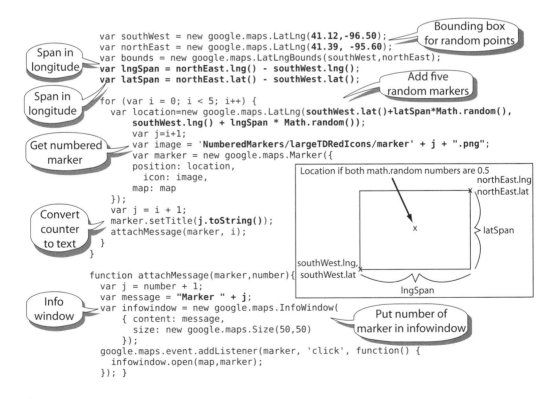

FIGURE 12.4a. The JavaScript *random* function is used to place markers in a random pattern. The files for the numbered markers are located in the NumberedMarkers folder. Copyright 2013 Google.

FIGURE 12.4b.

each marker must be constructed before the attachMessage function is called. This is done by defining a variable *j* to be one more than the loop counter *i* so that the value for *j* ranges from 1 to 5. This number and the .PNG file designation is then concatenated to the name of the file with the "+" operator:

```
var image = 'NumberedMarkers/largeTDRedIcons/marker' + j + ".png";
```

A similar procedure is used to increment the number that is written inside each infowindow.

12.3.2 Mapping Points from Arrays

Arrays are efficient mechanisms to store a large number of points. Separate arrays could be defined to map multiple points, with the first array holding all of the latitude values and the second all of the longitude values. The third array would store the HTML code for each infowindow that could include an img reference to an external picture. The last array would store the hover text.

The two maps in Figure 12.5 shows the hover text and infowindow for a particular point of interest for the city of Wayne, Nebraska. The code defines the four arrays using the new Array statement. The for (var i = 0; i < 16; i++) statement starts the loop that processes the 16 points. The following statements define the points, infowindow content, and the hover text. The this object stores the infowindow content.

12.3.3 Mapping Points from an XML File

To map an even larger number of points, it is best to place the latitude and longitude of the points and any associated information in a file. In Figure 12.6, the file has

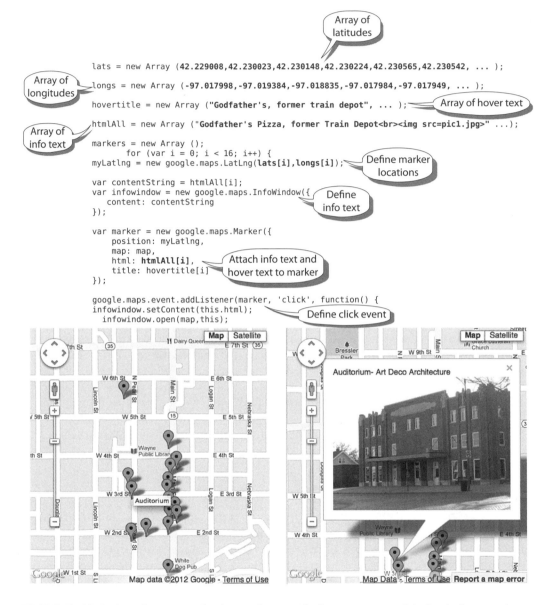

FIGURE 12.5. Displaying multiple markers with hover text and infowindows using a series of four arrays. The lats and longs arrays store the latitude and longitude values. The htmlAll array stores the text and links to external pictures. Finally, hovertext defines the 16 text strings that are presented when the mouse is hovered over a point. Copyright 2013 Google.

been formatted in extensible markup language (XML) and read using the downloadUrl function that is located in an external .js file (Williams 2011). In this case, the example.xml file contains three points defined as latitude/longitude. Text for the bubble is also defined in the file using HTML coding. The label text appears as a tooltip when the mouse is positioned over the marker. The createMarker function

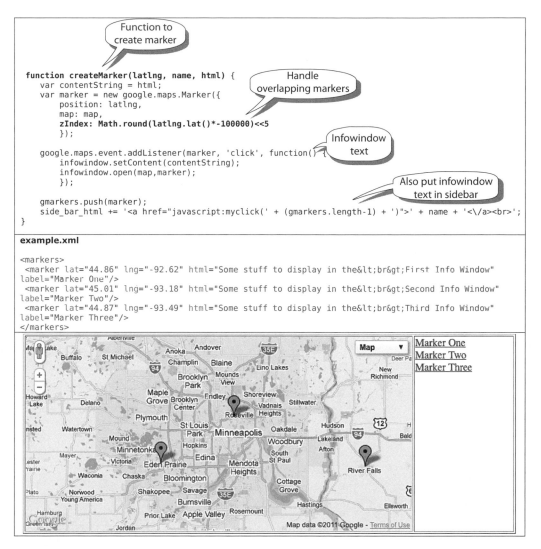

FIGURE 12.6. Displaying multiple markers with a sidebar. The position for each marker along with the associated bubble text and tooltip text is defined in an XML file. Copyright 2013 Google.

places the markers on the map. The zIndex option controls the marker's stacking level. This option can be used to control symbol overlap when a large number of symbols are being mapped.

12.3.4 Mapping Points from a JSON File

JavaScript Object Notation (JSON) is a data file format for representing simple data structures called objects. Based on JavaScript, it can be read quickly and is completely language independent. JSON is an ideal interchange format between languages as it can be used to read data in a variety of formats, including number, string, value, array, and object. Most web browsers have native JSON encoding/

decoding that improves security and speed by eliminating the need for a separate parsing function. Figure 12.7 depicts a reference to a JSON file called SW.json, a part of this JSON file, and the resulting map. GeoJSON is a JSON file that has been specifically designed for encoding geographic data structures. Spatial data types supported by GeoJSON are points, polygons, multipolygons, features, geometry collections, bounding boxes, and their associated attributes.

12.4   Mapping Points through GeoRSS and KML

In cases where data is frequently updated, a web format called RSS is often used. An RSS feed is composed of content and metadata. The metadata describes the

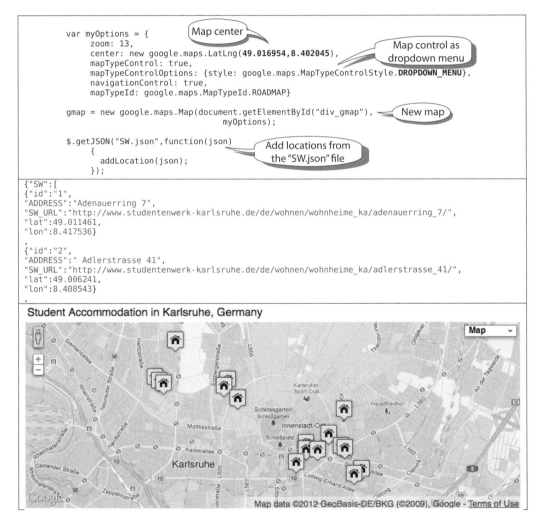

FIGURE 12.7. Displaying points with a JSON file. The position for each marker along with the associated bubble text and tooltip text is defined in the SW.json file, part of which is listed above the map. Copyright 2013 Google.

data and includes such attributes as date and author. Publishers of RSS feeds benefit by syndicating content automatically, while consumers benefit through timely updates from favored websites. A standardized XML file format allows the information to be published once and viewed using many different programs.

Keyhole Markup Language (KML) is an XML-based format used for describing two- and three-dimensional space that was originally developed for Google Earth. It is now an open standard officially named the OpenGIS® KML Encoding Standard (OGC KML) and is maintained by the Open Geospatial Consortium (OGC). The format specifies features such as placemarks, images, polygons, and 3D models. Places are always specified with latitude and longitude.

The `google.maps.KmlLayer` function reads a KML-formatted RSS feed specified with an http address (see Figure 12.8). The address in this example comes from flickr.com, an image hosting website. In this case, a user has geocoded a series of photos and has made the feed available for syndication. RSS is an efficient form of data distribution. Maps made in this way are usually displayed faster than those read from an XML file. The trade-off is that there is less control over the look of the actual map.

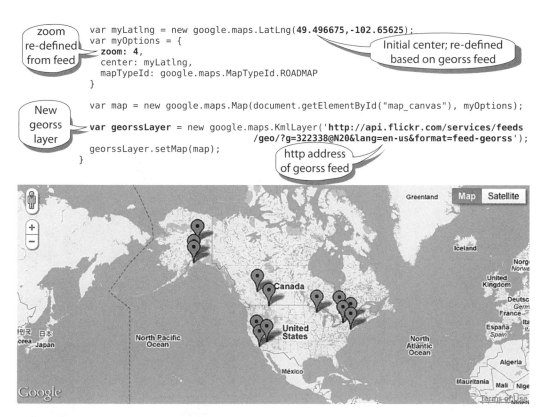

FIGURE 12.8. Displaying an RSS feed defined in KML format from the flickr website. Copyright 2013 Google.

12.4.1 Mapping Current Earthquake Locations

The example in Figure 12.9 shows an application of an RSS feed for the depiction of earthquakes. Notice that the code in this example is the same as that from the last (Figure 12.8) with the exception that the http address in the google.maps. KmlLayer line has been changed. This particular KML feed is updated every day and shows earthquakes in the past seven days. Each bubble is clickable to provide more information about the earthquake event.

```
var georssLayer| = new
google.maps.KmlLayer('http://earthquake.usgs.gov/earthquakes/catalogs/eqs7day-
M2.5.xml');
```

Earthquakes in the past week

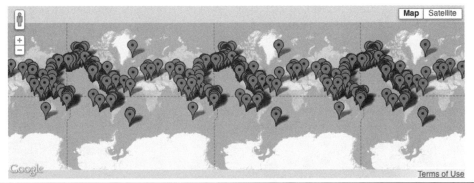

```
var ctaLayer = new   var ctaLayer = new
google.maps.KmlLayer('http://earthquake.usgs.gov/earthquakes/catalogs/eqs7day-
M2.5.xml',{preserveViewport:true});
  ctaLayer.setMap(map);
```

Earthquakes in the past week

FIGURE 12.9. Displaying an RSS feed defined in KML format from a USGS website. Each icon pinpoints the earthquake and, when clicked, describes the event. The top map is displayed using a basic call to a KML layer that ignores the defined center and zoom level and therefore duplicates most of the world. The bottom map is displayed with the {preserveViewport:true} option and ctaLayer that applies the user-defined center and zoom level. Copyright 2013 Google.

12.4.2 GeoRSS Feed Examples

Many organizations provide GeoRSS feeds, and some update their feeds throughout the day. Figure 12.10 shows four different examples of available KML feeds. Wave heights on the Great Lakes are continuously monitored and made available through a feed from the US National Atmospheric and Atmospheric Administration (NOAA). NOAA also makes weather station information available for Hawaii and the entire country through a number of continuously updated feeds. Current volcanic activity is made available through the Global Volcanism Program at the Smithsonian National Museum of Natural History. Nuclear power plants in Japan are made available through a KML file created in Google Earth.

12.4.3 Making Your Own KML File

KML files can be created in Google Earth or with any text editor. The code provided in Figure 12.11 places three points on the map. Each point is defined with the KML `<Placemark>`. The accompanying `<description>` tag defines text for display in the `infowindow`. The `google.maps.KmlLayer` tag is used to display the KML after

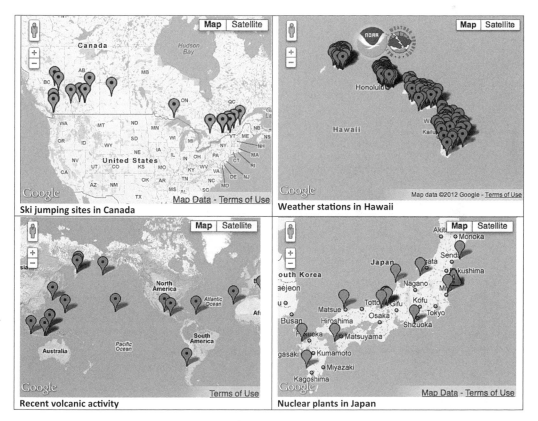

FIGURE 12.10. Four examples of geoRSS feeds distributed in the KML file format. Copyright 2013 Google.

KML File

Standard format of KML file

```
<?xml version="1.0" encoding="UTF-8"?>
<kml xmlns="http://www.opengis.net/kml/2.2">
<Document>
    <Placemark>
        <description>Nome, Alaska</description>
        <Point>
            <coordinates>-165.399409, 64.503877</coordinates>
        </Point>
    </Placemark>

    <Placemark>
        <description>Fairbanks, Alaska</description>
        <Point>
            <coordinates>-147.708316, 64.839314</coordinates>
        </Point>
    </Placemark>

    <Placemark>
        <description>Anchorage, Alaska</description>
        <Point>
            <coordinates>-149.898806, 61.211768</coordinates>
        </Point>
    </Placemark>
</Document>
</kml>
```

Attribute of point

Point as longitude, then latitude

http address of KML file

```
var kmlURL = "http://maps.unomaha.edu/OnlineMapping/Ch12/three_points.kml" +
             "?" + Math.round(Math.random() * 10000000000);

var georssLayer = new google.maps.KmlLayer(kmlURL);
georssLayer.setMap(map);
```

Associating KML layer with map

Adding random number to filename to circumvent Google's caching of KML files

FIGURE 12.11. A simple KML file with three points is displayed using the kmlLayer option. The text in the description is placed in the bubble for each point. The KML file can only be accessed through its http address. A random number must be added at the end of the filename to circumvent the caching of KML files at the server level. Copyright 2013 Google.

the file has been placed on a server. The default Google marker is used to indicate the three points on the map.

KML files present some major differences compared to previous methods of input. For example, the order of coordinates is reversed. Normally, the order is latitude and then longitude; with KML, the order is longitude, then latitude. KMLs also cannot be simply referenced on your local disk. Rather, a complete URL needs to be provided with an address like:

```
http://maps.unomaha.edu/Cloud/ch12/three_points.kml
```

KML files are also cached on the Google server. If a change is made to a KML file, refreshing the browser page will not force a refresh of Google's cache and the old version of the file will be mapped. At some point, perhaps after a few hours, the old version of the cache will be deleted and the new KML will be mapped properly. A solution to this problem is to change the name of the KML file. This would also require changing the reference to the file from within the JavaScript code. To avoid having to constantly rename the KML file and the reference to it, an arbitrary parameter can be added at the end of the URL, such as "?123" to make Google cache think that the file name has changed. The following line of code adds a random number at the end of the filename to fool the Google cache to think that it is a different file:

```
var kmlURL = "http://maps.unomaha.edu/Cloud/ch12/three_points.kml" + "?"
+ Math.round(Math.random() * 10000000000);
```

The chances are infinitesimal that the random number generator would produce the same number in two successive calls.

Another difference is that the formatting of the map is defined in the KML file as well, not in the API that displays the KML. Changing the icon involves changing the KML code. Figure 12.12 shows the code from 12.11 that has been modified to reference a specific icon file at:

```
http://maps.unomaha.edu/Cloud/ch12/icon49.png
```

This icon is then referenced before each point using the *<styleUrl>* tag.

12.5 Mapping Data from a Fusion Table

Another method for storing and displaying point data is Google Fusion Tables. A Fusion Table is a Google service for storing and managing tabular data in the cloud. A certain amount of data can be uploaded and shared with collaborators at no charge. Once uploaded, data may be filtered, aggregated, merged, and exported in multiple formats. Part of a Fusion Table of crime in Chicago is shown in Figure 12.13.

A Fusion Table is accessed in two different ways. In the initial version of Google Fusion Tables, a numeric code would access the table through Google Maps:

```
google.maps.FusionTablesLayer(139529)
```

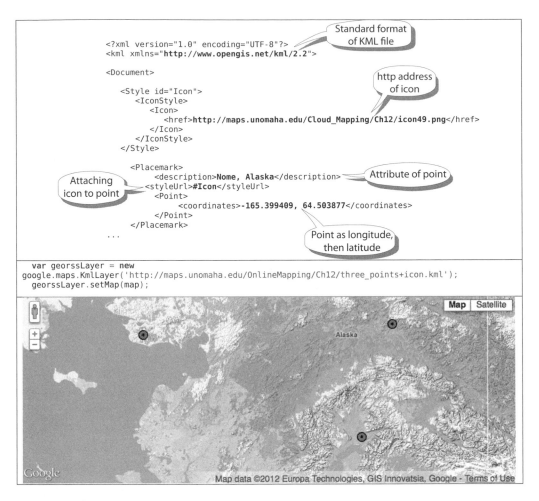

FIGURE 12.12. A different icon is referenced within the KML file. In this example, an icon is defined with the Style tag. StyleURL associates the icon with the point. Copyright 2013 Google.

where 139529 is the name of the Fusion Table. A large number of Fusion Tables were created with this type of code. In a subsequent version, this numeric code was changed to a much longer encrypted ID such as:

```
("1S4EB6319wWW2sWQDPhDvmSBIVrD3iEmCLYB7nMM")
```

These codes may be used to examine the tables directly through a web page in this way:

```
http://tables.googlelabs.com/DataSource?dsrcid=139529
```

or

```
http://www.google.com/fusiontables/DataSource?docid=1KRjg1Wofws9KsEvAB
Uun-xDPX1MoFdn9H0tWypc
```

Notice the use of "dsrcid" with the numeric ID and "docid" with the newer encrypted version.

Fusion Tables are managed using the Google Fusion Tables website. This site has tools to search for existing tables, upload new tables, query, and visualize the data. Query functions are based on a subset of the SQL query language—a standardized method for querying data. Visualization functions include charts and maps. Data can also be synchronized with an offline repository.

In the second Fusion Table example, public transport ridership information is provided at different stops (see Figure 12.14). A query can be performed to limit the results to a certain level of ridership, creating alternative views of the data. Here, two maps are made showing stops that are used daily by more than 2000 and 5000 people.

Fusion Table 424076 includes the largest 100 cities in the world. Figure 12.15 illustrates how the table can be used to create maps with all data points. The table

```
function initialize() {

  var chicago = new google.maps.LatLng(41.850033, -87.6500523);

  map = new google.maps.Map(document.getElementById('map_canvas'), {
    center: chicago,
    zoom: 11,
    mapTypeId: 'roadmap'
  });

  layer = new google.maps.FusionTablesLayer(139529);
  layer.setMap(map);
}
```

Map center

New map

Add Fusion Table layer

Fusion Table data and map may also be viewed in this way:
http://tables.googlelabs.com/DataSource?dsrcid=139529

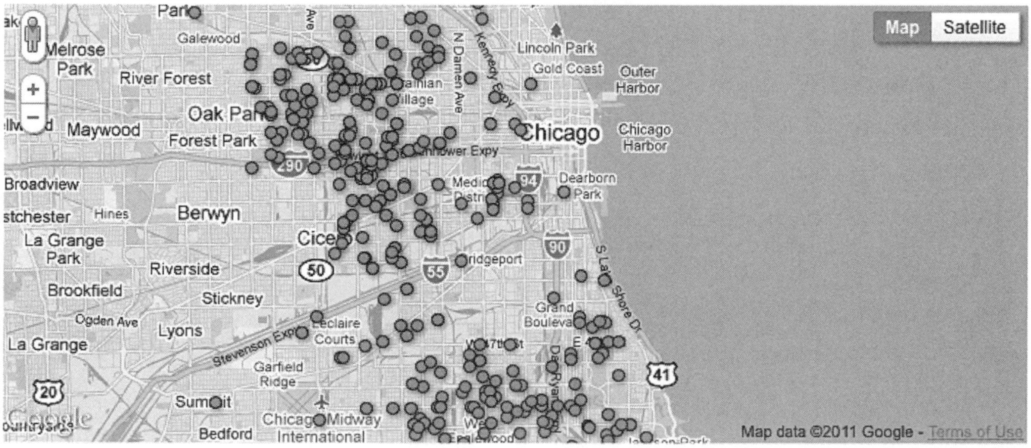

FIGURE 12.13. Displaying data from a Google Fusion Table. Copyright 2013 Google.

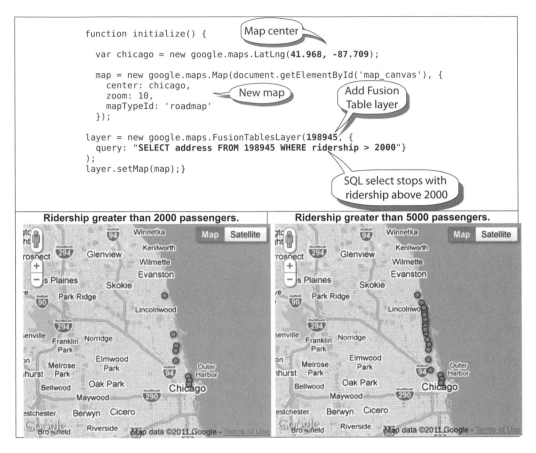

FIGURE 12.14. Displaying data from a Google Fusion Table on public transport ridership in Chicago. The two maps show the result of a query for different levels of daily ridership at different stops. Ridership of 2000 and above is shown on the map at left. Changing 2000 to 5000 results in the map on the right. Copyright 2013 Google.

can also be queried for more specific, user-defined results. The bottom map is a result of a query of the cities with over 5 million people.

A large number of tables are available through the Google Tables search page (see Figure 12.16). Searching on a particular topic returns tables that have been made public. The contents of these tables may be examined, and any table that contains either latitude/longitude values or street addresses can be displayed as a map. Selecting About under the File menu will provide the ID of the table for display through Google Maps.

The easiest way to create a new Fusion Table is to upload an existing table of data such as an Excel file. In Figure 12.16, a series of locations have been placed in an Excel file. These can be expressed either in latitude and longitude or simply as street addresses (in the parts of the world where geocoding is supported). Once uploaded, the Fusion Table number is accessed through File/About menu. Data formats accepted by Google Fusion Tables include:

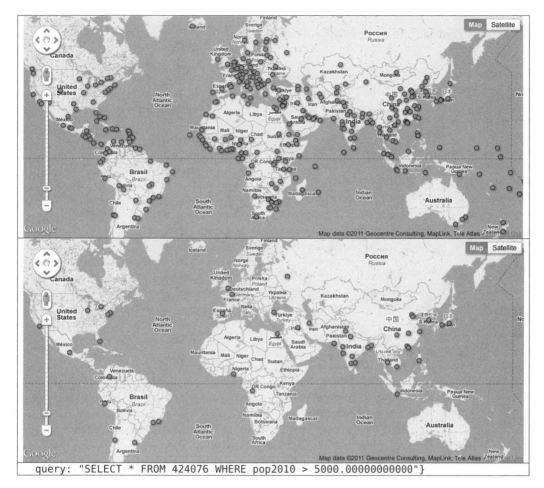

query: "SELECT * FROM 424076 WHERE pop2010 > 5000.00000000000"}

FIGURE 12.15. The top map presents Fusion Table 424076 which shows the 100 largest cities in the world. The bottom map is a result of a query specifying cities over 5 million. Numbers in the table are in thousands. Copyright 2013 Google.

- Comma-separated files (.csv)—Up to 100 mb
- Microsoft Excel files (.xls, .xlsx)—Up to 1 mb
- OpenDocument Spreadsheet (.ods)—Up to 1 mb
- Data already in a Google Spreadsheet

If spreadsheets in Excel or Open Office are larger than 1 mb, they can be saved as comma separated values (CSV) files to take advantage of the higher 100 mb limit for this file type.

KML files may be imported into a Fusion Table. The map in Figure 12.17 is the result of importing the KML file of three points for cities in Alaska used in Figure 12.11. A column is created in the Fusion Table that contains the KML code, in this case the location of the cities.

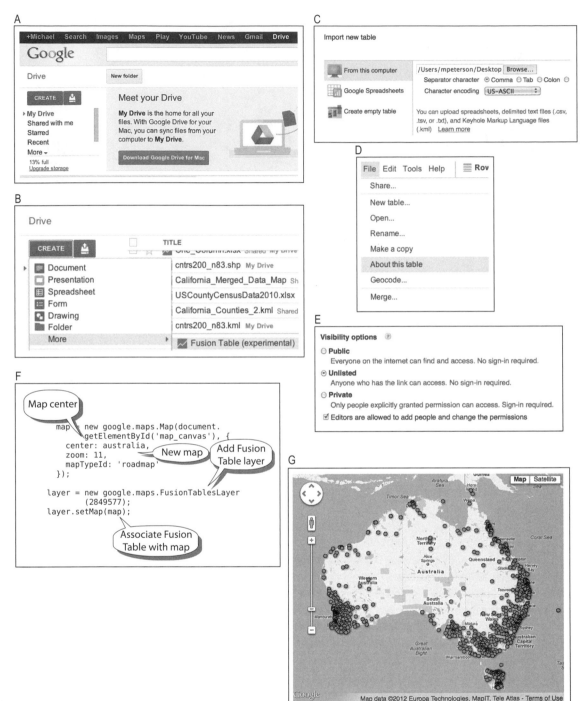

FIGURE 12.16. Making a map with a Fusion Table of addresses. A fusion table is created under Google Documents by selecting Table. An Excel file is imported that contains separate columns for latitude and longitude. The About menu item provides a numeric address of the table, in this case, 2849577. This number is inserted into the FusionTableLayer in Google Maps. In order for this reference to work, the table must be made public or unlisted to be accessed by Google Maps. Copyright 2013 Google.

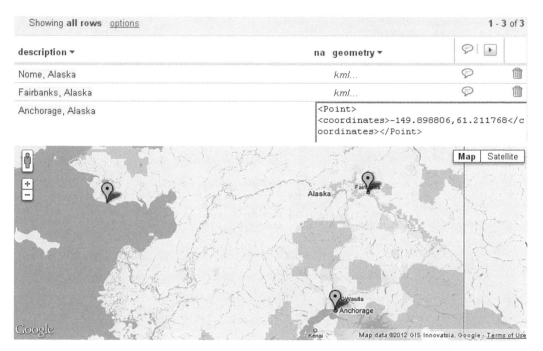

FIGURE 12.17. The KML file of three points from Alaska after being imported into Fusion Table 2950479. The table view on top shows the geometry column with the coordinates defined in KML. Copyright 2013 Google.

12.6   Geocoding

12.6.1   Marker Based on a Street Address

One of the most powerful functions of Google Maps is the geocoding of addresses. Figure 12.18 shows how a single address is geocoded using the `geocoder.geocode` function. This location is defined as the center of the map and depicted with a marker. The example also shows the use of an input dialog box within the body of the HTML to enter the address.

12.6.2   Mapping Multiple Street Addresses

When mapping the location of many points, such as crime incidents, it may be necessary to geocode several thousand addresses. This resource-intensive task can require a considerable amount of processing time. Further, it is not possible to expect the online mapping site to geocode the addresses every time the map is viewed. When mapping many addresses, it is necessary to pre-geocode all of the addresses and store the computed latitude/longitude locations into a separate file. Once the latitude/longitude positions have been calculated, it is a relatively easy task for the server to plot the locations. Several online services are available to geocode street addresses.

Figure 12.19 shows the address for each homicide in Omaha, Nebraska, for 2001 and the corresponding map. The addresses were first geocoded using an

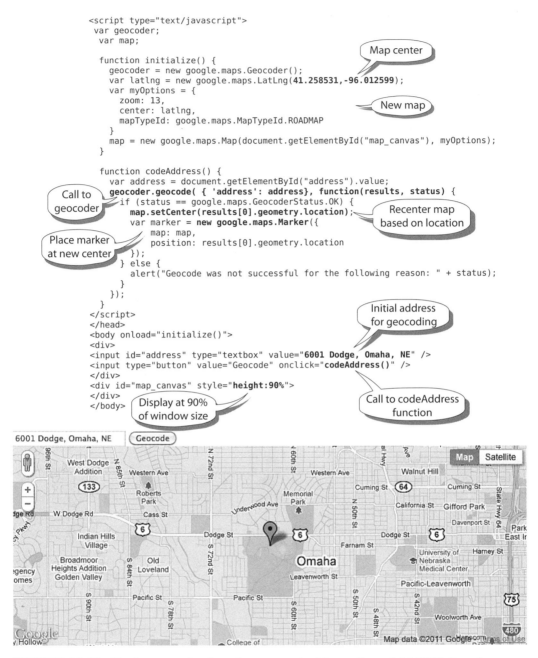

FIGURE 12.18. Placing a marker based on a street address. Copyright 2013 Google.

online service. The latitude and longitude values were then formatted into an XML file on the right side of the table. This formatting is done with the *concatenate* function in Microsoft Excel by merging the text strings and numbers in the specified format. The code described in 12.6 is used to plot the locations on the map, but no bubble info text or tooltip text is defined. Some online geocoding services output the resulting data file of coordinates in the KML format. In this case, the code introduced in Figure 12.9 can be used to map the results after the KML file has been placed on a server and accessed through its URL address. The easiest way to map a series of addresses as shown in Figure 12.19 is to import into a Fusion Table a single column of addresses stored in an Excel or a CSV (comma separated values) file.

12.7 Graduated Point Symbols

12.7.1 Circles

Another type of point map depicts quantities at locations. In this case, graduated symbols are used to show the magnitude of a variable at a given point. The symbols

Addresses of Homicides in Omaha 2001			Locations of Homicides defined in XML
Address	City	State	`<markers>`
4315 LEAVENWORTH ST	Omaha	NE	`<marker lat="41.252221" lng="-95.978036" />`
3600 BURDETTE ST	Omaha	NE	`<marker lat="41.279305" lng="-95.966148" />`
2030 MARTHA ST	Omaha	NE	`<marker lat="41.238332" lng="-95.942881" />`
3411 N 106 PA	Omaha	NE	`<marker lat="41.29042" lng="-96.078653" />`
3411 N 106 PA	Omaha	NE	`<marker lat="41.29042" lng="-96.078653" />`
3411 N 106 PA	Omaha	NE	`<marker lat="41.29042" lng="-96.078653" />`
1512 HOWARD ST	Omaha	NE	`<marker lat="41.255599" lng="-95.936109" />`
0 J CREIGHTON/GRANT	Omaha	NE	`<marker lat="41.260645" lng="-95.940469" />`
3000 AMES AV	Omaha	NE	`<marker lat="41.299609" lng="-95.956666" />`
4907 S 121 ST	Omaha	NE	`<marker lat="41.209511" lng="-96.101254" />`
4617 S 33 ST	Omaha	NE	`<marker lat="41.213203" lng="-95.962325" />`
1 PARK AV/LEAVENWORTH	Omaha	NE	`<marker lat="41.260645" lng="-95.940469" />`
6800 S 27 ST	Omaha	NE	`<marker lat="41.192012" lng="-95.954053" />`
1300 JEFFERSON ST	Omaha	NE	`<marker lat="41.195794" lng="-95.933739" />`
3054 PARKER ST	Omaha	NE	`<marker lat="41.276619" lng="-95.957739" />`
2400 AMES AV	Omaha	NE	`<marker lat="41.299674" lng="-95.947084" />`
4111 MIAMI ST	Omaha	NE	`<marker lat="41.282891" lng="-95.974279" />`
1607 EVANS ST	Omaha	NE	`<marker lat="41.291293" lng="-95.937481" />`
5501 N 35 ST	Omaha	NE	`<marker lat="41.308632" lng="-95.964784" />`
2437 CAMDEN AV	Omaha	NE	`<marker lat="41.305551" lng="-95.947853" />`
7200 BLONDO ST	Omaha	NE	`<marker lat="41.277939" lng="-96.024139" />`
15429 ADAMS ST	Omaha	NE	`<marker lat="41.195241" lng="-96.155312" />`
2907 Q ST	Omaha	NE	`<marker lat="41.205361" lng="-95.956316" />`
			`</markers>`

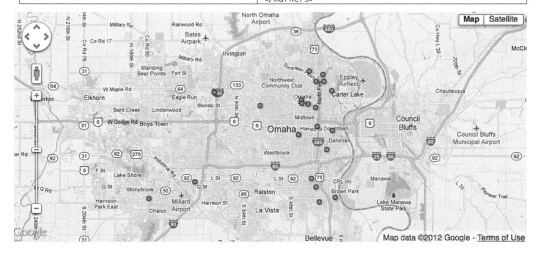

FIGURE 12.19. A geocoded list of addresses produced by geocode.com. The latitude and longitude values on the right in the top of the illustration have been formatted into an XML file using the concatenate function in Microsoft Excel. The data may also be mapped through a Fusion Table by simply importing a column of comma-delimited addresses. Data courtesy of Omaha Police Department; copyright 2013 Google.

are made progressively larger to show increases in quantity, such as population. Although any type of symbol can be graduated in size, geometric shapes like circles and squares are most often used. Circles are often used to show population for cities.

Data depicted at points may be nominal, ordinal, interval, or ratio. Mapping qualitative or nominal data may involve features such as restaurants, movie theaters, and bus stations. Here, a different icon would be used for each feature. No legend is needed if the icon is pictographic in the sense that it is recognizable as the feature it is representing. If geometric symbols such as a circle, square, or star are used, then a legend is needed to define the association between the symbol and the feature it is representing.

For ordinal, interval, and ratio quantitative data, size is used to convey value. If mapping ordinal data with categories of low, medium, and high, only three symbols of different sizes may be used. Interval and ratio data may either be placed into categories—effectively resulting in ordinal data—or mapped directly by making symbols proportional in size to the data values.

A variety of ways can be used to create graduated symbols with Google Maps. One method would be to define a series of icons with different size symbols. Ten icons could be defined from a small 4×4 grid to a size of 256×256 pixels or more. The data to be mapped would then be classified into ten categories and assigned to one of the symbol sizes.

An alternative method is to draw a symbol in relative proportion to the data value (Williamson 2011). In Figure 12.20, a circle function computes the outline of the circle. The call to the circle function includes a number of parameters such as the color of the circle, the thickness of the line, and its opacity.

Color is added to the graduated symbol using a *hexadecimal code* specified with a value such as #009900. The code represents a three-byte number for the three primary additive colors of red, green, and blue (RGB). Each primary color is coded with one byte or 8 bits resulting in a possible 16.7 million colors (2^{24}). The primary colors are represented in hexadecimal, with #FF0000 for red, #00FF00 for green, and #0000FF for blue. White is #FFFFFF and black is #000000. A web search for hexadecimal color codes will return many sites that define these colors on associated color charts.

The level of opacity is defined for the line that outlines the circle as well as for the color fill. Opacity is defined with a value between 0 and 1. If the value is 1, the color is completely opaque and the map will be completely covered by the color. If the value is less than 1, the underlying map will be visible. If the opacity is 0, the circle will no longer be visible.

The circle function also allows setting the number of points that define its outline. In the digital world, the circle is always only an approximation. It is created by drawing a series of straight lines between points along the outline of the circle. The more points used, the more it will look like a circle. A value of 6 for the number of points will create a hexagon, whereas a value of 4 will produce a square.

Multiple circles may be drawn at different locations to represent a distribution such as the population of cities, as shown in Figure 12.21. The size of each circle is proportional to the population it represents.

Some symbols such as circles are perceptually underestimated in size. In order

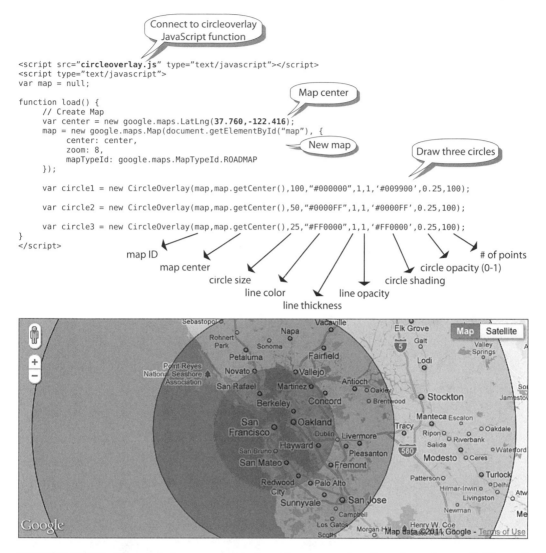

```
                            Connect to circleoverlay
                            JavaScript function

<script src="circleoverlay.js" type="text/javascript"></script>
<script type="text/javascript">
var map = null;

function load() {                         Map center
    // Create Map
    var center = new google.maps.LatLng(37.760,-122.416);
    map = new google.maps.Map(document.getElementById("map"), {
        center: center,
        zoom: 8,                         New map              Draw three circles
        mapTypeId: google.maps.MapTypeId.ROADMAP
    });

    var circle1 = new CircleOverlay(map,map.getCenter(),100,"#000000",1,1,'#009900',0.25,100);

    var circle2 = new CircleOverlay(map,map.getCenter(),50,"#0000FF",1,1,'#0000FF',0.25,100);

    var circle3 = new CircleOverlay(map,map.getCenter(),25,"#FF0000",1,1,'#FF0000',0.25,100);
}
</script>
```

map ID   map center   circle size   line color   line thickness   line opacity   circle shading   circle opacity (0-1)   # of points

FIGURE 12.20. Three circles are centered on the same point, from large to small. The call to the circle function includes the following attributes: latLng of center, radius of the circle, strokeColor of bounding line, strokeWidth of bounding line, strokeOpacity of bounding line, fillColor of circle, fillOpacity of circle, and numPoints (number of points that define the outline of the circle). Copyright 2013 Google.

to most effectively and accurately communicate the information, the mapmaker needs to compensate by adjusting the size of the circles. For example, a circle may need to be made three times bigger than another circle in order for it to be perceived as being twice as large in area. This compensation is usually accomplished with an exponent, and a variety of different exponents have been proposed to compensate for the underestimation in circle size (Chang 1977). Figure 12.22 shows a graduated symbol map of the larger cities in the United States with circles that have been

Object containing LatLng and population

```
<script>

    var citymap = {};
    citymap['chicago'] = {
        center: new google.maps.LatLng(41.878113, -87.629798),
        population: 2842518
    };
    citymap['newyork'] = {
        center: new google.maps.LatLng(40.714352, -74.005973),
        population: 8143197
    };
    citymap['losangeles'] = {
        center: new google.maps.LatLng(34.052234, -118.243684),
        population: 3844829
    }
    var cityCircle;

    function initialize(){
        var mapOptions = {
            zoom: 4,
            center: new google.maps.LatLng(37.09024, -95.712891),
            mapTypeId: google.maps.MapTypeId.TERRAIN
        };

        var map = new google.maps.Map(document.getElementById('map_canvas'), mapOptions);

        for (var city in citymap) {
            var populationOptions = {
                strokeColor: '#FF0000',
                strokeOpacity: 0.8,
                strokeWeight: 2,
                fillColor: '#FF0000',
                fillOpacity: 0.35,
                map: map,
                center: citymap[city].center,
                radius: citymap[city].population / 20
            };
            cityCircle = new google.maps.Circle(populationOptions);
        }
    }
</script>
```

Map center

Loop for each value in citymap

Circle parameters

Scale population by 20

Call Google circle function

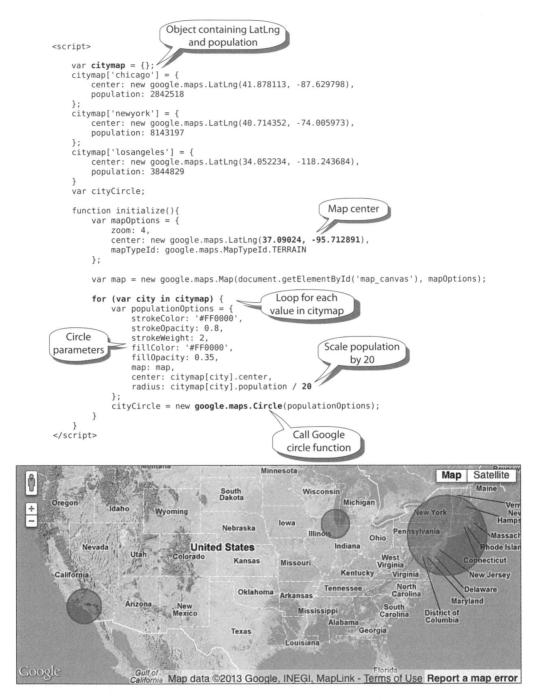

FIGURE 12.21. Population of three cities depicted with circles. Rather than using an array to store the point and population, the example uses an object called citymap. Copyright 2013 Google.

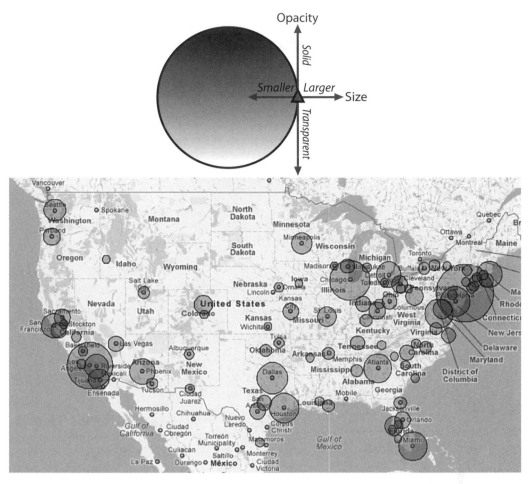

FIGURE 12.22. A graduated circle map showing the larger cities in the United States. In this implementation, the size of the circles can be proportionally altered by clicking and dragging on a circle outline. Moving up or down while dragging changes the shading of the circle; search: graduated circle map unomaha; copyright 2013 Google.

perceptually adjusted. The example also allows the interactive changing of circle sizes and opacity by clicking on and dragging on the side of any circle (Paziak 2012).

12.7.2 Pseudo Circles

The map in Figure 12.23 displays graduated symbols that depict scratch lottery ticket sales from convenience stores in the city of Omaha. The symbols appear to be circles, but they are actually short line segments of different thicknesses oriented at a 45-degree angle. Line thicknesses are relative to the *maxsize* symbol size. Although the symbols are not quite circular, they are drawn much faster than the circles in the previous examples.

12.8   Problem of Symbol Overlap

Severe symbol overlap can occur when mapping many points or graduated symbols. In mapping the 3404 burglaries in Omaha in 2001, a great deal of overlap is apparent in the eastern part of the city, particularly at this scale (see Figure 12.24). A smaller symbol size could be used to show the distribution. The use of Google's Marker Manager function allows different icons to be used depending upon the current zoom level.

12.9   Summary

The mapping of points, along with their qualitative or quantitative value, is a primary application of cloud mapping. There are a wide variety of methods for specifying the location of points. The methods have advantages and disadvantages.

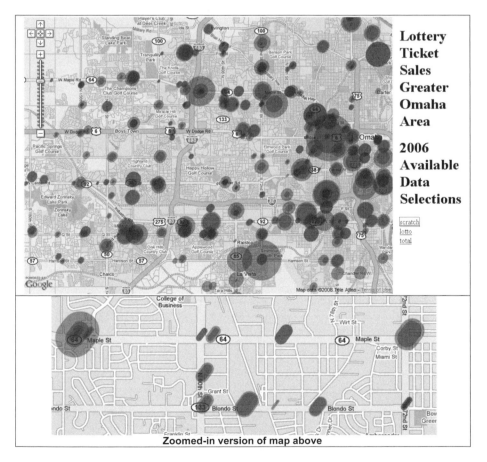

FIGURE 12.23. Graduated symbol map for the Omaha area showing scratch and lottery ticket sales for different time periods. Although the symbols appear to be circles, they are actually short line segments with different thicknesses and shadings, as shown in the zoomed-in version. Copyright 2013 Google.

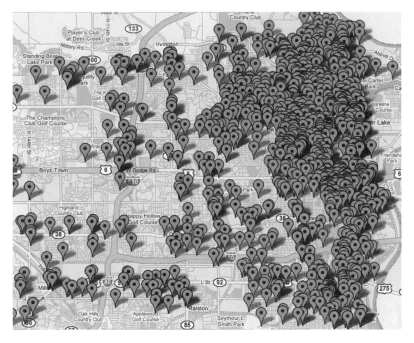

FIGURE 12.24. Burglaries in Omaha during 2001. A large number of symbols results in severe symbol overlap. The use of a smaller symbol would have alleviated some of the overlap problem. Data courtesy of Omaha Police Department; copyright 2013 Google.

In general, the location of the points can either be inserted directly into the JavaScript code or referenced through an external file. When mapping only a few points, it is easiest to place the points into the code—either individually or within an array. External file options include an XML-coded file, a JSON file, a geoRSS feed with KML, or a Fusion Table. The speed of map display generally increases in this order. Fusion Tables have the additional advantage of automatically geocoding street addresses in parts of the world where geocoding by street address is supported.

The making of graduated symbols is not well supported by mapping APIs. One option is to create a series of icons at different sizes. A second option is to draw out the symbols, such as circles, at different sizes. This approach makes it possible to make symbols proportional to the data values rather than classifying the data.

12.10 Exercise

Upload the code12.zip file to your server and make the suggested changes to the code examples.

12.11 Questions

1. Describe an application of the three mouse events—mouseout, mouseover, and mouseclick.

2. Describe one way to implement the numbering of markers.

3. Describe the different ways of providing coordinate values for the mapping of points.

4. Describe some advantages and disadvantages of the KML file format.

5. Describe the relationship between GeoRSS and KML.

6. What would be the best way to create an atlas of KML map offerings?

7. What are some alternatives for creating KML files?

8. Describe some advantages of Google Fusion Tables for mapping.

9. Is the display of thematic information on a general reference type of Google Map a good idea? Does the method convey the spatial patterns effectively?

10. Describe some problems in placing circles on maps.

11. Explain the following code after each letter A, B, C, and D.

```
var georssLayer = new google.maps.KmlLayer('http://api.flickr.com/
    services/feeds/geo/?g=322338@N20&lang=en-us&format=
    feed-georss');
georssLayer.setMap(map);
```

12. Insert a descriptive comment in each of the three places.

```
google.maps.event.addListener(map, 'click', function() {
        infowindow.close();
        });
    //A)
    //
    //
    //
        downloadUrl("xample.xml", function(doc) {
        var xmlDoc = xmlParse(doc);
        var markers = xmlDoc.documentElement.
getElementsByTagName("marker");
        for (var i = 0; i < markers.length; i++) {
          //B)
          //
          //
          //
          var lat = parseFloat(markers[i].getAttribute("lat"));
          var lng = parseFloat(markers[i].getAttribute("lng"));
          var point = new google.maps.LatLng(lat,lng);
          var html = markers[i].getAttribute("html");
          var label = markers[i].getAttribute("label");
          //C)
          //
          //
          //
          var marker = createMarker(point,label,html);
        }
        //D)
        //
        //
        //
        document.getElementById("side_bar").innerHTML = side_bar_html;
      });
    }
```

12.12 References

Chang, K. T. (1977) Visual Estimation of Graduated Circles. *The Canadian Cartographer* 14(2): 130–138.

Google (2011) *Google Maps JavaScript V3: The Solution for Maps Applications for both Desktop and Mobile Devices.* [http://code.google.com/apis/maps/documentation/javascript/]

Paziak, Douglas (2012) User Scalable Graduated Circles with Google Maps, in Michael Peterson (Ed.), *Online Maps with APIs and WebServices.* Berlin, Heidelberg: Springer Verlag.

Svennerberg, Gabriel (2010) *Beginning Google Maps API 3.* New York: Springer.

Williams, Mike (2011) *Google Maps API Tutorial.* Community Church Javascript Team. [http://econym.org.uk/gmap/basic3.htm]

Williamson, Matt (2011) *CircleOverlay extension for Google Maps.* App Delegate Inc. [http://appdelegateinc.com/samples/Google-Maps-Circle-Overlay/]

CHAPTER 13

The Online Map

You can use the Internet to find out, from anywhere on the planet: exactly how much coffee is in a certain coffee machine at Cambridge University in England; exactly how many sodas are available in certain vending machines at certain major universities; and much, much more.

—DAVE BARRY

13.1 Introduction

It is important to step back and consider that all maps based on an application programming interface (API) can only be made with an Internet connection. Servers provide both the tiles for the background map and the API tools to display the added points, lines, and areas. To put it simply, if a connection to the Internet is not available, these maps do not appear. The online map, as well as everything else online, is dependent on a sophisticated system of distributed computers—a system to which we are increasingly connected and dependent.

Underlying distributed computing is the client–server architecture. In this system, the user's client computer communicates with a server through a specific protocol. Most often, the server resides in a data center that houses a cluster of computers that respond to user requests. Some servers store maps while others store the programs that manipulate the maps. A client computer requests these services using the Internet as the medium of communication. Maps may also be distributed through the Internet using the peer-to-peer (P2P) model in which every computer is both server and client. The P2P architecture is especially useful for large files.

The online map is a product of the web. Although maps had been distributed through the Internet before the introduction of the World Wide Web, it was the web that made it possible for large numbers of people to access the available static and interactive maps. By the end of the 1990s, it was estimated that 200 million user-defined maps were being embedded within web pages on a daily basis. The tile-based system of map distribution, introduced by Google in 2005, makes it impossible to determine this number because the maps are no longer distributed as single entities but as 256×256 pixel tiles.

Internet mapping is based on Internet standards and web publishing guidelines. We have already seen that data definition formats such as extensible markup language (XML) are being used for mapping. Web standards and publishing guidelines provide a consistent and useful framework for formatting maps within web pages.

13.2   Client–Server Architecture

13.2.1   Server Data Centers

Figure 13.1 depicts the typical client–server architecture, the major distributed computing model. In this system, clients request services that are provided by servers. The server may be viewed as a dominant computer connected to several other client computers with fewer resources.

The distributed model provides an open and flexible environment in which a wide variety of client applications can be distributed and used by large numbers of computing devices, including mobile phones. Client–server computing has already reshaped the way computers are used and is affecting nearly every facet of our lives. The popularity of the Internet is driving its development, and some predict that all computer applications in the future will be *in the cloud* on a system of distributed computers. An example is Google's GDocs, an online application for

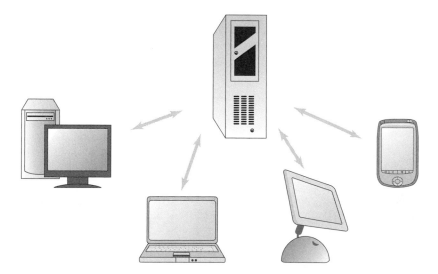

FIGURE 13.1. The client–server model. The client–server architecture is an example of a centralized architecture where the whole network depends on a central server to provide services.

the creation and distribution of word processing, spreadsheet, and presentation documents. Amazon Web Services is another commercial example of distributed computing.

There are many different types of servers, including those designed for the web, mail, and video. Each of these servers is usually a dedicated computer or cluster of computers that respond to client/user requests. Any computer, even a laptop, can be a server. To be available to client computers, it is best that the server be assigned a static IP, a nonchanging number that indicates the computer's address, and be registered through a domain name server (DNS). A DNS maintains a database of network names and associated IP addresses for all web servers. It allows a client to access a particular server by name. For example, a server at the University of Nebraska at Omaha is named: http://maps.unomaha.edu. The IP address for this server is 137.48.16.54. The server may be accessed either through this numeric IP or the server name as registered in the DNS.

Data centers are specialized buildings, usually without windows, that house a large number of computers. Figure 13.2 depicts the Google data center in Council Bluffs, Iowa, and part of the associated power facilities behind the main building. Diesel generators provide emergency backup in case of a power failure. A lead-acid battery backup system is used to power the computers until the generators are up and running. In the case of a natural disaster such as an earthquake, Google data centers in California have contingencies to acquire diesel fuel by helicopter to continue operations. It would be difficult for any individual to implement these types of backup contingencies to maintain the operation of a server.

Power is a major concern for a data center. Each is estimated to use 10 megawatts of electricity, requiring about 10 large diesel generators. Google has calculated the amount of energy used for each user search submitted through its search engine. They estimate that each search requires 0.0003 kWh. In terms of greenhouse gases, 1000 search requests generates the equivalent CO^2 as that of a car driven 1 km (0.61 m) (search: Powering a Google Search). To reduce costs and greenhouse emissions, companies operating data centers have invested in renewable energy.

The main program for serving web content is Apache, begun in 1995 as a series

FIGURE 13.2. A Google data center in Council Bluffs, Iowa. Power generators, pictured on the right, are located behind the windowless main building. Photos by M. Peterson.

of "patches" (thus, the name "Apache") to an HTTP (HyperText Transfer Protocol) server running at the National Center for Supercomputer Applications (NCSA) at the University of Illinois. Apache is now open-source software maintained by the Apache Software Foundation (http://www.apache.org). The program is available on multiple platforms, including Windows, Unix/Linux, and Mac OS X. Approximately 60% of all web servers use the Apache web server, with most of the remaining 40% using a combination of commercial software from Microsoft and Google, and Nginx—also a free and open-source project (Netcraft 2013).

13.2.2 Peer-to-Peer Networks

The peer-to-peer (P2P) protocol was made popular by Napster, an early file-sharing system that circumvented copyright restrictions on music. In contrast to the client–server model, each computer in a P2P network may act as both client and server. Clients in a P2P network can interact freely with other clients without the intervention of a server. All computers are peers with each other. While computers have equal status in the network, they don't necessarily have equal capabilities. A P2P network can consist of anything from small mobile devices to large mainframes (see Figure 13.3). A mobile peer might not be able to act as a server due to its intrinsic limitations, but the network does not restrict it from doing so.

In P2P, each peer contributes certain resources to the network, such as storage space and computing power (CPU cycles). As peers join the network, the capabilities of the network increase. This characteristic is termed *scalability* and is more difficult to achieve with the client–server architecture unless the capabilities of the server or server cluster are drastically increased. While there is a potential to share CPU resources, the main use of a P2P system is file exchange. Rather than downloading files from a single server and competing for server resources, clients can download files from a large number of different client–servers simultaneously.

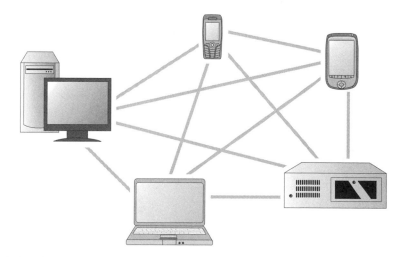

FIGURE 13.3. The P2P network model. In this networking system, each computer has a peer relationship with other computers and can act as both client and server.

Being less centralized and not dependent on a large data center, P2P is also less susceptible to server outages.

In actual implementation, a P2P network may rely on a directory server that stores IP addresses and other information about the computers in the network, including the locations of the files being shared. P2P systems vary in their reliance on directory servers. In the case of Napster, an early music file-sharing service that operated from 1999 to 2001, the server maintained a centralized directory of the location of all music files on registered computers around the world. The file transfer proceeded between two Napster-registered client computers. The Napster directory server did not store any music files. Through the coordinated efforts of the music industry, Napster was outlawed and its central server was shut down. Without the directory server, Napster users could no longer access the locations of music files, and the system of file exchange became immediately inoperable. The reliance on a directory server made Napster vulnerable.

Napster was soon replaced with the Gnutella open-source software that was not dependent on a directory server. A typical Gnutella session consists of a client connecting to a nearby client–server and sending a Gnutella packet. This "advertisement," also known as a ping, is forwarded to other connected client–servers called nodes (see Figure 13.4). All computers that receive the ping reply with a similar ping about themselves and send a list of nodes that it receives from other nodes. In theory, this packet would find its way to all Gnutella users, although, in practice, most never reach more than 50% of the network because of a time limitation placed on the ping responses. Searches for a specific file are performed at each Gnutella node. Searching and downloading files using this type of network is often unreliable because nodes are constantly disconnecting and reconnecting to the

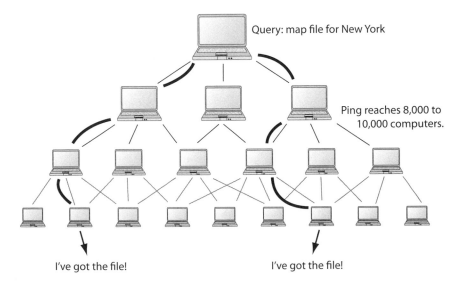

FIGURE 13.4. The Gnutella P2P network model. A client computer sends a ping for a specific file. This ping reaches a large number of computers that respond to the request. Downloading a file can be done from several computers simultaneously because the file is divided into packets.

network. Many different file-sharing programs use the Gnutella model, and a large number of music and movie files are distributed in this way, although it is impossible to determine an exact number.

13.2.3 *Thin versus Thick Clients*

In a typical client–server configuration, clients can have varying capabilities. The client may be slow and have limited storage, relying on the server to do most of the processing. Or the client may have a faster processor than the server and only rely on the server for some data. In almost all client–server interactions, the client's capabilities are underutilized. A concerted effort has been made to find a way to "download" processor-intensive applications to the client. These efforts have been largely futile because of the wide variety of client computers and operating systems and the difficulty servers encounter in responding separately to each.

Browser plug-ins have become the primary method of increasing the client's processing and display capabilities. Plug-ins such as Acrobat and Flash expand the functionality of the client's browser. In many cases, these plug-ins are added to the browser with minimal user action. As a browser collects more plug-ins, it becomes "thicker" and better able to process and display the wide variety of different file types that exist on the web. Specialized and proprietary plug-ins can even be created for certain applications. In designing online applications, a decision must be made about whether to design for a thin client with few browser plug-ins and limited processing power or for a faster client whose system has been augmented with such specialized plug-in software. The general trend has been to move away from plug-ins.

13.2.4 *Extending the Client–Server Model with Ajax*

Asynchronous JavaScript and XML (Ajax) is a technique that combines JavaScript and XML to create very interactive, server-client web applications (Garrett 2005). Ajax is not a programming language in itself, but a term that refers to the combined use of a number of different tools. The technique uses a combination of freely available tools such as cascading style sheets (CSS), document object model (DOM), and the extensible markup language (XML). Asynchronous communication is used to exchange data with the server while the user is idle so that the entire web page does not need to be reloaded each time the user makes a change (see Figure 13.5). The result is increased interactivity, speed, and an improved user interface.

Asynchronous communication is made possible by the Ajax engine, JavaScript code, which resides between the user and the server. Instead of loading the web page at the start of a web session, the Ajax engine is initially loaded in the background. Once loaded, the XMLHttpRequest object begins its work. This JavaScript code downloads data from the server without refreshing the web page. A user action that normally would generate an HTTP request to the server becomes instead a JavaScript call to the Ajax engine. If the engine can respond to a user action, a response from the server is not required. If the Ajax engine needs something from the server in order to respond to a user request—such as retrieving new data—the engine makes the request without interrupting the user's interaction

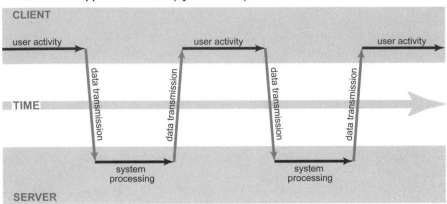

Classic Web Application Model (synchronous)

Ajax Web Application Model (asynchronous)

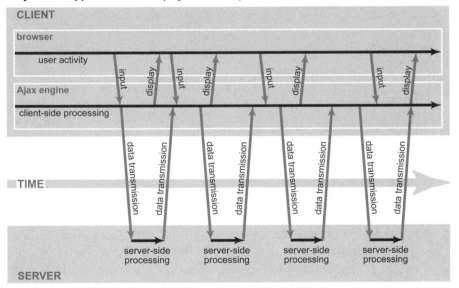

FIGURE 13.5. The typical client–server communication is synchronous (top illustration). Ajax uses the asynchronous communication between the client and the server. A connection is maintained to the server to speed interaction.

with the application. By eliminating the start-stop type of server interaction, Ajax transformed the online client–server experience.

Google Maps is a good example of an Ajax application. When the map is scrolled, additional map tiles are automatically downloaded. The tiles are added almost instantly because a connection is maintained to the server, allowing additional tiles to be quickly loaded. As the user scrolls, more map or satellite tiles are downloaded from the server without the user specifically asking for them. All major online map services, including Google, Bing, Yahoo!, OpenStreetMap, and Nokia, have implemented an Ajax interface. This new server/client method of communication has made MSP maps possible.

13.3 XML

A central part of Ajax is XML, developed as a markup language that would be less presentation-oriented. Rather than being concerned with how a page looks, like HTML, XML concentrates on what the elements within the file mean. It is extensible because new statements can be defined that in turn define the data, and it can be used to define other languages. Moreover, the meta-language is an open standard designed to share data and information between computers and computer programs as unambiguously as possible.

XML documents are composed of markup and content. The six types of markup that can occur in an XML document are shown in Table 13.1. The most important of these is the document type declaration (DTD), which creates new tag names, making it possible to define new languages. Other declarations allow a document to communicate meta-information relating to its content. There are four kinds of

TABLE 13.1. Six Different Types of XML Markup

XML Markup Type	Explanation

Elements	In the most common form of XML markup, elements identify the nature of the content they surround. It begins with a start-tag, <element>, and ends with an end-tag, </element>. In the following example, sex, firstname, lastname are subelements of the person element:
	
	`<person>`
	`  <sex>female</sex>`
	`  <firstname>Anna</firstname>`
	`  <lastname>Smith</lastname>`
	`</person>`

Attributes	Attributes provide additional information about elements. They are always surrounded by quotes. In the following example, sex is an attribute.
	
	`<person sex="female">`
	`  <firstname>Anna</firstname>`
	`  <lastname>Smith</lastname>`
	`</person>`

Entity References	An entity is a symbolic representation of information. Text such as "Hello, my name is" can be represented as an entity symbol. An entity must be created in the document type definition (DTD).
Comments	Comments begin with <!-- and end with -->. Comments are not part of the textual content of an XML document.
CDATA Sections	An XML document might contain characters that the XML parser would ordinarily recognize as markup (< and &, for example). In order to prevent this, a CDATA section can be used.
Document Type Declarations	The most important of the six different types of markup is the document type declaration (DTD) that makes it possible to create new tag names. Declarations allow a document to communicate meta-information about its content. There are four kinds of declarations in XML: element-type declarations, attribute list declarations, entity declarations, and notation declarations.

declarations in XML: element-type declarations, attribute list declarations, entity declarations, and notation declarations.

The primary benefit of XML is that it can encode the meaning of information. Programs can "understand" XML documents much better and therefore process the information in ways that are impossible using HTML. For example, unlike a simple listing of data, XML can explicitly state the meaning of each value. In making a database of pets, rather than presenting a table like this:

```
Shadow 12 cat Black
```

XML could define each element in this way:

```
<pets>
  <pet>
    <name>Shadow</name>
    <age>12</age>
    <type>cat</type>
    <color>black</color>
  </pet>
</pets>
```

Defining a markup language with XML is relatively simple and many new languages are based on XML. The three most important languages for maps are the geographic markup language (GML), scalable vector graphics (SVG), and keyhole markup language (KML). It is worthwhile to take a closer look at each.

13.3.1 *Geographic Markup Language*

The geographic markup language (GML) was defined by the Open Geospatial Consortium (OGC), a nonprofit, international, consensus standards organization. Released in 1999, GML encodes geographic space using geometry, features, and properties. The following would define a lake in GML with a feature ID of 155, a depth of 26 m, and a water quality type of 3. The outline of the lake is subsequently defined with seven coordinates.

```
<Feature fid="155" featureType="lake">
  <Description>Sawdust Lake</Description>
  <Property Name="Depth" type="Integer" unit="meters" value="26"/>
  <Property Name="Water Quality Type" type="Integer" value="3"/>
  <Polygon name="extent" srsName="epsg:27354">
    <LineString name="extent" srsName="epsg:27354">
      <CData>
        491888.999999459,5458045.99963358
        491904.999999458,5458044.99963358
        491908.999999462,5458064.99963358
        491924.999999461,5458064.99963358
        491925.999999462,5458079.99963359
        491977.999999466,5458120.9996336
        491953.999999466,5458017.99963357
      </CData>
    </LineString>
  </Polygon>
</Feature>
```

Georgraphic markup language (GML) incorporates a spatial reference system that embodies the main projection and geocentric reference systems. The encoding scheme defines coordinates and positions through a coordinate reference system (CRS) definition. It also defines transformations and conversions that allow coordinates to be changed from one CRS to another.

13.3.2 Scalable Vector Graphics

Also based on XML, scalable vector graphics (SVG) is a graphic description language that encodes features with lines and shapes in two dimensions. Although the format can also contain raster files, it is primarily designed for the display of vector information. With the ability to incorporate JavaScript, it is ideal for database-driven, user-defined graphics. Most of the major vector graphic editors, like Adobe Illustrator, are able to import and export SVG files.

Introduced in 1999, SVG has had a difficult history. Originally supported by a fully functional plug-in through the major graphic software company Adobe™, that support was dropped in 2008. Firefox and some other browsers have since added native support for the format that bypasses the need for a plug-in. Microsoft initially sought to support its own vector markup language (VML) with its Internet Explorer browser and did not display even basic SVG content until the release of version 9. Differing levels of support among browsers for the full SVG standard will continue to hamper the graphic format.

SVG allows for three types of graphic objects: vector graphic shapes (e.g., paths consisting of straight lines and curves), images, and text. Graphical objects can be grouped, styled, transformed, and combined into previously rendered objects. SVG drawings can be dynamic and interactive. Event handlers, such as *onmouseover* and *onclick*, can be assigned to any SVG graphical object. Because of its compatibility with other web standards, features like scripting can also be integrated.

Figure 13.6 depicts a series of graphic objects and the corresponding SVG code. The fill property specifies the fill color of the path, rect, or circle. Colors are defined by using named or hexadecimal color formats such as fill:red or fill:#ffff00. The stroke property defines the thickness, color, and individual points that constitute its shape. The path creates a curved line by calculating a series of points around centrally defined points. SVG can produce very detailed and readable maps that rival the look of maps on paper. Figure 13.7 shows a portion of a USGS topographic map that was converted to SVG.

The WikiProject Maps initiative started by Wikipedia encourages map submissions in SVG. The project seeks to make map contributions possible in a similar way to encyclopedic entries. Outline maps are provided, and free software like InkScape can be used to edit and symbolize the maps. Instructions are provided to standardize their look. While maps are accepted in JPEG, GIF, and PNG, submissions in SVG are strongly encouraged because the maps can be scaled without the loss of detail. In addition, the underlying XML format means that the maps can be easily edited and converted by scripts or cascading style sheets (CSS).

Probably the major problem with SVG—as with any format based on XML—is the size of the resulting file. As a relatively verbose, text-based format, SVG files can be quite large, slowing download and display. A zipped format called SVGz is

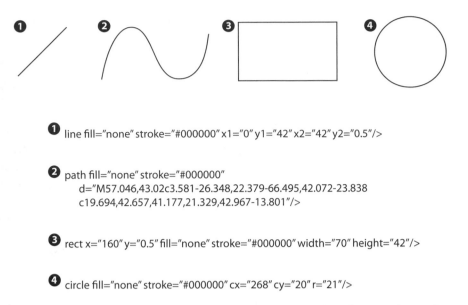

❶ line fill="none" stroke="#000000" x1="0" y1="42" x2="42" y2="0.5"/>

❷ path fill="none" stroke="#000000"
    d="M57.046,43.02c3.581-26.348,22.379-66.495,42.072-23.838
    c19.694,42.657,41.177,21.329,42.967-13.801"/>

❸ rect x="160" y="0.5" fill="none" stroke="#000000" width="70" height="42"/>

❹ circle fill="none" stroke="#000000" cx="268" cy="20" r="21"/>

FIGURE 13.6. The four graphic objects—line, path, rect, and circle—are shown here with their description in scalable vector graphics. All four shapes have both fill and stroke attributes.

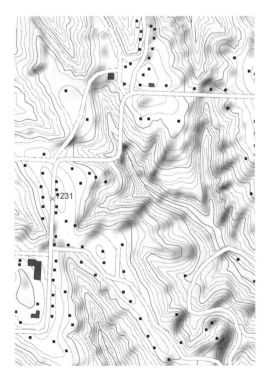

FIGURE 13.7. Part of a topographic map created with SVG. From Pavlicko and Peterson 2004. Reprinted by permission.

commonly used in an attempt to speed the delivery of the files. It has been found that user satisfaction with online maps is largely based on the speed of map display.

13.4 Web Mapping Services

A considerable amount of geographic information has been placed into GIS databases since these systems came into widespread use in the 1980s. In order for this information to be useful to more people, a method was needed for both "pulling" and disseminating the information from a database. In 1999, the Open Geospatial Consortium (OGC) defined a set of standards for distributing geographic data. The purpose was to facilitate the distribution of data and make layers of information more accessible. A series of standardized services were defined for supplying geodata universally across any platform. With standardization, map data services are able to interact with and display maps through an Internet-based interface.

Prior to standards being established, extracting information from a GIS database required interacting with a variety of large and complicated databases. The OGC streamlined this process by placing the burden for the extraction of data on the server. As defined by OGC, the web mapping service consists of two functions: (1) GetCapablites, which defines the capabilities of a server such as defining the supported file formats, the available map layers, and the method of display; and (2) GetMap, which tells the database what is needed. The database reads a request and creates the map-based data based on the requirements that have been defined by GetCapabilities. The requested data package is then sent to the web mapping service.

In addition to the two primary functions just discussed, most web mapping services support a handful of other functions as well. For example, GetFeatureInfo sends specific information about locations on the map, such as the name of a road or the altitude at a location. Another function, GetLegendGraphic, provides information about symbols used on the map.

The OGC standard has led to the development of a variety of services, including:

Web Map Service (WMS)—Georeferenced map images typically in the form of raster tiles (PNG, GIF, or JPG). The tiles can also be delivered in a vector format. Requests are made using a standard web URL address.

Web Coverage Service (WCS)—A geographical area that can be overlaid on a map but cannot be edited or analyzed. WCS is used to transfer coverages that consist of objects such as data points, pixels, or paths defined with vectors.

Web Feature Service (WFS)—A service that allows requests for geographical features, essentially providing the information behind the map. The WFS web service allows features to be queried, updated, created, or deleted by the client. This data is usually provided in an XML format like GML.

The open-source GeoServer application is the reference implementation of a server for the WMS, WFS, and WCS standards.

13.5   Web Publishing

The great majority of maps distributed through the Internet are embedded in web pages. Web publishing techniques therefore have a major influence on how maps are presented to users. Guidelines specify how information can be best formatted and how interactive elements can be integrated. They also address how to make web pages accessible to all people under a variety of viewing conditions.

The development of web pages has progressed through at least three stages. In the beginning, the pages were simply long documents. Frames were then used to separate content on multiple pages. An index would commonly appear in the left frame and would be used to select pages to view in the right frame. The use of frames became outdated with the development of cascading style sheets (CSS) that standardized the appearance of a series of web pages. CSS has now pervaded most web page designs.

13.5.1   Cascading Style Sheets (CSS)

Cascading style sheets (CSS) establish a common format for a set of web pages. A link can be made to a single CSS document that contains information for how the pages are presented in a browser. CSS makes it possible to change the appearance of a series of pages without having to edit every single page. In addition to controlling fonts and colors, the CSS file can also control the positioning of items on the page, whether images are viewable, and the language used. Changes to a CSS file can have a dramatic effect on the look of the pages that use it. In the example in Figure 13.8, the same HTML content is presented with four alternative CSS files. The illustration clearly shows how different presentation styles can have a major influence on the delivery of a message.

A CSS definition can be incorporated in a web page or reside in a separate file with the .css extension. The file contains two basic elements: the selector and the declaration. The selector chooses the HTML element that will have the style assigned to it. The declaration defines the style that will be applied to the selected elements. In the following example, h1 (largest heading text) is the selector. Next, the declaration defines the size of the font, its height, and its color—with the color given in hexadecimal format. All h1 text in the document would then be formatted in this way.

```
h1 { font-family: Times New Roman, Times;
    font-size: 40px;
    height: 200px;
    color: #333333; /* hexadecimal for a dark grey shading */
}
```

A link is established to the stylesheet from within the HTML file. In this case, the stylesheet is named style.css.

```
<link rel=StyleSheet href="style.css" TYPE="text/css">
```

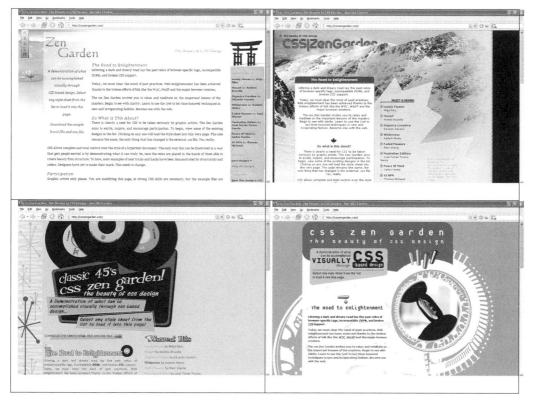

FIGURE 13.8. All four of these web pages are based on the same HTML file. The only difference is the underlying CSS file that alters the look of the page. Copyright 2012, Dave Shea.

A major benefit of CSS is the ability to inherit style properties between elements. The body selector references the entire HTML document. The following entry in the style sheet would set all text to Arial (Helvetica):

```
body {
      font-family: Arial, Helvetica;
   }
```

In CSS, HTML elements inherit their styling from parent elements, unless defined otherwise. All text elements in this example, including h1 header text, would be Arial. Table 13.2 summarizes other aspects of CSS.

13.5.2 Accessibility

While the concept of accessibility is normally associated with making web pages available to people with disabilities, the term is used here for any kind of issue that would limit the normal conveyance of information through the Internet. Among these issues are differences between web browsers whether desktop, mobile phone,

TABLE 13.2. Examples of Cascading Style Sheets

Explanation	CSS stylesheet definition
CSS can be used to set margins. The following would indent the margins by 10% on both sides.	`p { margin-right:10%; margin-left:10%;}`
Selectors and declarations can be grouped to save space.	`h1, h2, h3 { font-family:Arial }h1 { font-weight:bold; font-size:14pt;line-height:16pt; }`
A separate class attribute can be defined for a specific selector. In this case, `h1.artdeco` would be written with a different color when called in this way: `<h1 class="artdeco"> A loud color! </h1>`	`h1.artdeco { color: #00FFFF }`
Pseudo-elements are used for common typographic effects such as initial caps and drop caps: `<p>This sentence will have a drop cap</p>`	`p { font-size:14pt; line-height:16pt; font-family:Helvetica }` `p:first-letter { font-size:200%; float:left }`
This would place the selector 100 pixels from the top of the page and 100 pixels from the left. Percentage values can also be used.	`position:absolute; left:100px;top:100px`

or voice-activated, or constraints caused by the user's environment. In addition to blindness, accessibility concerns itself with language differences, color blindness, and the speed of the Internet connection. Given all of these issues, most web users will be confronted with accessibility barriers at one point or another, which will limit their use of the Internet. To address this problem, the World Wide Web consortium (W3C) has created guidelines for how to make web pages universally accessible.

Making websites for the blind is perhaps the most challenging of all accessibility requirements. Programs can read text automatically, and an "alt=" option in the img tag will describe a graphic, as in the following tag for a map depicting world population density:

```
<img src="WorldPopDensity.png" alt="Map showing world population
density. Darker shadings show areas of higher population density.
These are located in eastern China, northern India, western Europe,
isolated coastal areas of South America, and the eastern and western
coasts of North America.">
```

Such descriptions are helpful only if the person has become blind later in life. Visual mental conceptions are difficult to communicate effectively to those who have never had the ability to see.

Another accessibility issue comes from the ever-increasing need to design for different-sized displays. Display sizes are generally increasing but there will always be a major difference between desktop and mobile devices. Most web pages are designed for desktop displays with a standard size of 1024 × 768 or 1366 × 768 pixels. Initially, cell phone displays were on the order of 176×220, and the orientation was typically vertical rather than horizontal. Obviously, a web page designed for the larger display would not work well on such a mobile device. Automatically redesigning the web page to fit on all of these displays proved to be extremely challenging. The transformation needed was more than simply changing the format—it had to also change the content, analogous to the adjustments required to produce a smaller scale map. Most websites do not make these adjustments. Larger smartphone displays have now made it easier to view web pages on these devices. Meanwhile newer computers with resolutions over 300 dots per inch (dpi) have further complicated the display landscape. Images that have been down-sampled for presentation on lower resolution screens look pixelated on these high-resolution displays. Redesigning raster graphics to take advantage of higher resolution displays will be a never-ending problem.

The language barrier is perhaps the major accessibility issue. Automated methods have been developed to translate pages between different languages, but the quality of these pages is generally poor both from the standpoint of the translations and the resultant design. While automated translation services are getting better, they will likely never be able to properly translate between languages. Besides the actual translation, a number of other considerations are involved in designing multilingual pages. One of these considerations is the character set. Most Indo-European languages use one byte for a total of 256 possible letters and numbers. The Unicode system uses two bytes to represent characters, for a total of 65,535 possible characters—a number sufficient for any known language. Unicode is defined using the following tag in the header part of the html file:

```
<meta http-equiv="Content-Type" content="text/html; charset=utf-8">
```

In addition to an expanded character set, the page should also support languages that are written right to left, such as Arabic, Persian, and Hebrew. This may also influence the placement of menu bars that are typically on the left side in left-to-right languages. A final concern is the different amount of space required for text in different languages. The only way to effectively design web pages for multiple languages is to have separate pages for each language.

13.6 Summary

The online map is a relatively new entity. The first map to be incorporated within a web page appeared in 1993. Initially, scanned paper maps predominated as online maps. Database-driven maps appeared almost immediately but did not achieve their major adoption until the latter part of the 1990s.

Most online maps are based on browser-compatible, graphic file formats that are designed for the display of pictures. These raster file formats conform to the raster display of computer monitors. Increases in monitor resolution are making lower resolution raster files obsolete. Vector file formats, such as SVG, overcome this limitation. But, vector files may not appear as quickly because the files are larger and they must first be rasterized by the client computer before they can be displayed.

The major development for online maps is XML, a verbose markup language that allows for the definition of other markup languages. The promise of the XML-based languages, such as GML, SVG, and KML, is the development of web pages that are better able to share data and information. A great deal of work still remains to transform these promises into reality.

13.7 Questions

1. Describe different types of client computers.

2. What is a domain name server (DNS)?

3. What is Apache?

4. What is P2P and why is it important to the distribution of music and video files?

5. What is the difference between a thin and thick client?

6. How did Ajax change the client's interaction with a server?

7. Why was Ajax a particularly important development for the distribution of maps?

8. How has XML changed the online information landscape?

9. Describe some different map-based formats that are based on XML.

10. What are some advantages and disadvantages of SVG?

11. What is a web mapping service?

12. What are some accessibility concerns in the design of web pages?

13.8 References

Carto.net: A Place for Cartography on the Internet. (2012) Maintained by André M. Winter and Andreas Neuman, http://www.carto.net

Chisholm, Wendy, and Vanderheiden, Gregg (Eds.) (1999) *Web Content Accessibility Guidelines, W3C Recommendation.* [http://www.w3.org/TR/WAI-WEBCONTENT/]

Garrett, Jesse James (2005, Feb. 18). Ajax: A New Approach to Web Applications. AdaptivePath.com. [http://www.adaptivepath.com/ideas/essays/archives/000385.php]. Retrieved 2008-06-19.

Netcraft (2012) *Web Server Survey.* [http://news.netcraft.com/archives/web_server_survey.html]

Open Geospatial Consortium (2012) *OGC Standards and Specifications.* [http://www.opengeospatial.org/standards]

Pavlicko, Petr, and Peterson, Michael (2004, Sept. 7–10) Topographic Maps with SVG. *Proceedings of the SVG Open Conference*, Tokyo, Japan.

Line and Area Map Mashups

"What's color got to do with it?"

"It's got everything to do with it. Illinois is green, Indiana is pink. You show me any pink down here, if you can. No, sir; it's green."

"Indiana pink? Why, what a lie!"

"It ain't no lie; I've seen it on the map, and it's pink."

"Well, if I was such a numbskull as you, Huck Finn, I would jump over. Seen it on a map! Huck Finn, did you reckon the states was the same color out-of-doors as they are on the map?"

"Tom Sawyer, what's a map for? Ain't it to learn you facts? Well, then, how's it going to do that if it tells lies? That's what I want to know."

—*Tom Sawyer Abroad* by Mark Twain

14.1 Introduction

Like point symbols, line symbols can be used to show both qualitative and quantitative attributes. Road maps, for example, may be viewed as being both qualitative and quantitative. They are qualitative because they distinguish roads from the background. They are quantitative in the sense that roads are usually categorized into different ordinal widths to show their importance or size. Lines may also show the flow of a commodity such as oil, or depict elevation through the use of contour lines—also considered a line symbol.

In the same way, areas are shaded on a map to show both qualitative differences, as between land and water, and quantitative values such as population density. A map with shadings, either with grays or color, is referred to as "choropleth," from the Greek *choros* for area or region and *plethos* for value. A qualitative choropleth map would indicate nominal differences like zoning areas or soil types. Most maps are by definition qualitative choropleth maps because they use shadings to represent basic differences in land cover. For example, shadings are used to distinguish between land and water or urban and forested areas. In contrast, a quantitative choropleth map displays differences in the quantity of a variable as with a map that shows the percentage of votes received by a candidate by state or election precincts.

One of the main problems in line and area mashups is the amount of data that must be brought to the map. A single contour line, for example, may contain thousands of points. Polygon outlines may also be very detailed, especially when they follow a river or coastline. Reducing the number of points so that these types of maps can be effectively displayed is one way of solving this problem.

14.2 Lines

14.2.1 Simple Line

All mapping APIs incorporate an option to draw lines. The google.maps.Polyline is used to draw lines with the Google Maps API. In Figure 14.1, the Polyline function connects points that have been previously defined. Options for controlling the appearance of the line include strokeColor, strokeOpacity, and strokeWeight. As always, an appropriate center and zoom-level must also be defined. The center could be the midpoint of the line itself.

14.2.2 Flow Map

The easiest of all quantitative line maps to make is a flow map. Using different colors or line thicknesses, a flow map shows the flow of traffic, goods, or commodities between places. A traffic map, for example, uses colors of red, yellow, and green to depict the flow of traffic. In Figure 14.2, a thicker line weight is defined between Chicago and another city using the strokeWeight option. Modifying the strokeWeight value for the different lines based on a data value would result in a flow map.

14.2.3 Great Circle

Because of the projection of the spherical Earth, the shortest distance between two points on a map is rarely the shortest distance on the ground. Only one projection, the gnomonic, shows the great circle as a straight line as it would be on a globe. On all other projections, including the Mercator, the shortest distance on the ground is represented as an arc and seemingly a longer distance between two locations. The great circle is the shortest distance between two points and is supported in the Google Maps API through the geodesic: true Polyline option. Figure 14.3 shows two lines connecting Los Angeles and London. The curved line is the great circle, the route that best approximates a flight between the two cities. The straight line is depicted to highlight the difference between the two routes. Although the straight line looks shorter, it would represent a much longer distance.

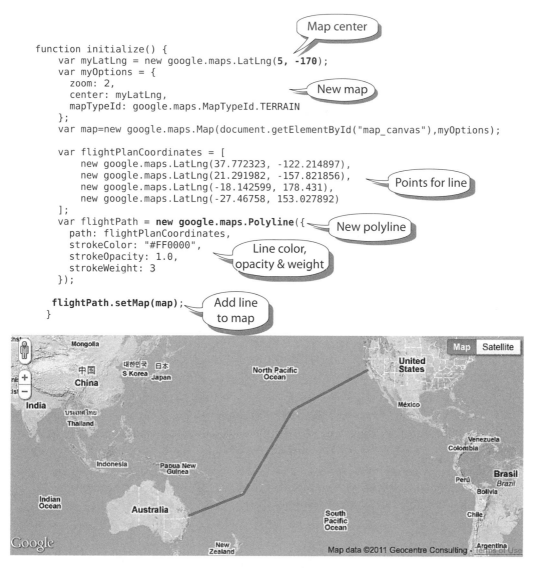

```
function initialize() {
    var myLatLng = new google.maps.LatLng(5, -170);          [Map center]
    var myOptions = {
        zoom: 2,                                             [New map]
        center: myLatLng,
        mapTypeId: google.maps.MapTypeId.TERRAIN
    };
    var map=new google.maps.Map(document.getElementById("map_canvas"),myOptions);

    var flightPlanCoordinates = [
        new google.maps.LatLng(37.772323, -122.214897),
        new google.maps.LatLng(21.291982, -157.821856),      [Points for line]
        new google.maps.LatLng(-18.142599, 178.431),
        new google.maps.LatLng(-27.46758, 153.027892)
    ];
    var flightPath = new google.maps.Polyline({              [New polyline]
        path: flightPlanCoordinates,
        strokeColor: "#FF0000",                              [Line color,
        strokeOpacity: 1.0,                                   opacity & weight]
        strokeWeight: 3
    });

    flightPath.setMap(map);                                 [Add line to map]
}
```

FIGURE 14.1. A line made from three individual segments and four points. Copyright 2013 Google.

14.2.4   *Line Input with XML and JSON*

The points that constitute a line may also be input through an external extensible markup language (XML) file. In Figure 14.4, points are imported using the same downloadUrl function as was demonstrated in Chapter 12 to read a series of points. The function parses each line in the XML file looking for data labeled with lat, lng, label, and html. The lat and lng variables are used to place the points, label becomes the mouseover text, and HTML is the text placed in the infowindow. Points are connected with the Polyline option and marked with createMarker. The

Second line

```
var flightPlanCoordinates = [
    new google.maps.LatLng(41.8, -87.7),
    new google.maps.LatLng(44.9, -93.3)
];

var flightPath = new google.maps.Polyline({
    path: flightPlanCoordinates,
    strokeColor: "#00FF00",
    strokeOpacity: .5,
    strokeWeight: 10
});

flightPath.setMap(map);
```

Endpoints of line

New polyline

Line color, opacity & weight

Add line to map

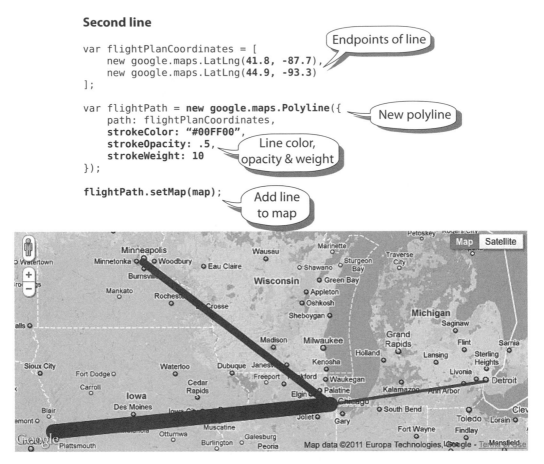

FIGURE 14.2. A simple flow map with three lines of varying thickness (strokeWeight). Color can also be used to show flow, as on maps of traffic that use green, orange, and red for the speed of traffic. The code shows how a single line is drawn. Copyright 2013 Google.

parsing of XML data is relatively slow, and the procedure works best for less than a few hundred points.

Figure 14.5 demonstrates how lines are drawn based on points read from a JSON file. While JSON is also based on XML, it can be input faster because it is recognized natively by JavaScript.

14.2.5 Lines from KML

KML (keyhole markup language) represents an even faster way of reading coordinates for lines. But, any formatting of the line must be defined within the KML file. The file is mapped with the `google.maps.KmlLayer` option in this way:

```
var georssLayer = new google.maps.KmlLayer('http://maps.unomaha.edu/
    Cloud/ch14/kml_lines.kml'); georssLayer.setMap(map);
```

In contrast to the XML and JSON methods of input, the full HTTP address must be used to specify the location of the KML file. This requires that the file be uploaded to a server before it can be mapped.

In the KML file, the line is defined with `<linestring>` (see Figure 14.6). LineStyle is used to change the color and thickness of the line. KML files include a z-value or elevation that is ignored by Google Maps.

14.2.6 Lines from a Fusion Table

Lines may also be drawn on a map using a Google Fusion Table. The example in Figure 14.7 depicts all recorded tornado tracks since 1950 in the major tornado-prone areas of the United States. All tracks are stored as KML in a Fusion Table. Any KML file can be imported to a Fusion Table, even large KML files with many thousands of points.

In addition to short line segments defined with only two points, a Fusion Table can also quickly display lines containing many thousands of points. The maps in Figure 14.8 show 500 ft and 200 ft contour lines for the Hawaiian island of Kauai. These were derived from a Shapefile, a GIS-type map file that was converted to KML using the MyGeodata Converter website and input into a Fusion Table. The 200 ft contour map requires over 60,000 latitude/longitude coordinates to produce the highly detailed contour lines. As can be seen when the map is plotted, the lines are placed onto each individual tile before being displayed.

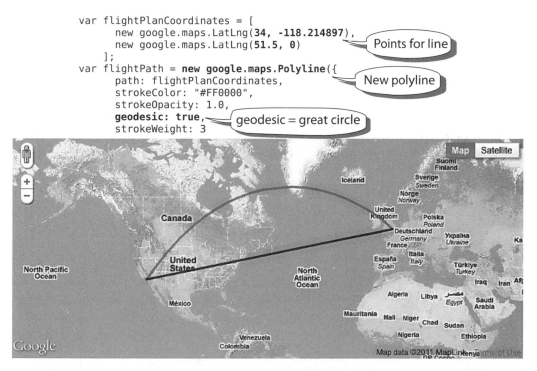

FIGURE 14.3. The *geodesic:true* Polyline option connects two points through the great circle, the shortest distance between two locations on the spheroid. It appears as a longer line on this map because of the projection. Copyright 2013 Google.

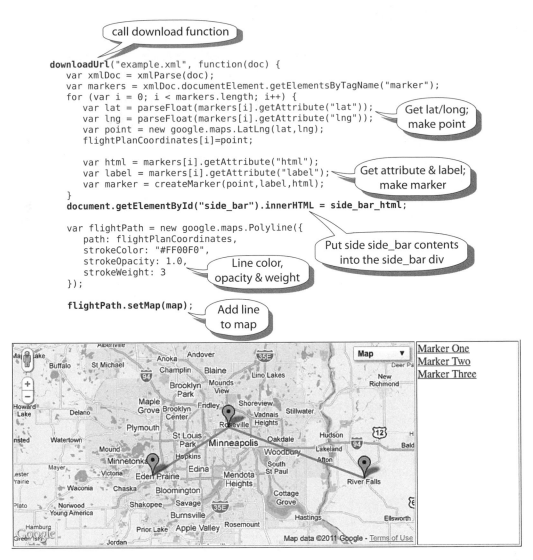

```
                    call download function

downloadUrl("example.xml", function(doc) {
    var xmlDoc = xmlParse(doc);
    var markers = xmlDoc.documentElement.getElementsByTagName("marker");
    for (var i = 0; i < markers.length; i++) {
        var lat = parseFloat(markers[i].getAttribute("lat"));
        var lng = parseFloat(markers[i].getAttribute("lng"));       Get lat/long;
        var point = new google.maps.LatLng(lat,lng);                 make point
        flightPlanCoordinates[i]=point;

        var html = markers[i].getAttribute("html");
        var label = markers[i].getAttribute("label");           Get attribute & label;
        var marker = createMarker(point,label,html);                make marker
    }
    document.getElementById("side_bar").innerHTML = side_bar_html;

    var flightPath = new google.maps.Polyline({
        path: flightPlanCoordinates,
        strokeColor: "#FF00F0",                     Put side side_bar contents
        strokeOpacity: 1.0,                            into the side_bar div
        strokeWeight: 3           Line color,
    });                       opacity & weight

    flightPath.setMap(map);            Add line
                                        to map
```

FIGURE 14.4. Lines input through an XML file. The file contains three coordinates, which would be sufficient to draw two lines. A third line could be drawn by connecting the third point to the first. Copyright 2013 Google.

```
function drawLines(gmap, marker, location)
{
      var flightPlanCoordinates = [
        new google.maps.LatLng(49.016805,8.391109),
        location,];
      var flightPath = new google.maps.Polyline({
      path: flightPlanCoordinates,
      strokeColor: "#FF0000",
      strokeOpacity: 1.0,
      strokeWeight: 2
    });
   flightPath.setMap(gmap);
  }
```

FIGURE 14.5a. Lines input through a JSON file. Lines are drawn between a midpoint and the points in the JSON file. Copyright 2013 Google.

272

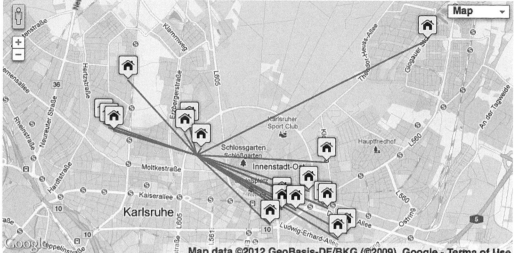

FIGURE 14.5b.

KML File

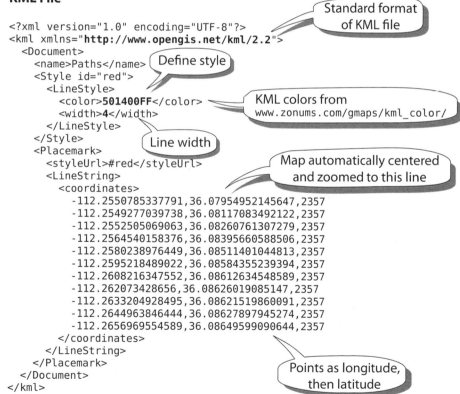

```
<?xml version="1.0" encoding="UTF-8"?>
<kml xmlns="http://www.opengis.net/kml/2.2">
  <Document>
    <name>Paths</name>
    <Style id="red">
      <LineStyle>
        <color>501400FF</color>
        <width>4</width>
      </LineStyle>
    </Style>
    <Placemark>
      <styleUrl>#red</styleUrl>
      <LineString>
        <coordinates>
          -112.2550785337791,36.07954952145647,2357
          -112.2549277039738,36.08117083492122,2357
          -112.2552505069063,36.08260761307279,2357
          -112.2564540158376,36.08395660588506,2357
          -112.2580238976449,36.08511401044813,2357
          -112.2595218489022,36.08584355239394,2357
          -112.2608216347552,36.08612634548589,2357
          -112.262073428656,36.08626019085147,2357
          -112.2633204928495,36.08621519860091,2357
          -112.2644963846444,36.08627897945274,2357
          -112.2656969554589,36.08649599090644,2357
        </coordinates>
      </LineString>
    </Placemark>
  </Document>
</kml>
```

Standard format of KML file

Define style

KML colors from www.zonums.com/gmaps/kml_color/

Line width

Map automatically centered and zoomed to this line

Points as longitude, then latitude

FIGURE 14.6a. A series of line segments input through a KML file. Copyright 2013 Google.

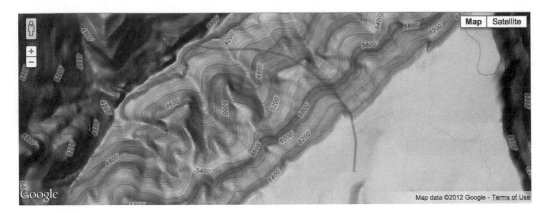

FIGURE 14.6b.

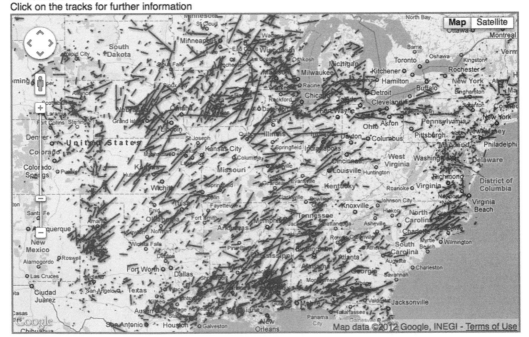

FIGURE 14.7. All recorded tornado tracks since 1950. Each track is stored as a coordinate pair in the KML format within a Fusion Table. Copyright 2013 Google.

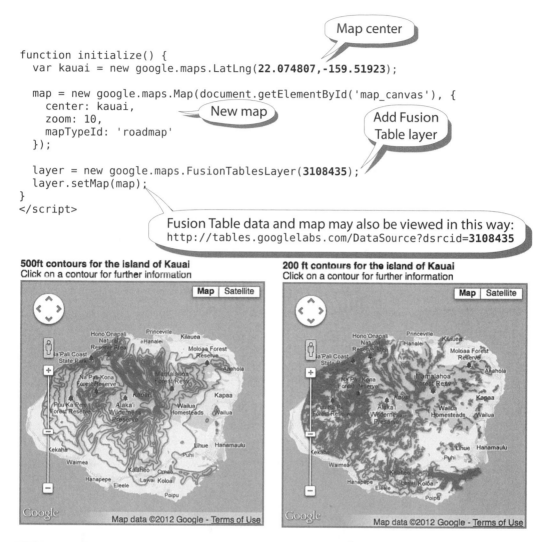

FIGURE 14.8. 500 ft and 200 ft contour lines for the island of Kauai in Hawaii. The lines were derived from a Shapefile that was converted to a KML and imported into a FusionTable. The 200 ft contour map contains over 60,000 latitude and longitude coordinates. Copyright 2013 Google.

As with KML, the map must be formatted before it is displayed with the Google Maps API. The map-styling option in Fusion Tables makes it possible to change attributes such as line width and color.

14.2.7   *Interactive Line Input*

Rather than inputting lines solely from a file, points for a line can also be input by simply clicking on the map. In the example Figure 14.9, a window to the right of the map displays the latitude and longitude of the most recently clicked point. Points

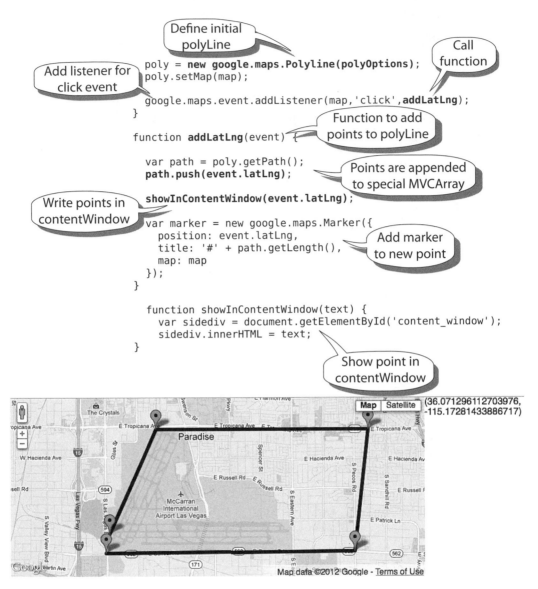

FIGURE 14.9. Lines are input by clicking on the map. The latitude and longitude of the last point clicked are shown in the upper-right corner. Copyright 2013 Google.

are "pushed" into a model-view-controller type of array. The so-called MVCArray differs from a normal array by incorporating special methods to retrieve, remove, and insert new objects. It may be thought of as a smart array that consists of more than a series of numbers.

14.2.8   Cross-Section Line

In Figure 14.10, the Google Maps API has been combined with Google's Visualization API to display the cross section of a line. This demonstrates how multiple APIs

```
function plotElevation(results, status) {
  if (status == google.maps.ElevationStatus.OK) {
    elevations = results;

    var elevationPath = [];
    for (var i = 0; i < results.length; i++) {
      elevationPath.push(elevations[i].location);
    }

    var pathOptions = {
      path: elevationPath,
      strokeColor: '#0000CC',
      opacity: 0.4,
      map: map
    }
    polyline = new google.maps.Polyline(pathOptions);

    var data = new google.visualization.DataTable();
    data.addColumn('string', 'Sample');
    data.addColumn('number', 'Elevation');
    for (var i = 0; i < results.length; i++) {
      data.addRow(['', elevations[i].elevation]);
    }

    document.getElementById('elevation_chart').style.display = 'block';
    chart.draw(data, {
      width: 700,
      height: 200,
      legend: 'none',
      titleY: 'Elevation (m)'
    });
  }
}
```

Extract elevations and store in array

Display polyLine on map

Extract data for elevation chart

Draw the elevation chart

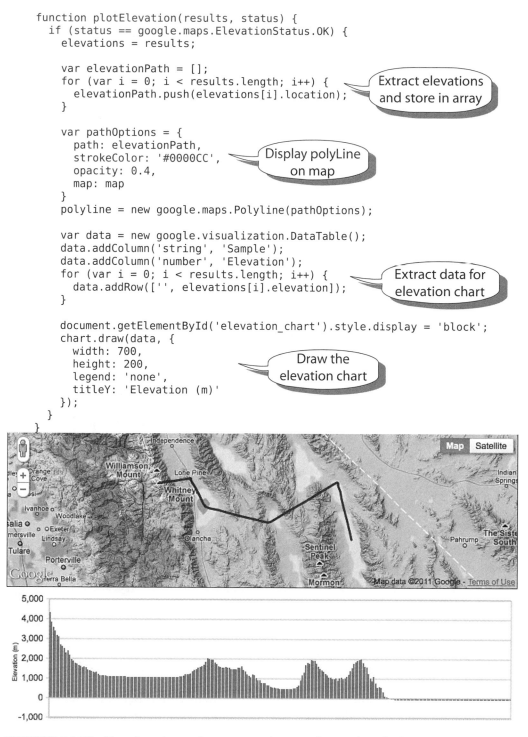

FIGURE 14.10. The elevations of a cross section are depicted with the Google Visualization API. Copyright 2013 Google.

can be used in conjunction with each other. In the example, six elevation points are stored in the `elevations` array and subsequently transferred to the `elevationPath` object. The visualization API constructs a cross section of the elevations below the map.

14.2.9 Encoded Polylines

To reduce storage and increase the speed of display, Google Maps implements a compact form of line encoding. This format is especially useful for lines with many points. A further benefit of the encoded `Polyline` is that a scale indicator can be attached to each point to define at what LOD the line would be visible. The inclusion of a point within a line can therefore be made dependent on the scale of the map. A line can be composed of more points on a large-scale map than on the corresponding smaller scale version.

Google's encoded Polyline consists of a series of text characters. An encoded line is shown in Figure 14.11. The encoding involves representing the numbers as binary and then converting to the ASCII equivalent characters. To save space, the encoding scheme uses relative coordinates for the second and third points. Once the first point is defined, only the difference needs to be provided for the second point. The third point is given relative to the second.

The points within a Polyline are made scale dependent by defining an additional character for each coordinate. The character defines the zoom level at which

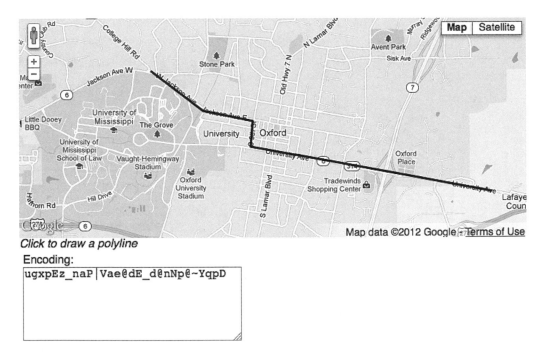

FIGURE 14.11. A line on a map is encoded as a Google encoded Polyline. Rather than representing the line with a series of latitude and longitude coordinates, an encoded Polyline converts a binary value into a series of characters. Copyright 2013 Google.

the point would be shown. By default, the zoom level for points is given at four different levels of zoom that range from 1–5, 6–10, 11–14, and 15–20.

Figure 14.12 shows an encoded Polyline on a map. A designator has been assigned to individual points in the line that correspond to the interstate indicating at what scale the point is shown. Some points in the line are therefore hidden in the smaller scale version of the line on the left. The points then appear when we zoom in to the map in the next higher zoom level on the right side. We can say that the line exhibits *scale dependence* because its depiction on the map varies by scale. This attribute is important for all features on a map. Fortunately, online utilities exist for converting a series of latitude and longitude values to encoded Polylines (search: Google Polyline encoder).

14.3   Polygons

A polygon may be defined as a particular type of line that closes upon itself. It consists of a series of points, with the first point always equal to the last. The two additional attributes that need to be defined for `google.maps.polygon` are the shading and opacity of the interior area. The main use of the polygon is to make a choropleth map showing either qualitative or quantitative data.

14.3.1   Single Polygon

Figure 14.13 shows the Bermuda Triangle in the Atlantic Ocean. Four points are defined, with the first point equal to the last. These points are loaded into an array named `triangleCoords`. This array is then passed to `google.maps.Polygon`. Other parameters include the `strokeColor`, `strokeOpacity`, `strokeWeight`, `fillColor`, and `fillOpacity`.

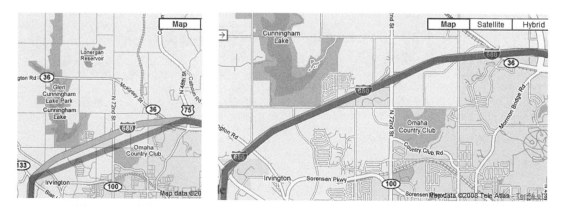

FIGURE 14.12. Scale dependence for an encoded Polyline. The compact encoded Polyline format assigns a scale factor along with the points. This scale dependence is illustrated here with a different line in the smaller-scale map on the left compared with the larger-scale map on the right. Copyright 2013 Google.

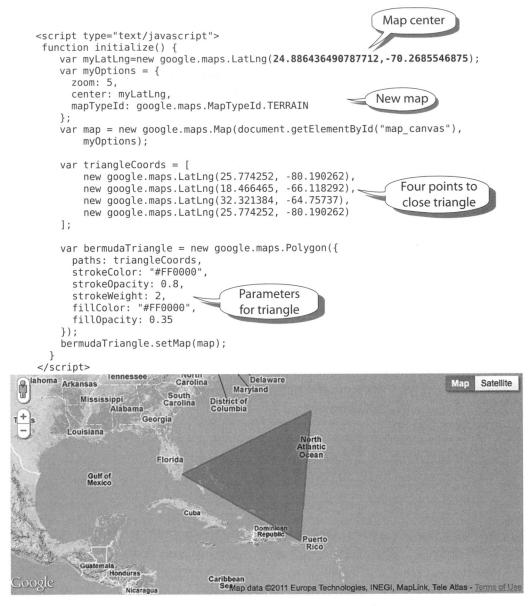

```
<script type="text/javascript">
 function initialize() {
    var myLatLng=new google.maps.LatLng(24.886436490787712,-70.2685546875);
    var myOptions = {
      zoom: 5,
      center: myLatLng,
      mapTypeId: google.maps.MapTypeId.TERRAIN
    };
    var map = new google.maps.Map(document.getElementById("map_canvas"),
        myOptions);

    var triangleCoords = [
        new google.maps.LatLng(25.774252, -80.190262),
        new google.maps.LatLng(18.466465, -66.118292),
        new google.maps.LatLng(32.321384, -64.75737),
        new google.maps.LatLng(25.774252, -80.190262)
    ];

    var bermudaTriangle = new google.maps.Polygon({
      paths: triangleCoords,
      strokeColor: "#FF0000",
      strokeOpacity: 0.8,
      strokeWeight: 2,
      fillColor: "#FF0000",
      fillOpacity: 0.35
    });
    bermudaTriangle.setMap(map);
  }
</script>
```

FIGURE 14.13. The Google polygon function draws a closed shape. Options include strokeColor, strokeOpacity, strokeWeight, fillColor, and fillOpacity. Copyright 2013 Google.

14.3.2 Polygon from XML File

The example in Figure 14.14 draws a polygon from a file of 17 points. The file is read in the same way as in the example defining the line in Figure 14.4. The last point is made equal to the first through the statement:

```
flightPlanCoordinates[counter+1]=flightPlanCoordinates[0];
```

In this case, each point is treated as a marker, and these markers are also

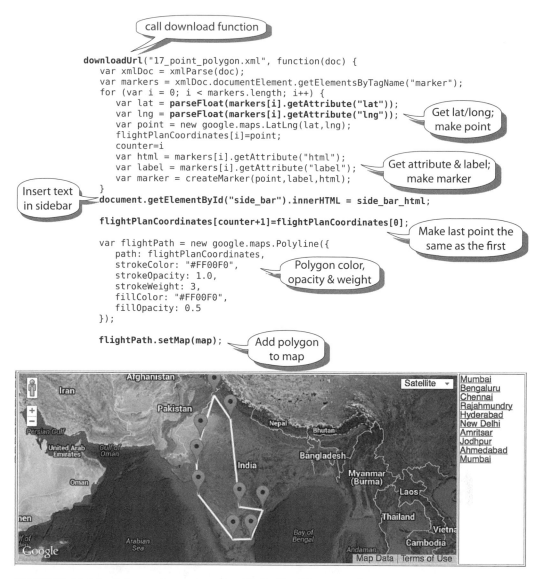

```
                    call download function

downloadUrl("17_point_polygon.xml", function(doc) {
    var xmlDoc = xmlParse(doc);
    var markers = xmlDoc.documentElement.getElementsByTagName("marker");
    for (var i = 0; i < markers.length; i++) {
        var lat = parseFloat(markers[i].getAttribute("lat"));
        var lng = parseFloat(markers[i].getAttribute("lng"));      Get lat/long;
        var point = new google.maps.LatLng(lat,lng);              make point
        flightPlanCoordinates[i]=point;
        counter=i
        var html = markers[i].getAttribute("html");
        var label = markers[i].getAttribute("label");       Get attribute & label;
        var marker = createMarker(point,label,html);           make marker
    }
    document.getElementById("side_bar").innerHTML = side_bar_html;

    flightPlanCoordinates[counter+1]=flightPlanCoordinates[0];
                                                    Make last point the
                                                    same as the first
    var flightPath = new google.maps.Polyline({
        path: flightPlanCoordinates,
        strokeColor: "#FF00F0",            Polygon color,
        strokeOpacity: 1.0,                opacity & weight
        strokeWeight: 3,
        fillColor: "#FF00F0",
        fillOpacity: 0.5
    });

    flightPath.setMap(map);          Add polygon
                                     to map
```

Insert text in sidebar

FIGURE 14.14. Polygon input from an XML file. The procedure is almost the same as that in the example in Figure 14.4 except that the last point is made equal to the first. Copyright 2013 Google.

drawn on the map. As the points are read from the file, they are put into the `flightPlanCoordinates` array. After defining the polygon with the strokeColor, strokeOpacity, and strokeWeight, setmap places the polygon on the map.

14.3.3 Donut Polygon

Many polygons are more complicated to define because interior areas may not be part of the polygon. In creating a so-called donut polygon, the outside polygon is defined first and then "inside" polygons are subtracted. Figure 14.15 shows how

```
function initialize() {
    var myLatLng = new google.maps.LatLng(48.209088,16.370959);
    var myOptions = {
        zoom: 12,
        center: myLatLng,
        mapTypeId: google.maps.MapTypeId.HYBRID
    };
    var map = new google.maps.Map(document.getElementById("map_canvas"),
        myOptions);

    var ring = [
        new google.maps.LatLng(48.217482,16.370337),

        new google.maps.LatLng(48.217482,16.370337)
    ];
    var gurtel = [
        new google.maps.LatLng(48.24614,16.381946),

        new google.maps.LatLng(48.24614,16.381946)
    ];

    var innerDistricts = new google.maps.Polygon({
        paths: [gurtel, ring],
        strokeColor: "orange",
        strokeOpacity: 0.8,
        strokeWeight: 4,
        fillColor: "orange",
        fillOpacity: 0.35
    });

    innerDistricts.setMap(map);
}
```

Map center

New map

All points for inside polygon

All points for outside polygon

Make the donut polygon

Add polygons to map

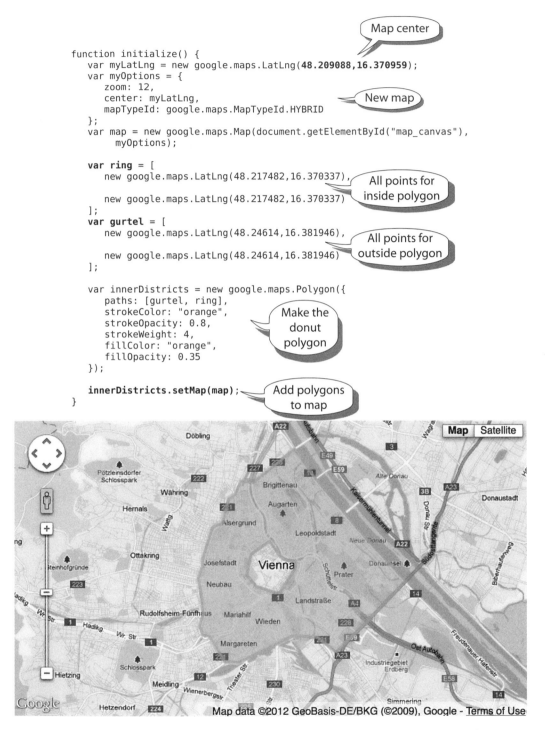

FIGURE 14.15. Definition of a polygon with a hole. The outside area is defined first. Subsequent polygons subtract from this initial polygon. Copyright 2013 Google.

an outside and inside polygon are defined for a map showing a district of Vienna, Austria.

14.3.4 *Donut Polygon with KML*

A donut polygon can also be defined using a KML file. Figure 14.16 shows the Pentagon building in the United States with the interior depicted as a separate polygon. The KML tags `<outerBoundaryIs>` and `<innerBoundaryIs>` distinguish between the two polygons.

14.3.5 *Multiple Polygons*

With potentially thousands of points, entering polygon outlines can be a very time-consuming task. Fortunately, many polygon files are available that can be converted for use by Google Maps. Once the polygons are defined, shadings can be assigned to indicate the attribute or value of each area.

The most common format for map files containing polygons is the Shapefile (.shp), a format developed by ESRI—the major distributor of GIS software. Introduced in the early 1990s, many Shapefiles are now available through the Internet from libraries and other online portals. Online repositories include diva-gis and dyngeometry. Tools are also available to convert these binary files for use by Google Maps.

Shapefiles can be very large with many points. For example, a Shapefile containing the boundaries for the counties of Nebraska, a state with many rectilinear outlines, has over 25,000 x, y coordinates, many more points than would be needed for a choropleth map. Most of the coordinates are along the eastern boundary of the state that coincides with the meandering Missouri River. Reading this many coordinates from a text file, like the XML files used here, would require a considerable amount of time. One option to speed the display of the map is to reduce the number of coordinates in the file.

14.3.6 *Line Coordinate Thinning*

The automated thinning of lines, a central aspect of map generalization, has been a major area of research and development in cartography since the 1970s. This development coincided with the computerization of maps. A number of algorithms have been introduced that determine which points can be removed from a line without changing its basic character. Essentially, the algorithms examine neighboring points and determine those that are important to preserve the shape of the line and those that are not. Through line simplification algorithms, unneeded detail in relation to map scale can be eliminated. More importantly, for purposes of mapping, they result in smaller file sizes and faster map display speeds.

Douglas and Peucker (1973) introduced one of the first algorithms for line coordinate thinning. The method has been used mainly as a scale-independent form of line generalization, particularly for thinning boundaries used on thematic maps. In the algorithm, two points in a line are connected by a straight-line segment. Perpendicular distances are then calculated from this segment to all intervening

KML File

Standard format of KML file

```
<?xml version="1.0" encoding="UTF-8"?>
<kml xmlns="http://www.opengis.net/kml/2.2">
  <Placemark>
    <name>The Pentagon</name>
    <Polygon>
      <extrude>1</extrude>
      <altitudeMode>relativeToGround</altitudeMode>
      <outerBoundaryIs>
        <LinearRing>
          <coordinates>
            -77.05788457660967,38.87253259892824,100
            -77.05465973756702,38.87291016281703,100
            -77.05315536854791,38.87053267794386,100
            -77.05552622493516,38.868757801256,100
            -77.05844056290393,38.86996206506943,100
            -77.05788457660967,38.87253259892824,100
          </coordinates>
        </LinearRing>
      </outerBoundaryIs>
      <innerBoundaryIs>
        <LinearRing>
          <coordinates>
            -77.05668055019126,38.87154239798456,100
            -77.05542625960818,38.87167890344077,100
            -77.05485125901024,38.87076535397792,100
            -77.05577677433152,38.87008686581446,100
            -77.05691162017543,38.87054446963351,100
            -77.05668055019126,38.87154239798456,100
          </coordinates>
        </LinearRing>
      </innerBoundaryIs>
    </Polygon>
  </Placemark>
</kml>
```

Extrude & Altitude for Google Earth

Map automatically centered and zoomed to outside polygon

Inside polygon

Points as longitude, then latitude

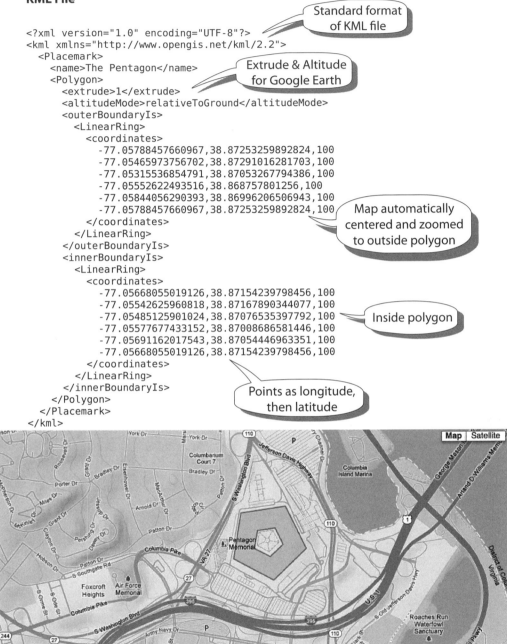

FIGURE 14.16. A KML file and the resulting display of a donut polygon in Google Maps. Copyright 2013 Google.

points. If none of these distances is greater than a tolerance value, the straight-line segment is considered suitable for describing the line. If the tolerance value is met, then the point with the greatest perpendicular distance from the line segment is selected as the new end-point and the process continues for the next point.

It should be remembered that a map is composed of more than lines, and line generalization is only one small part of the overall map generalization process. If it is done without considering other map elements, the line generalization algorithm may well affect the topology of the map. For example, a city located on one side of a river might end up on the other side after the line generalization operation (see Figure 14.17). Map generalization is therefore a very complicated process involving a consideration for all elements of the map, including point, line, and area features.

MapShaper.org is an online resource for the generalization of lines within shapefiles. Using a simple interface implemented with Adobe Flash, the website allows users to upload a Shapefile and then asks them to specify the amount of generalization. Once the map is simplified, the new Shapefile can be downloaded to the user's computer. MapShaper implements the Douglas–Peucker algorithm as well as more recent algorithms by Visvalingam and Whyatt (1993). Figure 14.18 shows the increasing simplification of U.S. state boundaries and the associated number of points. A variety of online tools are also available to generalize lines defined with KML.

14.3.7 Shapefile Conversion

Once a generalized Shapefile has been created, the file still needs to be converted for use by Google Maps. Various online utilities are available for converting a Shapefile to other formats including KML, CSV, and geoJSON. These utilities allow upload of the Shapefile and the conversion to an alternate format. Figure 14.19 shows the data conversion options for one of these online tools. Once converted to a comma separated value (CSV) file, additional conversion would be needed in a spreadsheet program like Microsoft Excel to change the coordinates to an XML format (see Figure 14.20).

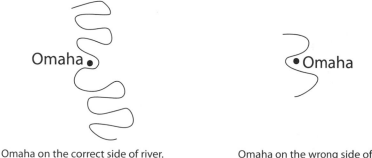

Omaha on the correct side of river.   Omaha on the wrong side of river.

FIGURE 14.17. The simplification of a river may cause a point feature, such as a city, to lose its correct topological relationship with the river.

FIGURE 14.18. Simplification of U.S. state outlines from 9407 points to 282 using MapShaper.org.

Run vector converter

List of acceptable input data formats

- ESRI Shapefile
- Arc/Info Binary Coverage
- Arc/Info .E00 (ASCII) Coverage
- Microstation DGN
- MapInfo File
- Comma Separated Value (.csv)
- GML
- GPX
- KML
- GeoJSON
- UK .NTF
- SDTS
- U.S. Census TIGER/Line
- S-57 (ENC)
- VRT - Virtual Datasource
- EPIInfo .REC
- Atlas BNA
- Interlis 1
- Interlis 2
- GMT
- X-Plane/Flighgear aeronautical data
- Geoconcept

List of available output data formats

- ESRI Shapefile
- Microstation DGN
- MapInfo File
- Comma Separated Value (.csv)
- GML
- GPX
- KML
- GeoJSON

FIGURE 14.19. Data conversion options from MyGeoDataConverter. Copyright 2013 MyGeoDataConverter.

12 Nodes	`<poly linecolor="#008800" linewidth="4" lineopacity = "1.0"`
	`  fillcolor= "#FFCC00" fillopacity = "0.5" html="State">`
-122.9679783 48.44379451	`<point lng ="-122.9679783" lat="48.44379451"/>`
-122.9679783 48.44379451	`<point lng ="-122.9679783" lat ="48.44379451"/>`
-123.0952329 48.47942282	`<point lng ="-123.0952329" lat ="48.47942282"/>`
-123.1597199 48.52184222	`<point lng ="-123.1597199" lat ="48.56256471"/>`
-123.1698993 48.56256471	`<point lng ="-123.1410538" lat ="48.62364712"/>`
-123.1410538 48.62364712	`<point lng ="-123.1037214" lat ="48.60837712"/>`
-123.1037214 48.60837712	`<point lng ="-123.0120949" lat ="48.55747774"/>`
-123.0120949 48.55747774	`<point lng ="-123.0086988" lat ="48.53371932"/>`
-123.0086988 48.53371932	`<point lng ="-122.9679800" lat ="48.52693332"/>`
-122.9679800 48.52693332	`<point lng ="-123.0222711" lat ="48.51335968"/>`
-123.0222711 48.51335968	`<point lng ="-123.0188829" lat ="48.48960517"/>`
-123.0188829 48.48960517	`<point lng ="-122.9679783" lat ="48.44379451"/>`
	`</poly>`

FIGURE 14.20. Conversion of latitude and longitude coordinates to a specific XML file. The points on the left were converted into the XML format on the right using the Excel concatenate function.

The map in Figure 14.21 shows the 93 polygons for Nebraska plotted with Google Maps. This overlay can be done fairly easily with a KML version of the Shapefile using the kmlLayer option. The bottom inset of the figure shows how well the border of Nebraska and Iowa match between the Shapefile and the corresponding Google map. It also shows that a rail line between the two states does not

FIGURE 14.21a. A Shapefile map of Nebraska by county mapped with Google Maps after line coordinate thinning with MapShaper and conversion to a KML file. The coordinates for a single polygon are listed. The bottom zoomed-in map (overleaf) shows the border between Nebraska and Iowa. The border between the states matches almost perfectly. The map also shows that a railroad line between Nebraska and Iowa does not connect, an error in the underlying Google Map database. Copyright 2010 Google.

```
<Polygon><outerBoundaryIs><LinearRing><coordinates>-98.29560321854305,41.914934699148525
-98.295469772942283,42.088814316934716 -98.300407260170289,42.088814316934716 -
98.300540705771056,42.303528288552499 -98.300140368968783,42.436973889309357 -
97.834415222327337,42.437774562913901 -97.835215895931881,42.232668674550609 -
97.834682113528856,42.089748436140013 -97.834148331125832,42.034768848628183 -
97.833214211920534,41.916402600756854 -98.064742329233681,41.915201590350037 -
98.29560321854305,41.914934699148525
</coordinates></LinearRing></outerBoundaryIs></Polygon>
```

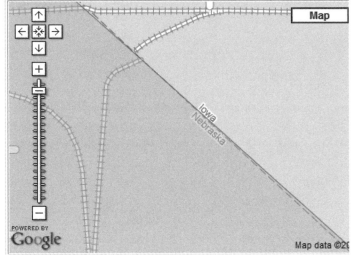

FIGURE 14.21b, c, & d.

match, highlighting a problem with the underlying map database. Non-connecting lines often occur when two state-based files are merged.

14.3.8 Mapping Population Data

Altering the shading of each polygon based on a data value would result in a choropleth map. In this case, we map population data for the state of Nebraska. The outline of the counties is input using an XML file. To simplify matters, we put the

93 population data values into an array called `popdata` (see Figure 14.22). Population values are listed in alphabetical order by county using the same order as the polygons in the XML file.

After the data have been assigned to the popdata array, they are converted to their natural log value using the Math.log function. The log conversion helps compensate for the extreme population values for the counties that contain the cities of Omaha and Lincoln in comparison to other counties in the state. The minimum and maximum values in the array are determined by sorting the data after the log 10 conversion. In the following step, the opacities for each county are calculated based on the maximum data value and the range of the data. The opacity of each county is then directly proportional to the log of the population. The counties with the highest populations are more opaque. The zoomed-in map in Figure 14.23 shows how the background map is visible for the less populated counties but not for those with a higher population.

The population map of Nebraska represents an unclassed choropleth map because the data values have not been put into categories. Usually, the data for choropleth maps are categorized into 4–7 classes although this classification may not be necessary. It can be difficult to create a proper sequence of colors for a classified map. The ColorBrewer website helps select a progression in different color schemes for a selected number of categories (search: ColorBrewer).

14.3.9 Polygons in Fusion Tables

KML files with polygons can be imported into a Fusion Table. In Figure 14.24, 2010 data for the percentage of the population under 18 is depicted as an unclassed gradient of shades (left) and in classified form with five categories (right). The

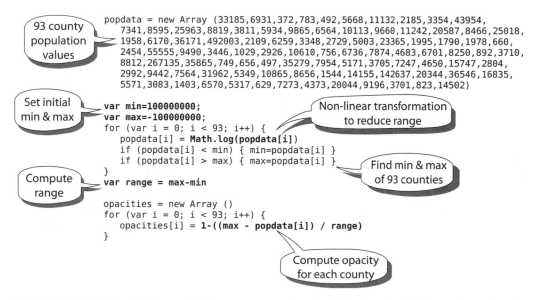

FIGURE 14.22. JavaScript code that computes the opacities for the 93 data values for the counties of Nebraska based on population.

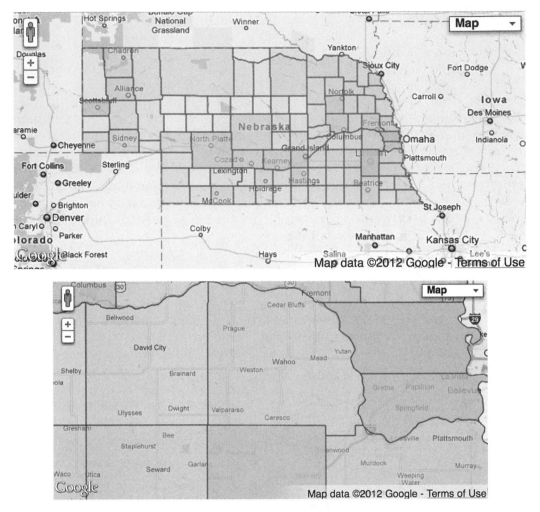

FIGURE 14.23. A population map of Nebraska. The opacity of the color assigned to each county is proportional to its population. The data have been converted to a log value to compensate for the skewed population values caused by the two largest cities, Omaha and Lincoln. The zoomed-in map on the bottom shows that city names are visible in the less populated counties. Copyright 2013 Google.

maps are based on a Shapefile that was converted to KML using the MyGeoData-Converter website, then uploaded to a Fusion Table. Data from the U.S. Census Bureau was also uploaded to a Fusion Table and merged with the map using corresponding county ID values in the two tables, GEOID10 and FIPS-0. After selecting "Map" under the "Visualize" menu, Configure styles is used to define the data to be mapped and the method of symbolization.

14.3.10   Heat Maps

The so-called heat map is a product of the mashup era. It shows the density of points where a higher density receives a darker color. As the density declines, the

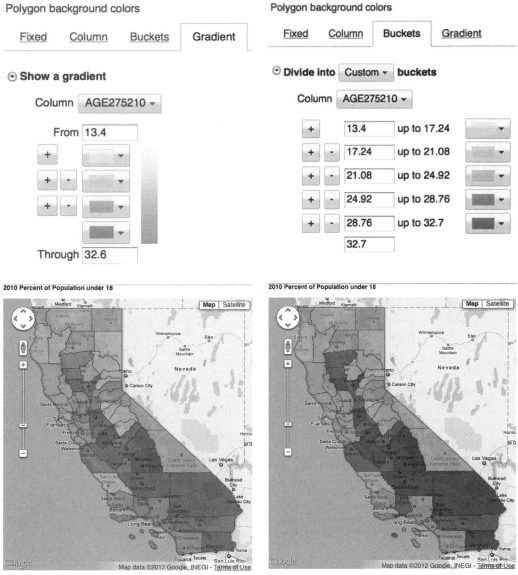

FIGURE 14.24. Two maps based on Fusion Table 3121331. The table includes polygons defined in KML. Another Fusion Table with 2010 U.S. Census Bureau data was merged with the county map Fusion Table and subsequently mapped with the Configure styles option. The map on the left depicts the data with a gradient of shadings from low to high. The map on the right has been categorized into five categories, called buckets in Google Fusion Tables. Copyright 2013 Google.

color changes to orange and then yellow. In cartographic terms, the heat map is an example of a shaded isarithmic map—analogous to a contour map that has been shaded. The darker colors on the heat map are essentially higher elevations. Figure 14.25 shows a density map of tornadoes in the United States. Data is provided through a Fusion Table. The `heatmap:true` option specifies the shaded representation.

The two heat maps in Figure 14.26 show coral bleaching in the Great Barrier Reef off the coast of Australia. Evidence indicates that elevated water temperature is the cause of large-scale bleaching events. Prolonged bleaching causes the coral to effectively die. The two maps differ because the density of bleaching events is a function of the area used in the calculation. The bottom, larger scale map has less bleaching per unit area, and therefore has fewer areas shaded red.

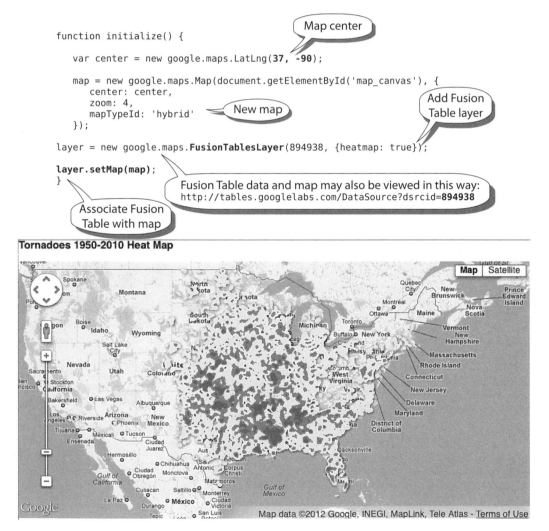

FIGURE 14.25. A heat map generated from the density of points defined in a Fusion Table. Copyright 2013 Google.

```
</body>

<b>Coral bleaching incidents in Australia between 1980 and 2006
depicted at two different scales</b>
  <div id="map_canvas" style="position: absolute; top: 30px;
width:700px; height:300px; color:#CEE3F6; border-style:solid;border-
width:3px;border-color:#7496C2"></div>
  <div id="map_canvas2" style="position: absolute; top: 340px;
width:700px; height:300px; color:#CEE3F6; border-style:solid;border-
width:3px;border-color:#7496C2"></div>

</body>
```

Coral bleaching incidents in Australia between 1980 and 2006 depicted at two different scales

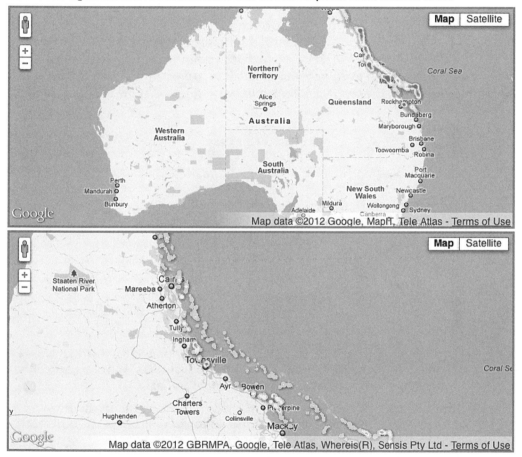

FIGURE 14.26. A heat map generated from the density of points defined in a Fusion Table at two different scales. The heat map presentation is scale dependent. Bleaching incident data were obtained from ReefBase.com as Excel files and imported into a Fusion Table. Copyright 2013 Google.

14.4 Summary

Line and area symbols generally require the definition of a large number of points. Many thousands of latitude and longitude points may be required to define a single polygon or contour line. The data volume problem as well as the representation at smaller scales may require generalization of the map data. Online utilities can both generalize a map and convert a map file to a different format. Some type of data conversion is usually necessary to make a map suitable for display with Google Maps.

KML files, whether stored separately or within a Fusion Table, represent the fastest way to bring lines and polygons to the map. The file is structured and organized in such a way to facilitate the overlay process. The limitation of the KML file is that its display cannot be modified without changing the KML file itself. The Fusion Table approach makes it possible to combine the map data with different attribute values, and even permits select alternative symbolization schemes. In both cases, however, the Google Maps API serves only to display the final product.

14.5 Exercise

Upload the code14.zip file to your server and make the suggested changes to the code examples.

14.6 Questions

1. What is a choropleth map, and what can it depict?

2. Why is the shortest distance between points over longer distances usually not a straight line with Google Maps and other online map providers?

3. Describe the different ways of entering points for display on a Google map.

4. Fusion Tables represent another promising development for distributing map data. Describe some advantages and disadvantages. What is needed to improve upon this method of map dissemination?

5. Compare and contrast the various methods for getting points, lines, and areas to a Google map. Your answer should incorporate all of the different tools that have been introduced to this point.

6. Google and other organizations maintain a free mapping service for way-finding based on business applications. Describe a model for a free mapping service that distributes thematic maps.

7. What is a donut polygon, and what makes it difficult to define?

8. What are some problems with line generalization?

9. What is the easiest way to create a choropleth map with an online mapping service? What method offers the most flexibility in classification and symbolization?

10. Describe how opaqueness values can be calculated for choropleth mapping.

11. Describe an organized and unified system that would serve updated maps through a KML feed.

12. What would be the best way to create an atlas of KML map offerings?

13. What are heat maps? What is the problem with their depiction at different scales?

14.7 References

Douglas, D. H., and Peucker, T. K. (1973) Algorithms for the Reduction of the Number of Points Required to Represent a Digitised Line or Its caricature. *The Canadian Cartographer* 10(2): 112–122.

Google (2011) *Google Maps V3 API Reference.* [http://code.google.com/apis/maps/documentation/reference.html]

Visvalingam, M., and Whyatt, D. (1993) Line Generalization by Repeated Elimination of Points. *The Cartographic Journal* 30(1): 46–51.

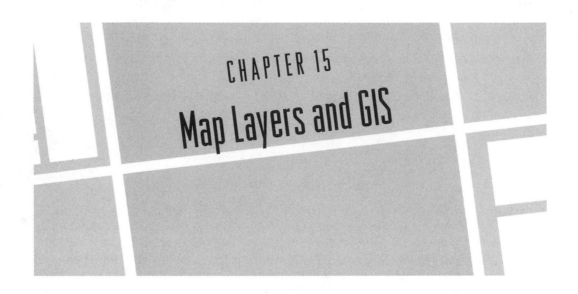

CHAPTER 15

Map Layers and GIS

Once a new technology rolls over you, if you're not part of the steamroller, you're part of the road.

—STEWART BRAND

15.1   Introduction

Geographic information systems (GIS) manipulate spatial data for purposes of recordkeeping, display, and analysis. As an integrated system, GIS includes the functions of spatial data input, storage, analysis, and output. Commercial GIS software has been the primary method of encoding, converting, and analyzing information about the spatial environment. Much of this work is now being done using free and open-source resources in the cloud.

GIS analysis is often characterized as an overlay of map layers as illustrated in Figure 15.1. In this model, GIS compares the relationship between the different maps layers to better understand the often complex relationships between features on the Earth's surface. Although most applications associated with GIS involve recordkeeping and cannot be characterized as an overlay, the overlay concept is a powerful metaphor for the analytical power of such systems. It provides GIS a lofty significance that goes beyond its primary database function.

The main application of GIS is in the urban environment. Cities use GIS to keep track of land ownership, property taxation, and road, water, sewer, electric, gas, and cable networks. Information on such networks would include not only their location but attributes such as their age, type of material, and maintenance records. Although the creation of such a GIS database represents a major investment, the benefits of maintaining information in an integrated form far outweigh the costs. For example, knowing the precise location of natural gas pipelines can save lives.

Physical Layers   **Human Layers**

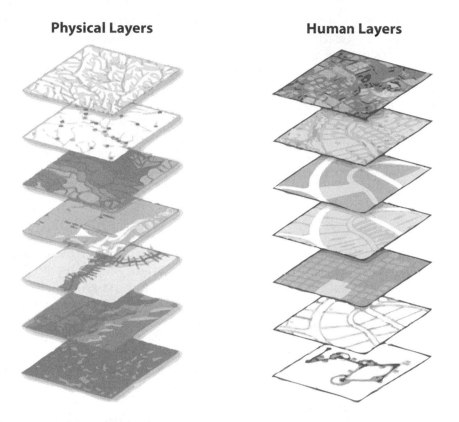

FIGURE 15.1. GIS as the overlay of map layers to demonstrate its analytical potential. In practice, most GIS systems are used for recordkeeping.

The ability to easily retrieve information from a GIS is a primary benefit of these systems. For example, the building of a new power line would impact both the physical and human environment. A GIS could be used to select the shortest route and the route with the least environmental impact. It can also be used for more mundane tasks such as generating a list of addresses of landowners along a proposed route so that the affected people can be informed and right-of-way permissions sought.

Common applications of GIS include agriculture, archaeology, criminology, demography, and forestry. The technology is young in the sense that new applications along with new analysis and manipulation functions are still being developed. In this chapter, we examine the concept of layering geographic information and the various data manipulation procedures that have developed in GIS. We divide our discussion based on the underlying data structure, raster and vector.

15.2 Raster

The overlay of maps is most easily accomplished with spatial information represented as a raster. This is because corresponding cells between raster files can be easily compared. The raster representation also makes it possible to analyze data in a variety of ways, especially those involving neighborhood analysis.

Although raster data is used to represent discrete entities, it is particularly suited for depicting continuous data. Most traditional maps code data by discrete areas, such as state, county, or census designations. In contrast, variables such as elevation or temperature change continuously over space and exhibit a great deal of variation between adjacent areas. This characteristic can be represented particularly well with the raster format where pixels can change values in a continuous fashion. The data structure is used mainly for spatial analysis functions in a wide variety of areas. Raster layers can also be easily integrated with remote sensing imagery.

15.2.1 Raster Overlay

Conceptually simple but analytically powerful, raster overlay involves placing two digital maps of the same area on top of each other, thereby creating a third "combined" map. An application might involve maps of land use and flood potential. The purpose of this overlay would be to find which types of land use are subject to flooding. The method is best explained by adding two maps coded in raster form as depicted in Figure 15.2. The combined map contains possible values between 51–54, 101–104, and 501–504, each indicating a particular combination of a flood event and land use. In this example, we are only interested in those areas with a value of 51 in the resulting map that would represent single family housing and the potential for a 50-year flood.

Additional functions are available to analyze the results of the overlay. For example, to determine the amount of area in single-family housing that would be subject to a 50-year flood event, you would count all cells with a value of 51 and multiply by the amount of area represented by each cell.

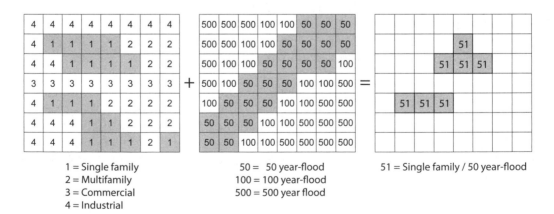

FIGURE 15.2. Overlay of maps in the raster format. Areas of single-family housing prone to a 50-year flood are identified by adding a map of land use and a map of flood potential. Areas of single-family housing more susceptible to flooding are identified with a value of 51 in the combined map.

15.2.2 *Map Algebra*

The use of mathematical operators in modeling map overlays suggests the great potential for application of more general algebraic expressions to the entire overlay process. The concept of map overlay was introduced in a comprehensive way by Dana Tomlin in the early 1980s (Tomlin 1990). In this system, maps may be viewed as variables in an expression to be manipulated by standard addition, subtraction, multiplication, and division operators. Once defined in this way, complex formulas can be devised to undertake almost any kind of map analysis.

Tomlin divides map algebra into different types of operators that combine maps or transform map data to create new maps. Relational operators involve the use of logical tests such as equal to or greater or less than. Boolean operators such as AND, OR, and NOT make logical tests (see Figure 15.3). They return values of true and false, usually expressed as 1 or 0.

The logical operators DIFF, IN, or OVER make logical tests on a cell-by-cell basis. The DIFF operator determines whether a cell in layer A is different from the corresponding cell in layer B. IN checks whether a cell value is in a list of acceptable values. With OVER, we can determine if a cell value in raster A is not equal to 0. If this is the case, the value of the cell in raster A is returned; otherwise, the value in B is returned.

Combinatorial operators work with multiple raster layers at the same time. They find unique combinations of values and assign a unique ID to each. Combinatorial operators include CAND, COR, and CXOR; CXOR, for example, evaluates whether a value is in A and B, but not C. Table 15.1 lists a variety of advanced map algebra functions that can be performed on single or multiple layers.

Functions can be grouped by operators to derive new values. Focal or local operators assign data values through a neighborhood analysis of a 3×3 matrix. Examples include a focal sum that adds up the values in a 3×3 matrix and assigns this value to the middle cell. Focal mean determines the average of the 3×3 matrix

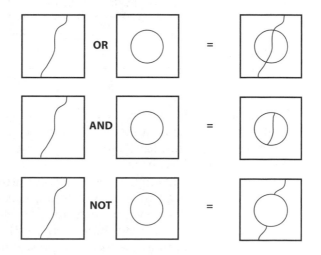

FIGURE 15.3. The result of OR, AND, and NOT overlay operators.

TABLE 15.1. Examples of Higher-Order Functions Based on Map Algebra

ASPECT	Identifies the direction of maximum rate of change in z value from each cell.
BOUNDARYCLEAN	Smoothes the boundary between zones by expanding and shrinking the boundary.
CON	Performs one or more conditional if/else evaluations.
EQUALTO	Evaluates, on a cell-by-cell basis, the number of times in an argument list that the input grid values are equal to the value specified by the first argument.
GREATERTHAN	Evaluates, on a cell-by-cell basis, the number of times in an argument list that the input grid values are greater than the value specified by the first argument.
HILLSHADE	Creates a shaded relief grid from a grid by considering the Sun's illumination angle and shadows.
INT	Converts input floating-point values to integer values through truncation.
ISNULL	Returns 1 if the input value is NODATA, and 0 if it is not.
LESSTHAN	Evaluates, on a cell-by-cell basis, the number of times in an argument list that the input grid values are less than the value specified by the first argument.
MEAN	Uses multiple input grids to determine the mean value on a cell-by-cell basis.
MERGE	Merges multiple, possibly nonadjacent input grids into a single grid based on order of input.
MOSAIC	Merges multiple adjacent continuous grids and performs interpolation in the overlapping areas.
NIBBLE	Replaces areas in a grid corresponding to a mask, with the values of the nearest neighbors.
REGIONGROUP	Records for each cell in the output the identity of the connected region to which it belongs. A unique number is assigned to each region.
SETNULL	Returns NODATA if the evaluation of the input condition is TRUE; if it is FALSE, returns the value specified by the second input argument.
SLICE	"Slices" (or changes) a range of values of the input cells by specified ranges, zones of equal area, or zones with equal intervals.
SLOPE	Identifies the rate of maximum change in z value from each cell.
ZONALAREA	Calculates the area of each zone in the input grid.

(see Figure 15.4). Buffers can also be defined by creating a layer of distances from a feature and then reclassifying to keep only a specific distance.

In contrast to focal operators, zonal functions assign values to all cells in a zone based on values supplied from another map, and global functions assign cell values by processing the entire grid. Examples include computing the distance from one cell to all other cells, or assigning a weight to all cells based on some impedance factor. Applications include modeling diffusion, siting facilities based on distance from selected points, and creating visibility surfaces that create a viewshed of points visible from certain locations. Another application would include finding the path of least resistance through a particular terrain.

The elegance of map algebra is that all operations result in a new map, making it easy to string a series of functions together to produce an increasingly sophisticated model. Tomlin shared all of the source code, documentation, and algorithms. Consequently, the ideas and source code were incorporated into many commercial and open-source software packages. Although there are different flavors of map algebra, the overall concept is used in every GIS system that supports raster calculations. In ESRI's ArcGIS, map algebra is implemented through the SpatialAnalyst extension referred to as the *raster calculator*.

15.2.3 Masking

Masking is a basic overlay operation that involves the multiplication of two matrices, in which one matrix consists of binary values of 1 and 0 (see Figure 15.5). The multiplication of cells by 0 essentially deletes the value, while the multiplication by 1 preserves it.

15.2.4 Siting Functions

Like the concept of map overlay, the siting of facilities is one of the defining applications of GIS. Here, a series of layers are combined to find the ideal location for siting a particular establishment. A weighting function can be applied to give some

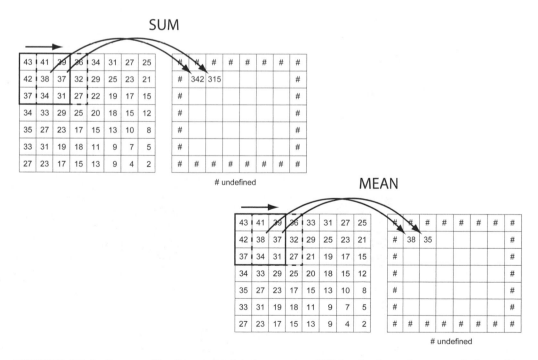

FIGURE 15.4. Two applications of a focal operator. SUM calculates the sum of the values in the 3 × 3 focal area while MEAN calculates the mean of the nine values. Edge pixels remain undefined.

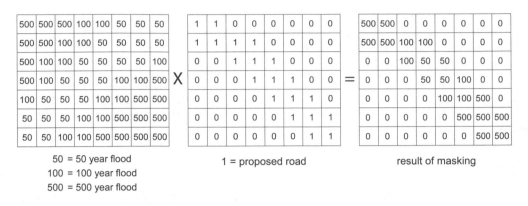

FIGURE 15.5. The masking operation uses a matrix of 1's and 0's to select specific cells.

variables more weight than others in the siting process. Figure 15.6 provides an example, showing the layers that might be used in siting a new school.

15.2.5 Cost of Movement Functions

These functions involve the calculation of "expense" when moving through the environment. In building a road, for example, one might want to minimize

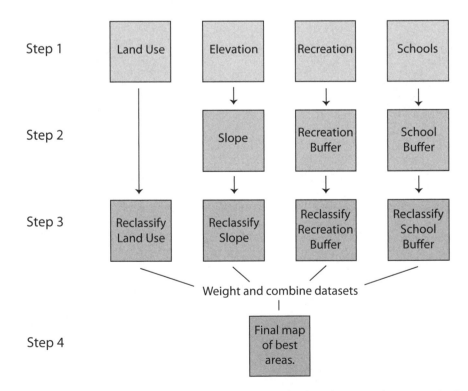

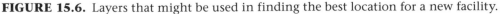

FIGURE 15.6. Layers that might be used in finding the best location for a new facility.

elevation change and the destruction of forests. Figure 15.7 summarizes some of the considerations involved in deciding where to build a road and how these considerations might be combined to find the ideal location.

15.2.6 Topographic Functions

Topographic functions operate on a digital elevation model (DEM), a raster representation of the elevation at grid locations. Using DEM data, a series of other layers can be calculated, including slope and aspect. These calculations are performed with a 3 × 3 kernel as shown in Figure 15.8. All slopes are calculated from the middle of the 3 × 3 focal area, with the slope being determined as a rate of change in the x and y directions. Slope is rise over run and, in the example, the distance between cells is 10 m. The formula multiplies the values immediately adjacent to the middle cell in horizontal and vertical directions to place more emphasis on those closer cells. This computed slope value is placed in the corresponding position of the SLOPE matrix. At the same time, the direction of view of this slope, called the aspect, is calculated and placed in the ASPECT matrix. Following this step, the focal operator is moved one cell to the right, and the slope and aspect calculations are performed again.

15.2.7 Interpolation from a Sparse Matrix

A sparse matrix is a raster representation of a surface or variable in which values are known for only certain cells. Using Tobler's first law of geography, "Everything

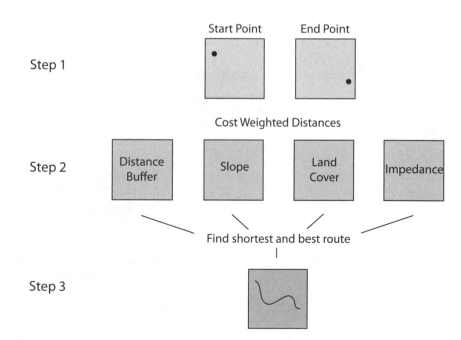

FIGURE 15.7. Finding the path of least "resistance" between two points to determine the location for a new road.

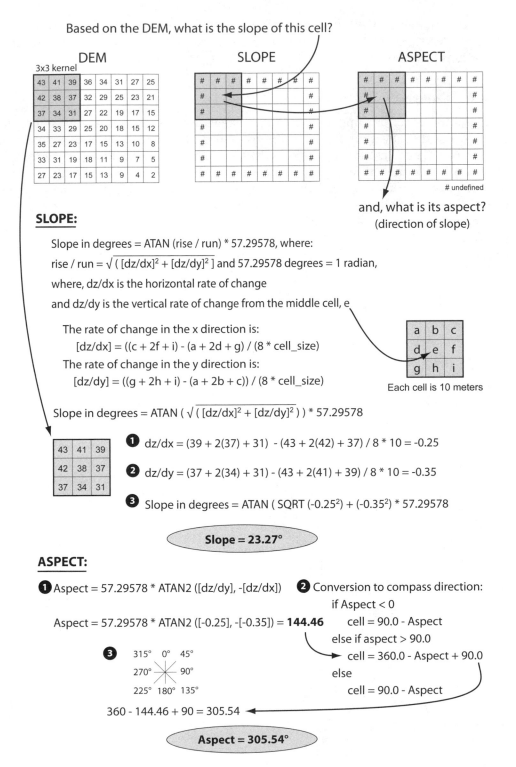

Based on the DEM, what is the slope of this cell?

DEM

3x3 kernel

43	41	39	36	34	31	27	25
42	38	37	32	29	25	23	21
37	34	31	27	22	19	17	15
34	33	29	25	20	18	15	12
35	27	23	17	15	13	10	8
33	31	19	18	11	9	7	5
27	23	17	15	13	9	4	2

SLOPE

ASPECT

undefined

and, what is its aspect?
(direction of slope)

SLOPE:

Slope in degrees = ATAN (rise / run) * 57.29578, where:

rise / run = $\sqrt{[dz/dx]^2 + [dz/dy]^2}$ and 57.29578 degrees = 1 radian,

where, dz/dx is the horizontal rate of change

and dz/dy is the vertical rate of change from the middle cell, e

The rate of change in the x direction is:
   [dz/dx] = ((c + 2f + i) - (a + 2d + g) / (8 * cell_size)
The rate of change in the y direction is:
   [dz/dy] = ((g + 2h + i) - (a + 2b + c)) / (8 * cell_size)

a	b	c
d	e	f
g	h	i

Each cell is 10 meters

Slope in degrees = ATAN ($\sqrt{[dz/dx]^2 + [dz/dy]^2}$) * 57.29578

43	41	39
42	38	37
37	34	31

❶ dz/dx = (39 + 2(37) + 31) - (43 + 2(42) + 37) / 8 * 10 = -0.25

❷ dz/dy = (37 + 2(34) + 31) - (43 + 2(41) + 39) / 8 * 10 = -0.35

❸ Slope in degrees = ATAN (SQRT (-0.25²) + (-0.35²) * 57.29578

> **Slope = 23.27°**

ASPECT:

❶ Aspect = 57.29578 * ATAN2 ([dz/dy], -[dz/dx])   **❷** Conversion to compass direction:
                                                              if Aspect < 0

Aspect = 57.29578 * ATAN2 ([-0.25], -[-0.35]) = **144.46**   cell = 90.0 - Aspect
                                                              else if aspect > 90.0
                                                                 cell = 360.0 - Aspect + 90.0
                                                              else
                                                                 cell = 90.0 - Aspect

❸   315° 0° 45°
        270° —✳— 90°
        225° 180° 135°

360 - 144.46 + 90 = 305.54

> **Aspect = 305.54°**

FIGURE 15.8. The calculation of slope and aspect based on a digital elevation model.

is related to everything else, but near things are more related than distant things" (O'Sullivan and Unwin 2003, p. 221), we can define a function that computes the values of unknown cells. This process is called interpolation.

Interpolation calculates the value for unknown cells based on the distance to, and the value of, a series of surrounding known points. The inverse-distance-squared formula, a common form of automated interpolation, is shown in Figure 15.9 where the value of an unknown point (z) in a matrix is calculated based on three surrounding points. By squaring the distance (D^2) to these points, the influence of the closest point, in this case 451, is magnified compared to the points that are farther away. For example, the use of a cubed function for distance would increase the influence of the nearer, known cells, and the resulting value would be even closer to the value of the nearest cell.

15.3   Vector

Most GIS systems are operated by cities or other local government entities. In terms of mapping, urban areas essentially consist of lines—roads, electricity, cable, water, sewer, and natural gas pipelines. The vector data structure is best suited for encoding and manipulating these types of linear features. A variety of considerations are involved in the structuring, scaling, and overlay of vector data.

15.3.1   Structuring the Vector Map

One of the major concerns in both cartography and geographic information systems relates to the efficient encoding and storage of maps. One source of inefficiency comes from the encoding of polygons (a multisided enclosed shape) as a complete series of individual vectors. The method results in a duplication of points between adjacent polygons (Figure 15.10).

In order to avoid the duplication of points between polygons that share a

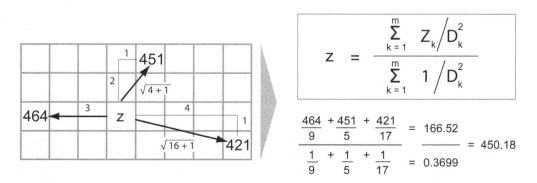

FIGURE 15.9. The inverse-distance-squared interpolation formula. Values for unknown cells are computed based on the distance to, and value of, a series of known cells, where Z_k represents the known values for grid cells, D_k is the distance to the unknown cell z, and m is the number of known data values to use in the calculation—in this case, three.

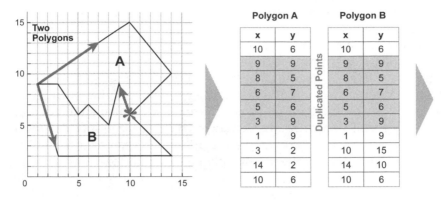

Polygon A

x	y
10	6
9	9
8	5
6	7
5	6
3	9
1	9
3	2
14	2
10	6

Polygon B

x	y
10	6
9	9
8	5
6	7
5	6
3	9
1	9
10	15
14	10
10	6

Total Number of Points: 20

FIGURE 15.10. Encoding adjacent shapes stores duplicate points. In this example, the two areas—A and B—share five coordinates that are included in the definition of each polygon.

border, a different method of encoding needed to be developed. In the arc-node system, polygons are stored in a more efficient fashion. In this approach, the polygon is subdivided into line segments called arcs. The individual arcs are delimited by "critical points" that are called nodes. A database is used to keep track of which arcs constitute a particular polygon (see Figure 15.11). When a map is produced, polygons are subsequently reassembled from the database.

In practice, the arc-node method reduces the storage of x, y coordinates by almost half through the elimination of duplicate points between adjacent polygons. The data structure also facilitates the overlay of maps and other complicated operations such as redistricting—the redrawing of political boundaries to ensure equal representation. These operations are made possible by defining the relationships

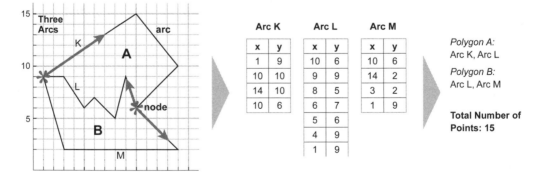

Arc K

x	y
1	9
10	10
14	10
10	6

Arc L

x	y
10	6
9	9
8	5
6	7
5	6
4	9
1	9

Arc M

x	y
10	6
14	2
3	2
1	9

Polygon A:
Arc K, Arc L

Polygon B:
Arc L, Arc M

Total Number of Points: 15

FIGURE 15.11. Arc-node encoding of polygons. In this example the two polygons are represented with three arcs—K, L, and M. Polygon A is composed of arcs K and L, while polygon B consists of arcs L and M. Fifteen coordinates are needed to represent the two polygons in contrast to 20 for the polygon method. In some implementations, arcs must be uniquely defined, and they are only unique if they have different nodes. In the illustration, all three arcs would be seen as identical. A so-called pseudo-node would need to be added to at least two of the arcs. The two polygons would then be represented with five distinct arcs but the same number of coordinates.

between arcs and the polygons they define. The definition of these relationships, as well as those between features, is referred to as topology.

A relational database stores information about the arcs that constitute the polygons. When the map is reconstructed, the database is consulted to determine which arcs constitute the polygons. The main advantage of the arc-node approach is that there is no duplication of points between adjacent polygons; its disadvantage is that a sophisticated database must be used to keep track of all the arcs and polygons. As a result, the map is more complicated to work with for simple tasks and more difficult to transfer between computers. The ESRI Shapefile provides a solution to these problems by encoding the complete polygons and often using a simpler dBase (.dbf) format to assign features to the spatial entities. The size of the file is greater because duplicate points between polygons are stored at least twice.

15.3.2 Scaling the Vector Map

The vector map can be processed in several ways, demonstrating the computer's real power in manipulating maps. With the vector approach, creating a map at a larger size would simply involve multiplying all x, y coordinates by the corresponding factor. Figure 15.12 presents the calculations involved in increasing the size of the map by a factor of 1.6. The minimum x and y values are subtracted from each value before multiplication by the scaling factor and are then added again so that the scaled figure is placed in the correct position relative to the origin. The addition of a constant to ensure coordinates is called translation, resulting in movement of the figure relative to the origin.

15.3.3 Rotating the Vector Map

The rotation of coordinates is a relatively simple process. The formula for rotation is:

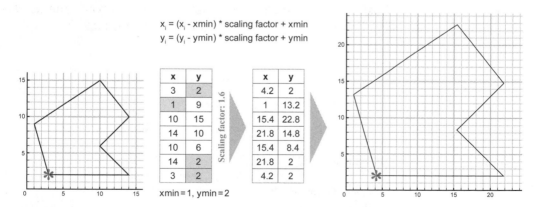

FIGURE 15.12. Enlarging a polygon. To scale a set of coordinates, the difference between each coordinate and the minimum x or y is multiplied by the scaling factor. The minimum x and y values are then added to the scaled value to maintain the same distance from the origin.

$$x_{i\_} = x_i * \cos(radang) + y_i * \sin(radang)$$
$$y_{i\_} = -x_i * \sin(radang) + y_i * \cos(radang)$$

where x_i is the set of x coordinates, y_i is the set of y coordinates, and radang is the angle of rotation in radians. The conversion of this formula into a computer program would be as follows:

```
u = cos (radang)
v = sin (radang)
for i = 1, n
    tempx = x(i) * u + y(i) * v
    y(i) = -x(i) * v + y(i) * u
    x(i) = tempx
next i
```

where u and v store the result of the cos and sin transformation of the angle, n is the total number of coordinates, and tempx stores the result of the rotated x coordinate. It is necessary to store the value of the x coordinate in variable tempx so that calculations of the rotated y coordinate can use the original, unrotated value of x.

15.3.4 *Overlaying the Vector Map*

The overlay of maps defined in vectors is much more complicated than the corresponding raster process (Burrough 1994, p. 85). Vector overlay finds the intersection between arcs in each layer (see Figure 15.13). Although this can be done using calculus where each line is described as a formula such as y = mx + b, it is still difficult to execute for each line segment between two polygons. Vector overlay is one area where commercial software still has the advantage over open-source GIS implementations.

15.4 Summary

GIS has become a big business and continues to grow. The various components of GIS, including computer hardware, software, consulting services, and data input

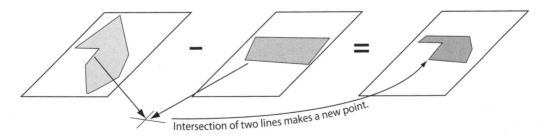

FIGURE 15.13. Polygon overlay is based on finding the intersection of lines between the two polygons. These points are determined by describing each line with a formula, y = mx + b and then solving an equation for their intersection. The points create the new vertices for the subtracted polygon.

are now a multibillion dollar enterprise. Governmental and private agencies use consulting services to help develop and implement a GIS. The initial investment in any GIS is the input of data, which alone can sometimes run into the millions of dollars.

The largest ongoing cost associated with GIS is human labor. Trained people are needed to operate the hardware, work with the often very sophisticated and complicated software, and perform the various data input, analysis, and output functions. GIS continues to be a major source of employment for people with a background in geography and cartography.

Government agencies, ranging from local entities to the federal government, are the major users of commercial GIS software. Many of the GIS procedures described in this chapter are possible using cloud-based tools. Budgetary concerns will inevitably lead to a search for less expensive technology, and many solutions will be found in cloud computing. APIs specifically designed for GIS and map overlay will also play a major role in this endeavor.

15.5   Questions

1. Why is GIS often described as the overlay of map layers? What are more common applications?

2. It could be argued that GIS software (ArcGIS, QGIS, etc.) has not properly adapted to modern methods of mapping, both desktop and mobile. To what extent is GIS software still in a "paper-thinking" mode that envisions map presentation as it was done on paper? Why is GIS behind in modern methods of mapping?

3. Why is it easier to overlay map information defined as rasters instead of vectors?

4. Describe the integration of algebra in the overlay of map layers.

5. Describe the masking, siting, and cost of movement functions in the raster environment.

6. Describe how slope and aspect are calculated.

7. What is a sparse matrix, and how can values for unknown cells be estimated?

8. What are the main advantages of the arc-node system for encoding spatial information compared to the polygon approach?

9. In rotating a series of coordinates, why must a temporary value be used for either the x or y coordinate?

10. Describe the general process for overlaying two polygons defined as a series of vectors.

15.6 References

Burrough, P. A. (1994) *Principles of Geographical Information Systems for Land Resources Assessment.* Oxford: Clarendon Press. 193 pp.

O'Sullivan, D., and David Unwin (2003) *Geographic Information Analysis.* Hoboken, NJ: John Wiley & Sons.

Tomlin, Dana (1990) *Geographic Information Systems and Cartographic Modeling.* Englewood Cliffs, NJ: Prentice-Hall.

CHAPTER 16
Map Layer Mashups

All models are wrong but some are useful.

—George E.P. Box

16.1 Introduction

In previous chapters, we overlaid points, lines, and areas that were defined as vectors in latitude and longitude. Here, we overlay a raster image that could be an air photo, satellite image, or scanned map. The advantage of overlaying an image is that the overlay can be done much more quickly. No major conversion or drawing is necessary to place the information because the underlying map is in the same format. Raster files can be overlaid as single entities or divided into tiles that precisely match the tiles of the base map. The overlay can cover only a small portion of the display or be as large as the entire display.

16.2 Layer Overlays

The examples in this chapter place a raster image, such as a scanned paper map or air photo, on top of the online map or satellite image. The procedure requires knowing the latitude and longitude of the southwest and northeast corners of the overlay. Some image file formats, such as GeoTIFF and JPEG2000, have this information coded in the header of the file. In most cases, the corners of the map will need to be determined by comparing the overlay to the map.

16.2.1 Basic Overlays

The three examples in this section overlay different types of files. The first example uses a map that has been scanned and saved in the JPEG format (see Figure 16.1). The corners of the image have been estimated and are then defined using image-Bounds, a variable that stores the lower-left and upper-right coordinates. These coordinates are combined with the address of the image in the oldmap object. Examining the streets in this example clearly shows that the map does not overlay properly. A variety of factors could be involved in making the two maps incompatible.

The example in Figure 16.2 uses a more sophisticated procedure to overlay a topographic map of Alaska on a Google satellite image. The initialize function again specifies the location of the map and the bounds. Here, the topographic map

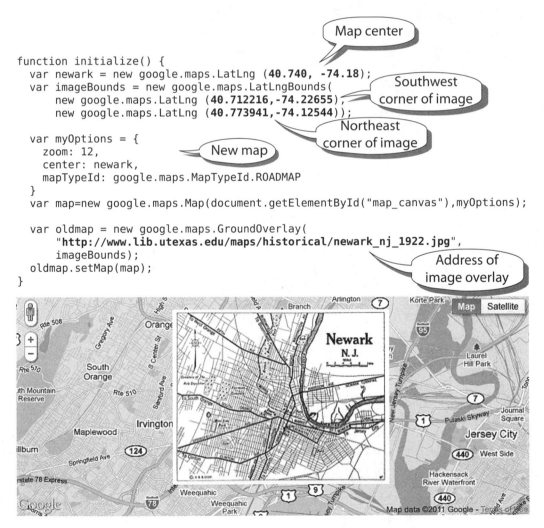

FIGURE 16.1. Overlay of a scanned map in the jpeg format. Map of Newark, NJ, courtesy of University of Texas Libraries, University of Texas at Austin; copyright 2013 Google.

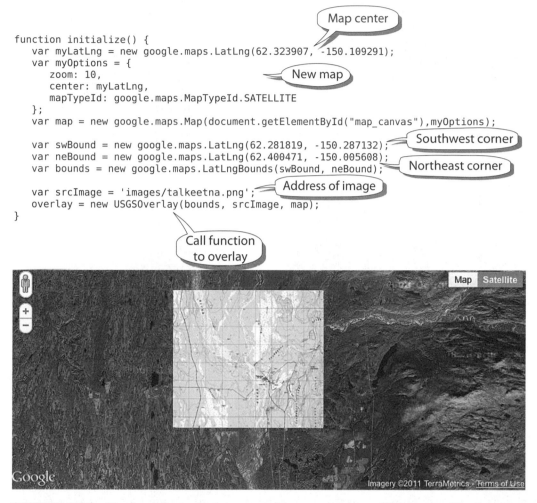

```
function initialize() {
    var myLatLng = new google.maps.LatLng(62.323907, -150.109291);
    var myOptions = {
        zoom: 10,
        center: myLatLng,
        mapTypeId: google.maps.MapTypeId.SATELLITE
    };
    var map = new google.maps.Map(document.getElementById("map_canvas"),myOptions);

    var swBound = new google.maps.LatLng(62.281819, -150.287132);
    var neBound = new google.maps.LatLng(62.400471, -150.005608);
    var bounds = new google.maps.LatLngBounds(swBound, neBound);

    var srcImage = 'images/talkeetna.png';
    overlay = new USGSOverlay(bounds, srcImage, map);
}
```

FIGURE 16.2. Overlay of scanned topographic map on a satellite image. The location is Alaska, and the map is more square than is typical of topographic maps in the lower 48 states because more than 7½ degrees of longitude is depicted. U.S. Geological Survey, Department of the Interior/USGS; copyright 2013 Google.

is provided as a PNG file at 920 × 762 pixels. The USGSOverlay function performs the overlay.

The example in Figure 16.3 toggles the view of the USGS quadrangle map with a toggle button. The toggling is done with a button defined in the body section. When the toggle button is clicked, a function is called that checks whether the div layer is visible. If it is visible, it turns the visibility off; if it is not visible, visibility is turned on.

An air photo or satellite image that is layered over the satellite view could provide more detail or a more current view. The map in Figure 16.4 displays a different satellite image. Clicking on the Map button overlays the same image on the map display.

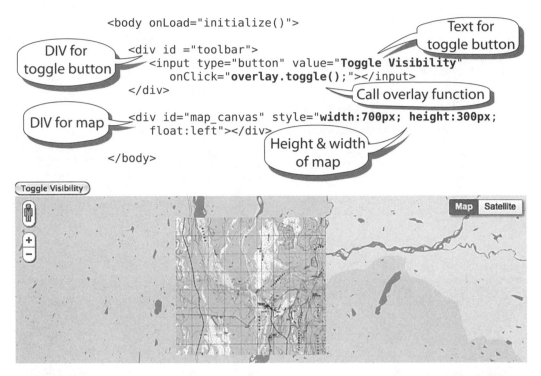

FIGURE 16.3. USGS map overlay with toggle button. The code shows how the button is defined. U.S. Geological Survey, Department of the Interior/USGS; copyright 2013 Google.

16.3 Standard Layer Overlays

The major online map providers offer many additional layers of information that are often defined as raster layers. Most of these layers are transparent, allowing for the display of the underlying map or satellite image. Viewed alone, the information on these layers would be meaningless. Table 16.1 shows the available layers provided by Google.

Many of these layers, notably the traffic layer, are frequently updated, whereas others are only occasionally updated. The traffic layer shows the speed of traffic for major streets in the larger cities of the world (see Figure 16.5) and is updated continuously throughout the day. The cities use traffic cameras or in-road sensors that measure the speed of traffic and upload these measurements to a central server. All of the major map providers can access this data and produce a traffic map at regular intervals. The maps are also tiled at multiple levels of detail for faster download. Only tiles that need updating are replaced.

Figure 16.6 shows how the traffic map can be displayed using the Google Maps API. The traffic tiles are made as a transparent PNG so that most of the map is still visible. Traffic is symbolized using a dual-level symbology based on the area. For highways, the colors are a general indication of the speed of traffic. Google shades its traffic map for highways as follows:

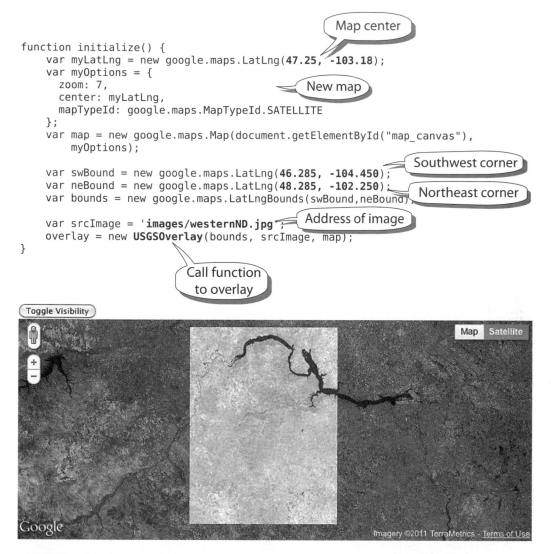

```
function initialize() {
    var myLatLng = new google.maps.LatLng(47.25, -103.18);
    var myOptions = {
        zoom: 7,
        center: myLatLng,
        mapTypeId: google.maps.MapTypeId.SATELLITE
    };
    var map = new google.maps.Map(document.getElementById("map_canvas"),
        myOptions);

    var swBound = new google.maps.LatLng(46.285, -104.450);
    var neBound = new google.maps.LatLng(48.285, -102.250);
    var bounds = new google.maps.LatLngBounds(swBound,neBound);

    var srcImage = 'images/westernND.jpg';
    overlay = new USGSOverlay(bounds, srcImage, map);
}
```

FIGURE 16.4. Image overlay based on a background satellite image. U.S. Geological Survey, Department of the Interior/USGS; copyright 2013 Google.

> **Green:** more than 50 mi per hour or 80 km per hour
>
> **Yellow:** 25–50 mi per hour or 40–80 km per hour
>
> **Red:** less than 25 mi per hour or 40 km per hour
>
> **Red/Black:** very slow, stop-and-go traffic
>
> **Gray:** no data currently available

Within urban areas, these traffic speeds no longer apply because they would not be realistic. Here, the colors simply indicate good, fair, and poor traffic conditions based on the average of previous conditions.

TABLE 16.1. Standard Layers Provided by Google—All Layers Are Tiled

Traffic	Current traffic conditions
Photos	Locations of available photographs
Labels	Street, city, and boundary names
Webcams	Locations of cameras with live imagery
Videos	Locations of YouTube videos
Wikipedia	Wikipedia pages for locations on map
Bicycling	Biking paths and trails
45°	45° angle bird's-eye view
Transit	Public transportation network

An example of a layer that is not updated as frequently is the bicycle layer showing bicycle paths (see Figure 16.7). In contrast to traffic, the bicycle layer is opaque and totally covers the underlying map at larger scales in urban areas, although (if the Internet connection is slow) one might be able to catch a brief glimpse of it before being covered by the overlay.

16.4   KML Overlays

Keyhole markup language (KML) has been used in the previous chapters to display vector data in the form of points, lines, and areas. A KML file can also contain a rasterized file. The raster data can make the KML very large, resulting in the need

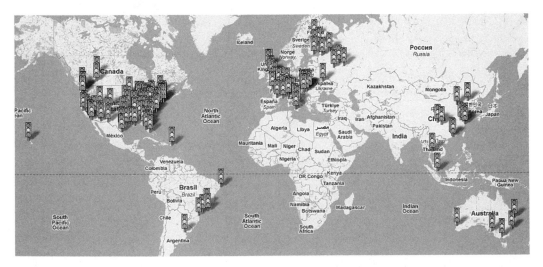

FIGURE 16.5. Available traffic maps for cities around the world. The system is based on cameras that capture the speed of traffic and send the data to a central server. Copyright 2013 Google.

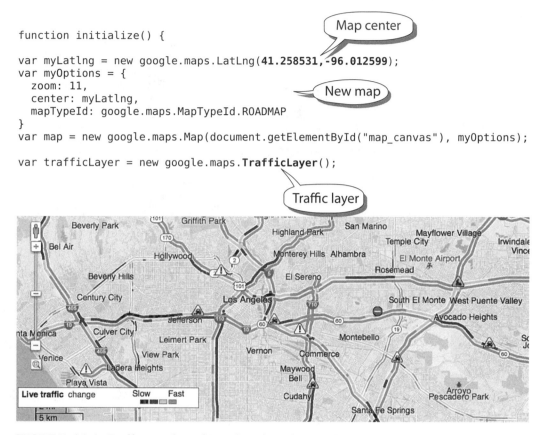

```
function initialize() {

var myLatlng = new google.maps.LatLng(41.258531,-96.012599);
var myOptions = {
  zoom: 11,
  center: myLatlng,
  mapTypeId: google.maps.MapTypeId.ROADMAP
}
var map = new google.maps.Map(document.getElementById("map_canvas"), myOptions);

var trafficLayer = new google.maps.TrafficLayer();
```

FIGURE 16.6. Traffic overlay of Los Angeles. The overlay is updated at regular intervals. Copyright 2013 Google.

for compression. The KMZ format is a zipped version of the KML with a smaller file size. KMZ files are automatically unzipped before they are displayed.

The map in Figure 16.8 overlays a KML file displaying a rasterized map of wave heights for the Great Lakes between the United States and Canada. The isoline map has been shaded. The zoomed-in version indicates the pixel size.

16.5   Image Tiling

Image tiling is the process of subdividing a raster file to match the tiles of online map servers. If the original file has a high enough resolution, tiles can be made at multiple levels of detail. One advantage of tiling the overlay is that only parts of the image need to be displayed based on the current zoom level. Tiling therefore reduces the amount of Internet traffic and increases the speed of display. The main advantage is the display of additional detail at the higher zoom levels.

Tiling an image is a good way to represent any type of high-resolution raster file. Photographs with a large number of pixels may be distributed in this way. Zooming further into the image makes the system switch to different tiles with

```
function initialize() {

var myLatlng = new google.maps.LatLng(41.258531,-96.012599);
var myOptions = {
  zoom: 11,
  center: myLatlng,
  mapTypeId: google.maps.MapTypeId.ROADMAP
}

var map = new google.maps.Map(document.getElementById("map_canvas"), myOptions);

var bikeLayer = new google.maps.BicyclingLayer();
bikeLayer.setMap(map);

}
```

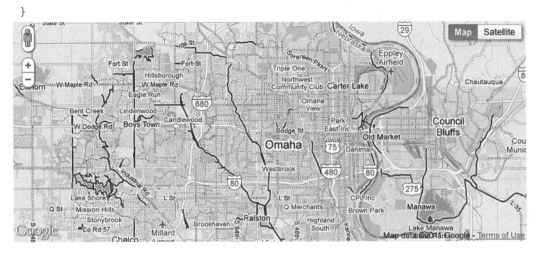

FIGURE 16.7. Bicycle path overlay from Google. In contrast to the traffic layer, the bicycle layer completely covers the underlying map in urban areas. Copyright 2013 Google.

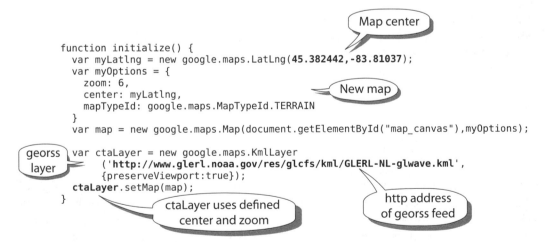

FIGURE 16.8a. A KML overlay of a rasterized file showing updated wave heights for the Great Lakes. The stair-step effect of the raster file is clearly visible in the bottom zoomed version. Copyright 2013 Google.

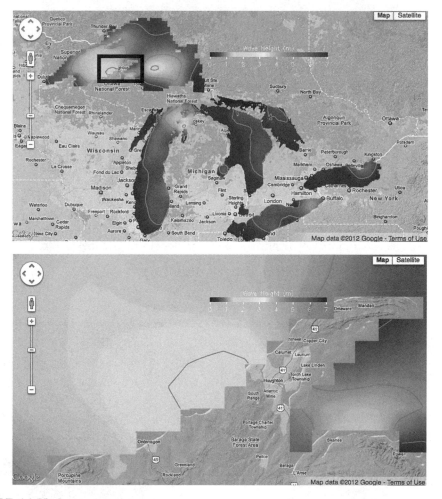

FIGURE 16.8b & c.

greater resolution. Using Google Maps to zoom into photos has been implemented in a variety of ways. The George Kremer art collection includes a series of old paintings that have been scanned and tiled for the Google Maps API (see Figure 16.9). The images can be smoothly zoomed, and fine details in the pictures can be examined that would not normally be visible (search: George Kremer collection). All types of historical documents could be distributed in this way.

Figure 16.10 displays a digital raster graphic (DRG) file. DRGs are high-resolution scans of 1:24,000 USGS topographic maps. Scanned at 250 dpi, the files are about 4600 × 5600 pixels and 8 mb in size. Georeferencing is accomplished using the GeoTIFF format that creates a link with an associated .tfw file containing the referencing information. DRG files are available through archive.org (search: USGS maps download internet archive).

One application for tiling images is MapTiler (Klokan 2011; search: MapTiler), which takes a single large image and cuts it into separate 256 × 256 pixel files. It begins by creating a smaller version of the original image with a 256-pixel-square

FIGURE 16.9. Example of a picture that has been made available for viewing using the Google Map API. The picture can be panned and viewed like a Google Map. Painting by Michael Sweerts, circa 1660. Copyright The Kremer Collection; Michael Sweerts, 1618–1664; Young Maidservant, c. 1660; Canvas, 61 × 53.5 cm.

Zoom into topo map to see tiles appear for the different zoom levels

Zoom into topo map to see tiles appear for the different zoom levels

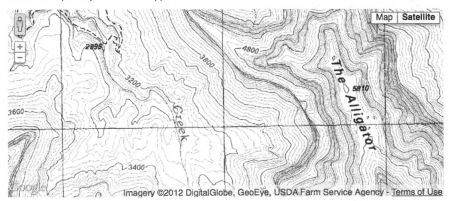

FIGURE 16.10a & b. Overlay of a digital raster graphic (DRG) as tiles. The upper map (this page) shows the entire DRG. The lower map (this page) at a higher zoom level completely covers the Google Map. The two tiles (upper opposite page) are from the 12th and 14th zoom levels. Search: MapTiler. U.S. Geological Survey, Department of the Interior/USGS; copyright 2013 Google.

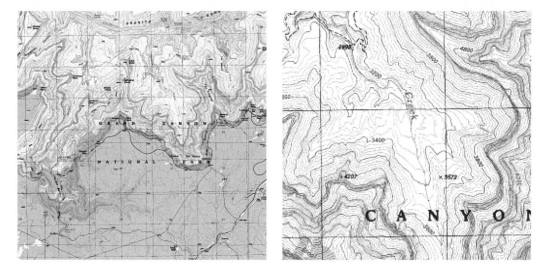

FIGURE 16.10c & d.

image. At the next level, there are 4 of the 256 pixel tiles, then 16, 64, and finally 256. This corresponds to overall sizes of 256, 512, 1024, 2048, 4096, etc., pixels when all tiles are combined. MapTiler automatically chooses the maximum zoom level to correspond to the original size of the image. Zooming in any further would cause the image to pixelate. It is also possible to input different images into MapTiler to create tiles for the different levels of detail.

The six-step process for tiling an image is shown in Figure 16.11. The first step presents a choice between creating Mercator tiles or an overlay for Google Earth. The next steps define the input file, in this case a geoTIFF of a scanned USGS 1:250,000 topographic map that is dragged onto the open file window in step three. This DRG file, listed in step four, has 4,600 × 5,600 pixels. The fifth step defines the type of tiles and the number of layers. The last step shows that the tile rendering process is complete.

MapTiler creates tiles for each zoom level and places them in multiple embedded folders. In addition to these tiles, MapTiler also creates a `tilemapresource. xml` file that includes the min/max box for the overlay. These values are used to specify the location of the tiles for display with Google Maps (see Figure 16.12).

Once a map or image has been tiled, the tiles are typically stored on a different server than those used for the map or satellite image. The map or satellite will always be displayed faster. It is difficult for an ordinary server to match the display speeds of commercial map tiles.

Another possibility in image overlay is the addition of a mechanism to control the visibility of layers. One method uses radio buttons to control which of several layers would be visible. Harris (2012) demonstrates how to add a slider bar to the display of the overlay (see Figure 16.13). Moving the slider bar to either side controls the visibility of the overlay. MapTiler outputs the necessary code to incorporate the Harris slider.

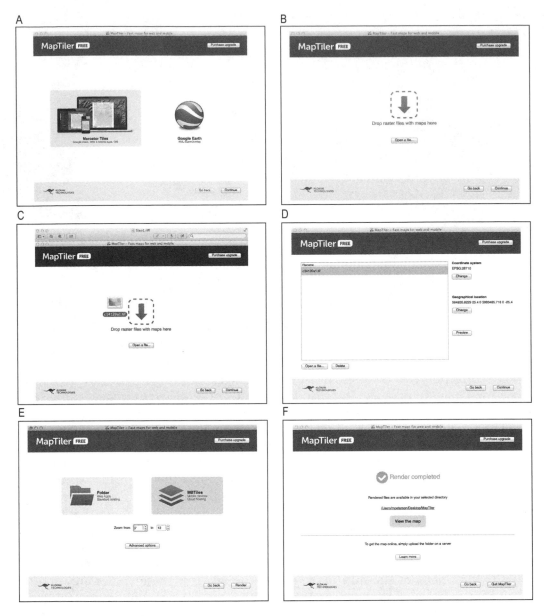

FIGURE 16.11. The six steps for creating tiles using MapTiler. Any georeferenced image with a sufficient number of pixels can be tiled. In this case, a USGS DRG file is used with approximately 4600 × 5600 pixels. Tiles are generated for the 10th through the 15th zoom level. U.S. Geological Survey, Department of the Interior/USGS.

16.6 Summary

Along with the ability to pan and zoom, the layering of map and image data is one of the most powerful benefits of the online map. The possibilities for adding layers are almost endless. Every existing map, air photo, or satellite image could be layered in a number of different ways. The ability to create images at multiple levels

XML File

```
<?xml version="1.0" encoding="utf-8"?>
    <TileMap version="1.0.0" tilemapservice="http://tms.osgeo.org/1.0.0">
      <Title>o36112a2.tif</Title>
      <Abstract></Abstract>
      <SRS>EPSG:900913</SRS>
      <BoundingBox minx="35.98263093762349" miny="-112.26055836038145"
      maxx="36.13311975190522" maxy="-112.11411831155569"/>
      <Origin x="35.98263093762349" y="-112.26055836038145"/>
      <TileFormat width="256" height="256" mime-type="image/png" extension="png"/>
      <TileSets profile="mercator">
        <TileSet href="10" units-per-pixel="152.87405654296876" order="10"/>
        <TileSet href="11" units-per-pixel="76.43702827148438" order="11"/>
        <TileSet href="12" units-per-pixel="38.21851413574219" order="12"/>
        <TileSet href="13" units-per-pixel="19.10925706787109" order="13"/>
        <TileSet href="14" units-per-pixel="9.55462853393555" order="14"/>
        <TileSet href="15" units-per-pixel="4.77731426696777" order="15"/>
      </TileSets>
    </TileMap>
```

Callouts: SW and NE corners · Map location · 256x256 tiles in PNG format · Zoom levels for tiles · Ground distance per pixel in meters

FIGURE 16.12. The `tilemapresource.xml` file created by MapTiler. The minx, miny, maxx, and maxy values represent the bounding box for the tiles.

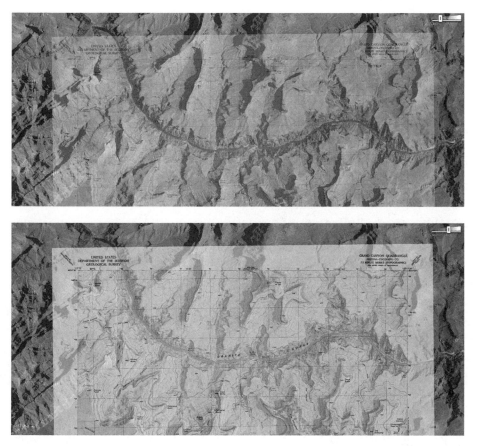

FIGURE 16.13. Topographic map of the Grand Canyon overlaid over a Google Maps satellite image with a slider bar. The overlay is just barely visible in the top display. The slider bar has been moved in the bottom illustration to show more of the map. U.S. Geological Survey, Department of the Interior/USGS; copyright 2013 Google.

323

of detail is another benefit. Future versions of online maps will make such overlays even more common.

As mapping APIs develop, they will certainly incorporate more analytical tools. For example, it should be possible to present a comparison between two image overlays. Other GIS-type comparisons should also be possible. A number of secondary APIs have already been developed demonstrating that analytical tools could be combined with the mapping API.

16.7   Exercise

Upload the code16.zip file to your server and make the suggested changes to the code examples.

16.8   Questions

1. What are some advantages and disadvantages of overlaying raster layers with Google Maps?

2. What is the minimum information required to overlay a map defined in an image file format?

3. Google Maps offers a number of additional layers that can be added to the map. Describe some other layers that could be offered.

4. How is the traffic layer constructed?

5. How can a raster KML overlay be distinguished from a vector overlay?

6. What is the advantage of presenting paintings with the Google Map interface?

7. Describe the operations performed by MapTiler.

8. What are the limitations of presenting a standard USGS topographic map through a Google Map interface?

9. Describe a system that would present all available USGS topographic maps at different available scales through a tiled mapping system.

10. Describe elements of a GIS API that could be used with Google Maps.

16.9   References

Harris, Gavin (2012) Image Slider for Google Maps v3. [http://www.gavinharriss.com/code/opacity-control]

James, Will (2011) Automatic Tile Cutter. MapKi—A Forum for Sharing Ideas for the Google Maps API. [http://mapki.com/wiki/Automatic_Tile_Cutter]

Klokan, Petr Predal (2012) *Map Tile Cutter—Map Overlay Generator for Google Maps and Google Earth*. Zurich: Klokan.

CHAPTER 17

Databases, MySQL, and PHP

The first thing we're gonna need is a lot of pictures.
Unfortunately, Harvard doesn't keep a public centralized
facebook so I'm going to have to get all the images from the
individual houses that people are in. Let the hacking begin.

First up is Kirkland. They keep everything open and allow
indexes in their Apache configuration, so a little w-get magic is
all that's necessary to download the entire Kirkland facebook.
Kids' stuff.

Dunster is intense. You have to do searches and your search
returns more than twenty matches, nothing gets returned. And
once you do get results they don't link directly to the images,
they link to a PHP that redirects or something. Weird. This may
be difficult. I'll come back later.

—JESSE EISENBERG AS MARK ZUCKERBERG IN *THE SOCIAL NETWORK*

17.1 Introduction

Everything on the Internet relies on a database. Defined as an organized body
of information that is arranged for ease and speed of access, a database is simply
the way information is stored and retrieved. In most cases, before anything is put
onto a web page, some type of information for that page has been extracted from a
relational database. The most commonly used relational database software for web
pages is MySQL (DuBois 2009).

MySQL, pronounced "my sequel" or "my ess cue el," is an open-source relational database management system (RDBMS). It is used for all types of database
applications including e-commerce, airline reservation systems, and online social
networks. MySQL (structured query language) is known for its performance,

reliability, and ease of use. It can handle large databases with billions of entries. MySQL continues to be adopted by organizations as they discover that it can handle their database needs at a fraction of the cost of commercial software. Any http address that ends with PHP has likely consulted a MySQL database before constructing the web page.

PHP, or PHP hypertext preprocessor, is a server-side scripting language that works with MySQL. It constructs dynamic web pages on the fly by combining HTML with information derived from a database. It is the most popular language for creating websites that compete against other commercial and open-source languages such as ASP.NET, Java, ColdFusion, Perl, Ruby, and Python. As of 2013, over 80% of all web servers use PHP as a server-side programming language (W3Techs 2013).

Like JavaScript, PHP is an interpreted language that needs to be compiled into executable code. In contrast, PHP code cannot execute locally on a client computer. Thus, PHP code is not viewable by choosing the source code option for a web page. The web page has itself been written by PHP, and what we see by viewing the source code is only the output.

Figure 17.1 shows the current usage of PHP compared to other server-side preprocessors. Although PHP is the most popular scripting language around the world, its use varies considerably in different areas. Over 90% of all server-side applications in Europe use PHP, but that rate is only 40% in China. Microsoft servers are much more common in China, prompting greater use of Microsoft's ASP.NET (W3Techs 2013).

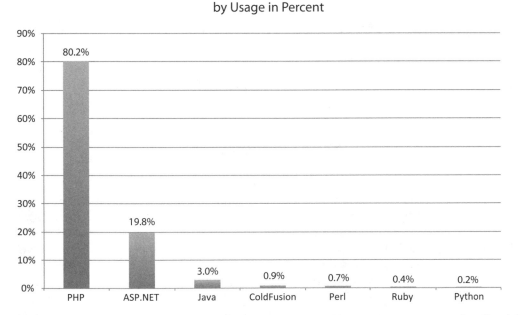

FIGURE 17.1. Relative use of server-side programming languages. PHP outranks all other languages combined. Based on June 2013 W3Techs.com server-side data.

Both MySQL and PHP are examples of free and open-source software (FOSS). FOSS presents a major challenge for commercial software companies because it makes software available for free that companies sell for a great amount of money. Although no one company is responsible for open-source software, there may be many thousands of installations and millions of users. In considering the long-term viability of any software, it is always best to stay with those that have the greatest number of users, whether commercial or open source. While still free, MySQL is technically no longer open source because it is now owned and supported by Oracle, the largest database company in the world.

This chapter provides an introduction to general database concepts, the MySQL database, the spatial extensions to MySQL, and the PHP server-side scripting language. Different types of databases are first examined. This discussion is followed by a more detailed description of relational databases through the underlying SQL language. Finally, Spatial SQL is introduced along with the PHP scripting language.

17.2 Types of Databases

Although many different ways of storing data have been proposed, the two most common in general usage are the flat file and the relational database. Spreadsheet programs such as Microsoft Excel™ are based on the flat file concept. All records in a flat file are stored in the same number of fields in a table format. A flat file is very good for summing columns of data or for sorting data with a certain theme, but cannot easily perform complex queries. In contrast, a relational database may be viewed as multiple tables of data tied together by a key field or fields (see Figure 17.2). The relational database results in less duplication of data and, more importantly, the ability to easily query the data.

Relational databases are the most common databases and the basis of almost

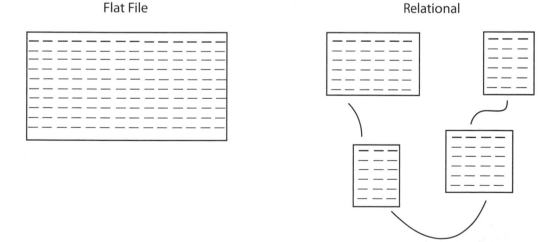

FIGURE 17.2. Flat file versus a relational database. Both consist of tables, but the relational model ties multiple tables together by relating key fields.

all information presented through web pages. All relational databases use SQL, often referred to as structured query language, a computer language designed to input, manipulate, query, and extract data. It is essentially a set of statements that result in certain actions on data stored in relational tables. SQL is what makes a relational database work.

Other types of databases include the *hierarchical database*, which organizes data in a tree-like structure. It defines a parent–child relationship in which each parent can have many children but each child has only one parent. The folder or directory structure for storing files on a computer is an example of a hierarchical data structure. *Object-oriented databases*, similar to object-oriented programming, carry out operations on items, called data objects, in which the operation is part of the definition of the object. Finally, *NOSQL* is an effort that promotes the use of nonrelational databases and does not require fixed table schemas as with the relational model.

17.3   Relational Database

Relational databases are often referred to as a *matrix of tables*. Each table contains records that are the horizontal rows in the table—sometimes called *tuples*. Fields refer to the columns of the table, called *attributes*. Domain refers to the possible values for a field. For example, if the field contains percentages, then the domain of values could only fall between 0 and 100.

Each table has a field that contains distinct identifiers, or keys, that specify records. The common field is used to relate any two tables. A key that uniquely identifies a record is called a *primary key*. A *foreign key* in another table links the two tables. The foreign key helps to ensure referential integrity, a property of a relational database that enforces valid relationships between tables, so that no foreign key can contain a value that does not match a primary key in the corresponding table. A value in the foreign key field must exist in the primary key field. In Figure 17.3, the fields `City` and `City_Name` would be the primary and foreign fields respectively.

City	Latitude	Longitude
Omaha	41.25	-96
Atlanta	33.755	-84.39
Lincoln	40.809722	-96.675278
Des Moines	41.590833	-93.620833
Memphis	35.1175	-89.971111
Minneapolis	44.95	-93.2
Washington DC	38.895111	-77.036667
Cincinnati	39.133333	-84.5
New York City	40.716667	-74
Salt Lake City	40.75	-111.883333
Chicago	41.881944	-87.627778
Milwaukee	43.052222	-87.955833
Cancun	21.160556	-86.8475

City_Name	Airport
Omaha	OMA
Atlanta	ATL
Lincoln	LNK
Des Moines	DSM
Memphis	MEM
Minneapolis	MIC
Washington DC	WAS
Cincinnati	CVG
New York City	JFK
Salt Lake City	SLC
Chicago	ORD
Milwaukee	MKE
Cancun	CUN

FIGURE 17.3. Two tables of a relational database. The `City` column is the *primary key*. It matches the `City_Name` column in the second table, defined as the *foreign key*.

17.4 SQL

SQL is the language to input, manipulate, and extract data from a relational database. Relational databases were first developed in 1970 by IBM and eventually led to a product called SEQUEL. The associated language to manipulate the data became known as SQL and was standardized by the American National Standards Institute (ANSI) in 1986. New features have been added to incorporate object-oriented functionality and extensions to handle spatial data. The latest standard defines a language called XQuery to query data in extensible markup language (XML) documents.

The SQL language can be categorized into statements that define how the data is structured, input, and manipulated. Figure 17.4 shows how the two tables from 17.3 would be created. The tables are named cities and airports. In the first table, the initial variable is city and has a maximum length of 30 characters. The second table names this identical `City_Name` variable with the same number of characters. These two columns bind these two tables together.

Once the structure has been defined, the data values can be entered into the database. Figure 17.5 depicts the SQL code that would insert the data into a MySQL database. Three fields are entered into the `cities` table. The first of these fields is the city name, limited to 30 characters. Following this field are the latitude (`lat`) and longitude (`long`) of the city.

To demonstrate database manipulation, a query could be constructed that would, for example, find all cities west of Minneapolis. First, the `SELECT` command would retrieve rows selected from one or more tables and create a new table, as shown in Figure 17.6. In this example, `city` indicates the column from the `cities` database from which elements that satisfy a condition will be retrieved. The `WHERE` clause indicates the condition or conditions that must be satisfied in order to be selected. In this case, the condition selected is being further west than Minneapolis with a longitude value that is less than –93.2.

17.5 Spatial SQL

The Open Geospatial Consortium (OGC) is an international group of companies, agencies, and universities participating in the development of conceptual solutions useful with all kinds of spatial data management applications. In 1997, the OGC published the *OpenGIS®* Simple Features Specifications for SQL, a document that proposed several conceptual ways for extending SQL to support spatial data. With

```
create table cities (           create table airports (
    city VARCHAR(30),               City_Name VARCHAR(30),
    lat FLOAT,                      Airport VARCHAR(3),
    long FLOAT,                     FOREIGN KEY (City_Name) references cities (city)
    PRIMARY KEY (city)          );
);
```

FIGURE 17.4. The definition of two tables using SQL. The *foreign key* is used to prevent any actions that would destroy the link between the two tables.

```
INSERT INTO cities (city, lat, long) VALUES ("Omaha", 41.25, -96);
INSERT INTO cities (city, lat, long) VALUES ("Atlanta", 33.755, -84.39);
INSERT INTO cities (city, lat, long) VALUES ("Lincoln", 40.809722, -96.675278);
INSERT INTO cities (city, lat, long) VALUES ("Des Moines", 41.590833, -93.620833);
INSERT INTO cities (city, lat, long) VALUES ("Memphis", 35.1175, -89.971111);
INSERT INTO cities (city, lat, long) VALUES ("Minneapolis", 44.95, -93.2);
INSERT INTO cities (city, lat, long) VALUES ("Washington DC", 38.895111, -77.036667);
INSERT INTO cities (city, lat, long) VALUES ("Cincinnati", 39.133333, -84.5);
INSERT INTO cities (city, lat, long) VALUES ("New York City", 40.716667, -74);
INSERT INTO cities (city, lat, long) VALUES ("Salt Lake City", 40.75, -111.883333);
INSERT INTO cities (city, lat, long) VALUES ("Chicago", 41.881944, -87.627778);
INSERT INTO cities (city, lat, long) VALUES ("Milwaukee", 43.052222, -87.955833);
INSERT INTO cities (city, lat, long) VALUES ("Cancun", 21.160556, -86.8475);
```

FIGURE 17.5. MySQL statements to input values into a table of a database.

version 5.0.16, MySQL supports a subset of spatial extensions to enable the generation, storage, and analysis of geographic features.

Spatial extensions refer to a SQL language that has been augmented with a set of commands that code geometry types such as points, lines, and areas. The specification describes a set of SQL geometry types, as well as functions on those types to create and analyze geometry values. The initial spatial extensions implemented by MySQL were only a subset of those proposed by OGC. More recent versions of MySQL incorporate more of these spatial extensions.

A geographic feature in SQL is anything in the world that can be specified by location. The following are all examples of geographic features:

- An entity, such as a mountain, a pond, or a city.
- A space, for example, a town district or the tropics.
- A definable location, such as a crossroad where two streets intersect.
- A location associated with a spatial reference system, which describes the coordinate space in which the object is defined.

The GEOMETRY spatial extension supports any type of point, line, or area feature. Other single-value types include POINT, LINESTRING, and POLYGON, which restrict their values to a particular geometry type. Still other data types hold collections of values, including MULTIPOINT, MULTILINESTRING, MULTIPOLYGON, and GEOMETRYCOLLECTION. GEOMETRYCOLLECTION can store a collection of objects of any type. The remaining collection types (MULTIPOINT,

Example of an SQL query:	Result of query:
SELECT city FROM cities WHERE long < -93.2	Omaha Lincoln Des Moines Salt Lake City

FIGURE 17.6. MySQL SELECT statement to manipulate values in a database. In this case, values of longitude that are west of Minneapolis, or less than –93.2, are selected.

MULTILINESTRING, MULTIPOLYGON) restrict collection members to those having a particular geometry type.

With the spatial extension, latitude and longitude are no longer input as separate fields, as in Figure 17.5. Figure 17.7 shows how these values are properly entered using the POINT spatial extension in MySQL.

17.6 PHP

After being born in Greenland and growing up in Denmark and Canada, Rasmus Lerdorf began writing his *Personal Home Page Tools* in the mid-1990s. At that point, he could not have imagined that his tools would eventually power so many websites. Nor could he have foreseen that PHP would compete so effectively with commercial solutions from Microsoft, Apple, and even the database giant, Oracle. His software is even used to run massive websites like Wikipedia and Facebook.

PHP is dependent on a server, and it is possible to convert almost any computer into a web server with PHP support (Ullman 2011). The major free and open-source (FOSS) software components such as the Linux operating system, the Apache web server, MySQL, and PHP are referred to by the acronym LAMP and can be downloaded and installed in unison. A similar package called WAMP, without an operating system, is available for Windows. MAMP is the corresponding version for the Macintosh, although the Apache web server, PHP, and a version of MySQL called SQLite are already integrated within the Macintosh operating system that is based on UNIX. SQLite is a popular database choice for small to medium-sized websites.

Like JavaScript, PHP can be embedded within an HTML file. However, in order to execute, the file with PHP code must reside on a server. Most commercial web servers offer PHP support, and some provide a limited service for no cost. Figure 17.8 uses PHP to write "Hello World." In contrast to client-side JavaScript, PHP generates HTML that is then sent to the client. The client's browser receives the results of running the script. The only indication that PHP generated the HTML are those three letters in the http address. When the web server is configured to process all HTML files with the PHP processor before they are sent to the client, then there is no indication that PHP is involved in creating the web page.

In parsing a file, PHP looks for opening ("<?php") and closing tags ("?>") that

```
INSERT INTO cities (city, pt) VALUES ("Omaha", GeomFromText('POINT(41.25 -96)'));
INSERT INTO cities (city, pt) VALUES ("Atlanta", GeomFromText('POINT(33.755 -84.39)'));
INSERT INTO cities (city, pt) VALUES ("Lincoln", GeomFromText('POINT(40.809722 -96.675278)'));
INSERT INTO cities (city, pt) VALUES ("Des Moines", GeomFromText('POINT(41.590833 -93.620833)'));
INSERT INTO cities (city, pt) VALUES ("Memphis", GeomFromText('POINT(35.1175 -89.971111)'));
INSERT INTO cities (city, pt) VALUES ("Minneapolis", GeomFromText('POINT(44.95 -93.2)'));
INSERT INTO cities (city, pt) VALUES ("Washington DC", GeomFromText('POINT(38.895111 -77.036667)'));
INSERT INTO cities (city, pt) VALUES ("Cincinnati", GeomFromText('POINT(39.133333 -84.5)'));
INSERT INTO cities (city, pt) VALUES ("New York City", GeomFromText('POINT(40.716667 -74)'));
INSERT INTO cities (city, pt) VALUES ("Salt Lake City", GeomFromText('POINT(40.75 -111.883333)'));
INSERT INTO cities (city, pt) VALUES ("Chicago", GeomFromText('POINT(41.881944 -87.627778)'));
INSERT INTO cities (city, pt) VALUES ("Milwaukee", GeomFromText('POINT(43.052222 -87.955833)'));
INSERT INTO cities (city, pt) VALUES ("Cancun", GeomFromText('POINT(21.160556 -86.8475)'));
```

FIGURE 17.7. MySQL statements to input latitude and longitude using the POINT SQL spatial extension.

```
<html>
 <head>
  <title>PHP Test</title>
 </head>
 <body>
 <?php echo '<p>Hello World</p>'; ?>
 </body>
</html>
```

FIGURE 17.8. Like JavaScript, PHP can be embedded within an HTML file but it can only execute on a server.

define the beginning and ending of the PHP code. Whatever exists between these delimiters is processed by PHP. These tags can be within an HTML document, as above, or they can define an entire file. In this case, PHP would write all of the HTML code.

Variable types are almost identical to JavaScript and similarly support booleans, integers, floating-point numbers, text strings, arrays, and objects. The major difference is that all variables in PHP begin with a $, as in $firstname. Figure 17.9 shows how variables are defined.

PHP includes similar control structures to JavaScript that support for, if, elseif, and while. Figure 17.10 demonstrates these different control structures with PHP.

PHP can be used to support multiple browser sessions in which data is preserved between subsequent accesses. It can upload single or multiple files. PHP even supports cookies, a mechanism to store data in a remote browser. Cookies can be set using the setcookie() and setrawcookie() functions. Setcookie() is done in the HTML header before anything else is sent to the browser. A time delay can be set in case the browser page needs to be adjusted based on what is found in the cookie data. This causes a delay in the display of the web page.

Both PHP and MySQL include comprehensive security measures that control access to data. Once access is established, controls can be set to allow only specific types of changes. User privileges can range from read-only to full editing capabilities. Access privileges can be restricted to the host on which the database resides.

```
<?php
  $myAge = 16;                    // a PHP Integer
  $myHeight = 6.5;                // a PHP Floating Point
  $haveHair = true;               // a PHP Boolean
  $greeting = "What's up?"        // a PHP String
  $carsOwned = new carsOwned();   // a PHP Array
?>
```

FIGURE 17.9. The definition of variables with PHP. All variables in PHP begin with a "$".

For	If
```for ($i = 1; $i <= 10; $i++){ echo $i;}```	```<?phpif ($a > $b) echo "a is bigger than b";?>```
Elseif	While
```<?phpif ($a > $b) { echo "a is bigger than b";} elseif ($a == $b) { echo "a is equal to b";} else { echo "a is smaller than b";}?>```	```<?php$i = 1;while ($i <= 10): echo $i; $i++;endwhile;?>```

FIGURE 17.10. Examples of PHP control structures of for, if, elseif, and while.

17.7   PostgresSQL and PostGIS

Oracle, the largest database company in existence, now owns MySQL. While the company has stated that it will keep MySQL open and free, many open-source developers are switching to another database package called PostgreSQL or Postgres for short (PostgreSQL Global Development Group 2010). In many ways, the two databases are very similar as both are based on SQL. At one time, MySQL was considered to be faster but with fewer features, while PostgreSQL was seen as slower but with more features. With improvements in both databases, these characterizations are no longer accurate. MySQL has added functions, such as spatial queries, and PostgreSQL has dramatically improved its speed. PostgreSQL with the PostGIS extension is more SQL compliant, while MySQL has greater support by Internet service providers. Perhaps the most important consideration is that while PostgreSQL has many sponsors and developers, it is not controlled by any one company—especially one as influential as Oracle.

PostGIS adds support for geographic objects to PostgreSQL (PostGIS 2011). In effect, PostGIS "spatially enables" the PostgreSQL server, allowing it to be used as a backend spatial database for geographic information systems (GIS). PostGIS follows the OpenGIS *Simple Features Specification for SQL* and has been certified as compliant with the Types and Functions profile (PostGIS 2011). As a more specialized service, PostGIS is not commonly offered by web-hosting sites, especially those offering minimal services at no cost.

17.8   Summary

Databases and database queries are the foundation of nearly everything on the Internet, including maps. Every time we use the Internet, we are conducting a query of a database on a remotely located server. The Internet may be seen as simply a medium to support interaction with remote databases.

The relational database is the most often used type of database. Based on work done by IBM in the 1970s, a language developed to input, manipulate, and extract information from multiple related tables. This language, SQL, is the way information is processed with a relational database. MySQL can be freely downloaded and has emerged as the most commonly used database. It has been adopted by major online sites like Facebook and Amazon, although these companies have modified the program for their own purposes.

Spatial extensions proposed by the Open Geospatial Consortium (OGC) have been incorporated within more recent versions of MySQL. The PostGIS extension to the PostgreSQL database has a more complete implementation of the OGC spatial primitives.

17.9   Questions

1. What are the relative advantages and disadvantages of PHP and JavaScript?

2. What is the relationship between MySQL and PHP?

3. Consult the W3Techs website for the current usage of server-side languages such as PHP compared to other preprocessors.

4. What are some advantages and disadvantages of the free and open-source software movement?

5. What are the origins of SQL? Describe the basic SELECT statement.

6. How is the WHERE clause used in manipulation of the database?

7. Describe primary and foreign keys in SQL.

8. What is OGC and Spatial SQL? Describe different spatial extensions to SQL.

9. How does Spatial SQL change the way latitude and longitude values are handled?

10. What are cookies, and how does PHP support their use?

17.10   References

DuBois, Paul (2009) *MySQL* 4th Ed. Upper Saddle River, NJ: Addison-Wesley.

PostGIS (2011) What Is PostGIS? [http://postgis.refractions.net/]

PostgreSQL Global Development Group (2010) PostgreSQL 9.0 Reference Manual— Volume 2: Programming Guide. UK: Network Theory Limited.

Ullman, Larry (2011) *PHP for the Web: Visual QuickStart Guide* 4th Ed. Berkeley, CA: Peachpit Press.

W3Techs (2013) Usage of server-side programming languages for websites. [http://w3techs.com/technologies/overview/programming_language/all]

CHAPTER 18

Mapping from a Database

A lot of that momentum comes from the fact that Linux is free.

—Nat Friedman

18.1 Introduction

Most websites are based on the Linux operating system (W3Techs 2013). The Linux movement began in the early 1990s as an alternative to the dominant and proprietary Windows operating system from Microsoft. In August 1991, a 21-year-old Finnish programmer by the name of Linus Torvalds posted the following message on a computer bulletin board:

> Hello everybody out there! I'm doing a (free) operating system . . . (just a hobby, won't be big and professional) for 386(486) AT clones. This has been brewing since april, and is starting to get ready. I'd like any feedback. (Deek and McHugh, p. 90)

The resulting operating system grew to eventually dominate the server computer industry. Known for its security and reliability, Linux is ideal for uninterrupted operation. A major advantage of Linux is that it is a free and an open source, and not tied to a particular company.

Today, Linux is available in many versions including Ubuntu (Thomas and Channelle 2009), Fedora (Tyler 2007), OpenSuse (Whittaker and Davies 2008), Debian (Kraft 2005), and FreeBSD (Hong 2008). Any of these versions can be downloaded without cost and set up to run on almost any computer, even a laptop. The process may require the creation of a CD or DVD used for installing a minimal system that then downloads and installs all of the other components.

Linux is based on a command-line interface (CLI) where the user types commands rather than pointing at icons using a mouse or selecting items from menus.

Although a variety of graphical user interfaces (GUIs) have been added to accommodate those who prefer the point-and-click interface, Linux is intended more for programmers than for ordinary computer users. Linux operating systems dominate server installations, but it is not widely used as a desktop operating system except in parts of Europe and South America where governments have actively sought to limit the use of proprietary operating systems like Windows. The most common version of Linux, Ubuntu, accounts for less than 2% of all desktop computers (search: Ubuntu market share). Europe has the largest percentage of users.

In contrast, almost all online web-hosting sites are based on Linux and other open-source software, including PHP and MySQL. These utilities are made available through another open-source project called the cPanel. The most important of these utilities for the administration of the MySQL database is phpMyAdmin, an open-source utility written in PHP. MySQL, PHP, and all of the utilities available through cPanel are programmed and maintained by a loosely associated group of programmers in different parts of the world.

This chapter describes how a MySQL database is created, how data is entered and queried, and how the results of a query can be mapped. We begin by examining the PHP scripting language.

18.2   HTML and JavaScript with PHP

PHP is primarily used as a server-side scripting language for creating HTML pages. Up to this point, we have hard-coded HTML pages; that is, the pages are created in some type of editor and subsequently made available to users. The pages remain unchanged until we decide to edit them. Most web pages that we view are created on the fly as a specific piece of information is requested. PHP is used both to extract information from a database and write the HTML code that is served to the client.

The example in Figure 18.1 uses a series of echo statements to produce an entire HTML file that places text in the window header and writes a line of text.

The second example, Figure 18.2, writes both the HTML and the JavaScript function that calculates the square of a number.

The third example demonstrates the ability of PHP to concatenate strings and a variable. PHP uses a period to combine strings and variables (see Figure 18.3). The text string ($text) and a variable ($number) are put together in the combined $msg string.

A loop is implemented with PHP in Figure 18.4. The variable $j is initially defined with a value of 10. A loop adds $j to itself five times, resulting in a final value of 320.

The next example demonstrates the use of the while statement to create a loop (see Figure 18.5). The variable $j is repeatedly added to $myCount until its value is less than or equal to 50. A while loop is used when a particular condition needs to be met before the loop stops.

The example in Figure 18.6 uses an if statement to evaluate a condition, in this case, whether $a is greater than $b. The else condition handles the alternate condition.

Figure 18.7 calls the PHP random number generator. The rand function is set

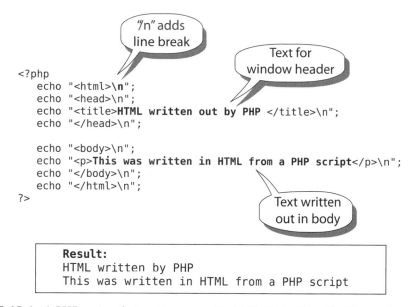

```php
<?php
    echo "<html>\n";
    echo "<head>\n";
    echo "<title>HTML written out by PHP </title>\n";
    echo "</head>\n";

    echo "<body>\n";
    echo "<p>This was written in HTML from a PHP script</p>\n";
    echo "</body>\n";
    echo "</html>\n";
?>
```

Result:
HTML written by PHP
This was written in HTML from a PHP script

FIGURE 18.1. A PHP script that writes an entire HTML file. The "\n" is used to add a line feed to make the code more readable when viewing the source code.

to return a random number between 0 and 10. An if statement then evaluates whether the number is greater than or equal to 5. Next, a loop executes 1000 times and adds the value of the random number to the variable $total. This number is then divided by 1000. The number should be close to 5, and the more times the random number is called, the closer it should get to 5.

```php
<?php
    echo "<html>";
    echo "<head>";
    echo "<SCRIPT LANGUAGE=\"JavaScript\">";
    echo "function square(number) { ";
    echo "    return number*number} ";
    echo "    </SCRIPT> ";
    echo "</head>";
    echo "<body>";
    echo "<SCRIPT> document.write('The square of 16 is ', square(16),'.') </SCRIPT>";
    echo "</body>";
    echo "</html> ";
?>
```

Result:
The square of 16 is 256.

FIGURE 18.2. A PHP script that writes an entire HTML file with an embedded JavaScript function. The "\n" is no longer used to add a line feed. The resulting source code listing does not separate the code on separate lines.

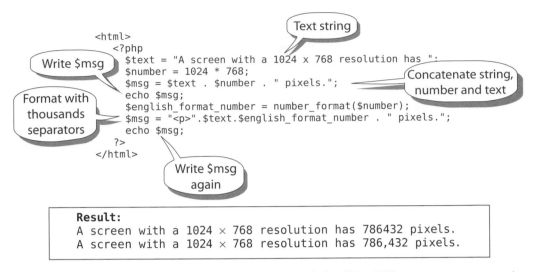

```
<html>
    <?php
    $text = "A screen with a 1024 x 768 resolution has ";
    $number = 1024 * 768;
    $msg = $text . $number . " pixels.";
    echo $msg;
    $english_format_number = number_format($number);
    $msg = "<p>".$text.$english_format_number . " pixels.";
    echo $msg;
    ?>
</html>
```

Text string

Write $msg

Concatenate string, number and text

Format with thousands separators

Write $msg again

```
Result:
A screen with a 1024 × 768 resolution has 786432 pixels.
A screen with a 1024 × 768 resolution has 786,432 pixels.
```

FIGURE 18.3. The example demonstrates the use of the "." in PHP to concatenate a series of strings and variables into the single $msg string variable.

The example in Figure 18.8 calls the PHP date function in two different ways. The date function has many different parameters that control how the date is returned (search: php date function parameters). The function is commonly used to insert the current date on a web page.

18.3   Creating the Online Database

Creating an online database is a two-step process. The first step is to define a database using MySQL. Second, the phpMyAdmin utility is used to add data to the new database. Both tools are available through the cPanel (see Figure 18.9).

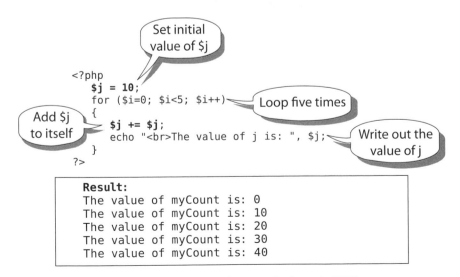

Set initial value of $j

```
<?php
    $j = 10;
    for ($i=0; $i<5; $i++)
    {
        $j += $j;
        echo "<br>The value of j is: ", $j;
    }
?>
```

Loop five times

Add $j to itself

Write out the value of j

```
Result:
The value of myCount is: 0
The value of myCount is: 10
The value of myCount is: 20
The value of myCount is: 30
The value of myCount is: 40
```

FIGURE 18.4. The example demonstrates the use of a loop in PHP.

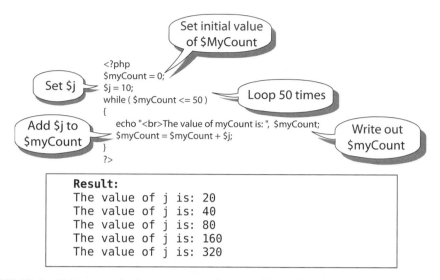

FIGURE 18.5. This example demonstrates the use of the while condition. The loop runs as long as $myCount is less than or equal to 50.

A new database is defined in MySQL by providing a name and a password (see Figure 18.10). Each MySQL database name begins with the account (in this case, a4768098), followed by the user-specified name. The account owner is the only person with access to the database. Providing the account and password to another user allows that user to access the database. PHP also provides a four-line summary of information about the database. These four lines create the connection for PHP to interact with the database.

Administering the MySQL databases is done using phpMyAdmin (see Figure

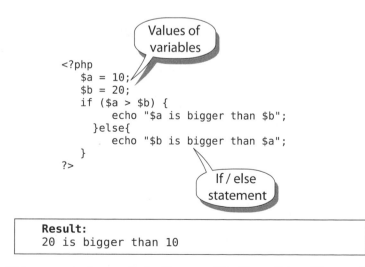

FIGURE 18.6. This example demonstrates the use of the if condition. The condition evaluates whether the variable $a is greater than $b.

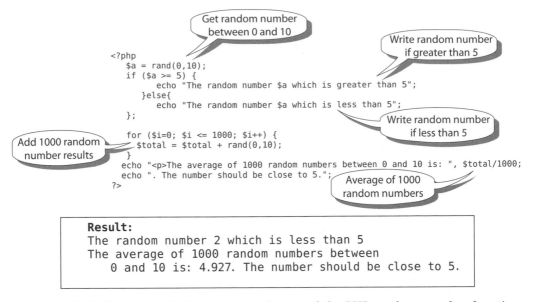

FIGURE 18.7. This example demonstrates the use of the PHP random number function. The if statement evaluates the variable $a to determine if it is greater than 5. The results from 1000 random calls are then averaged.

FIGURE 18.8. Two PHP statements within HTML that provide the current date and time in different formats.

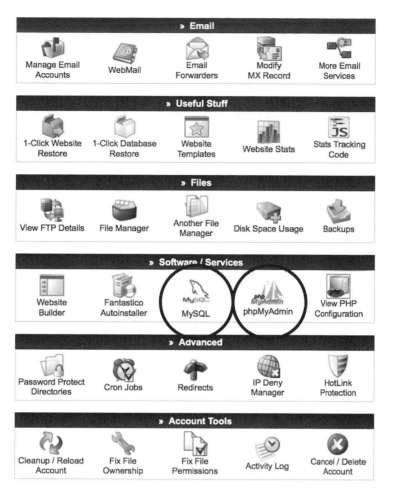

FIGURE 18.9. A standard web-hosting control panel, called cPanel, gives access to the different tools, including MySQL and phpMyAdmin. Copyright 2013 First-Class Web Hosting.

18.11). Written in PHP, this open-source program supports all types of database administration, including the importing of data and the creation, renaming, and copying of files. phpMyAdmin can also be used to query the database.

Although data can be entered directly using phpMyAdmin, a better alternative is to import data from an external file defined in SQL. Figure 18.12 shows a partial listing of an SQL file containing the location of all 50 U.S. capital cities. The top lines create a table called us_capitals, which contains four variables for each capital—name, usstate, population, and location. Location is defined with two variables, latitude and longitude, but is input as a single point entity. This is an example application of an OGC MySQL spatial extension. The following lines use the SQL command INSERT INTO to enter the data for all 50 cities into the table. The SQL file is imported by simply browsing to and selecting the file.

Once imported, various phpMyAdmin options can be used to examine the data and structure of the table. Figure 18.13 presents a partial listing of the data

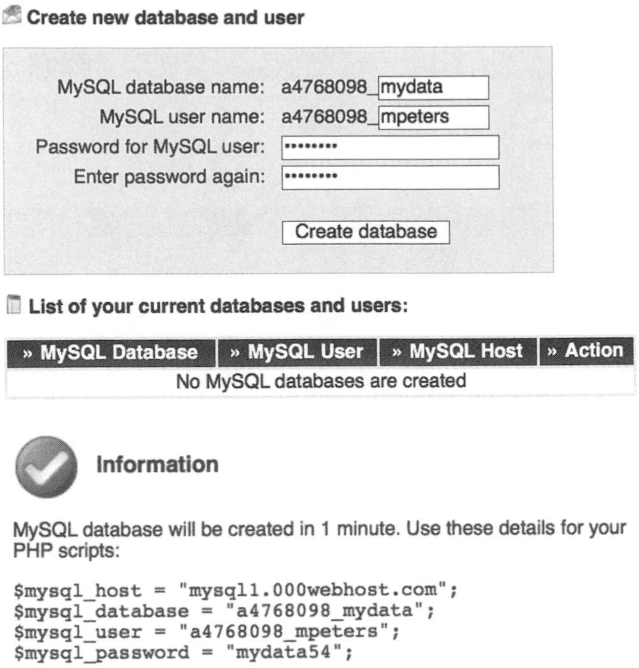

Create new database and user

MySQL database name:	a4768098_mydata
MySQL user name:	a4768098_mpeters
Password for MySQL user:	●●●●●●●●
Enter password again:	●●●●●●●●

Create database

List of your current databases and users:

» MySQL Database	» MySQL User	» MySQL Host	» Action
No MySQL databases are created			

Information

MySQL database will be created in 1 minute. Use these details for your PHP scripts:

```
$mysql_host = "mysql1.000webhost.com";
$mysql_database = "a4768098_mydata";
$mysql_user = "a4768098_mpeters";
$mysql_password = "mydata54";
```

FIGURE 18.10. The top illustration shows the creation of a MySQL database called mydata. The bottom illustration is the information to access the database through PHP. Copyright 2013 First-Class Web Hosting.

List of current databases:

» MySQL Database	» MySQL User	» phpMyAdmin
a4768098_mydata	a4768098_mpeters	Enter phpMyAdmin

Server: localhost ▶ Database: a4768098_mydata

Structure | SQL | Search | Query | Export | Import | Operations

Import

File to import

Location of the text file _____ Browse... (Max: 2,048 KiB)

Character set of the file: utf8

Imported file compression will be automatically detected from: None, gzip, bzip2, zip

Partial import

☑ Allow interrupt of import in case script detects it is close to time limit. This might be good way to import large files, however it can break transactions.

Number of records(queries) to skip from start 0

Format of imported file

⊙ SQL

Options

SQL compatibility mode NONE

Go

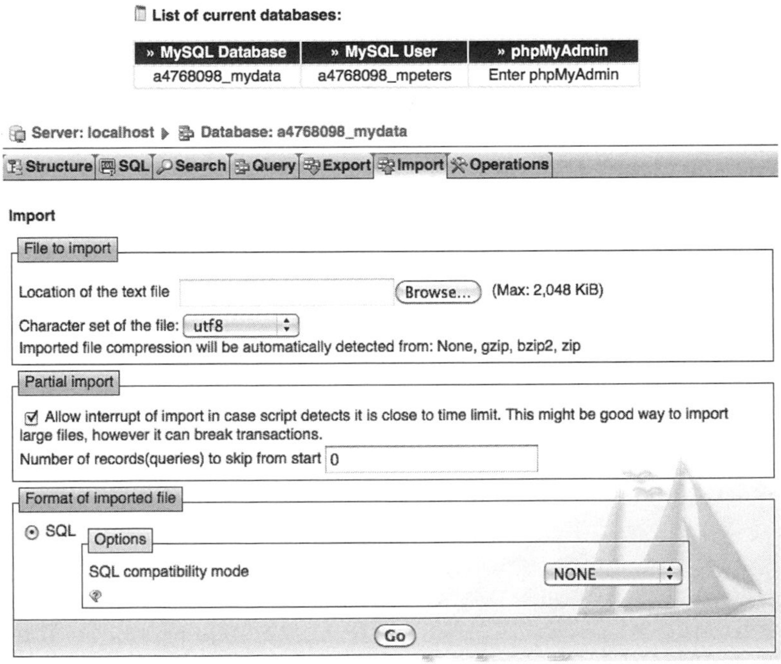

FIGURE 18.11. The phpMyAdmin listing of the current databases (top) and import window (bottom). Copyright 2013 First-Class Web Hosting.

342

```
create table us_capitals (
    name VARCHAR(30) PRIMARY KEY,
    usstate VARCHAR(5),
    population INT(10),
    location Point NOT NULL,
    SPATIAL INDEX(location)
);

INSERT INTO us_capitals (name, usstate, population, location) VALUES
    ( 'Montgomery', 'AL', 469268 , GeomFromText('POINT(32.377447 -86.300942)'));
INSERT INTO us_capitals (name, usstate, population, location) VALUES
    ( 'Juneau', 'AK', 30987 , GeomFromText('POINT(58.302197 -134.410467)'));
INSERT INTO us_capitals (name, usstate, population, location) VALUES
    ( 'Phoenix', 'AZ', 4039182 , GeomFromText('POINT(33.448097 -112.097094)'));
...

...
INSERT INTO us_capitals (name, usstate, population, location) VALUES
    ( 'Charleston', 'WV', 305526 , GeomFromText('POINT(38.333056 -81.613889)'));
INSERT INTO us_capitals (name, usstate, population, location) VALUES
    ( 'Madison', 'WI', 543022 , GeomFromText('POINT(43.074444 -89.384722)'));
INSERT INTO us_capitals (name, usstate, population, location) VALUES
    ( 'Cheyenne', 'WY', 85384 , GeomFromText('POINT(41.140278 -104.819722)'));
```

Create new MySQL table with these fields

Insert data into fields

Lat and long put into OGC point

FIGURE 18.12. SQL commands to create a database on US capitals. Four variables are being placed in a single table, including the name of the city, the corresponding state, its population, and its location in latitude and longitude. The POINT variable is an OGC spatial construct and consists of both numbers.

using the Browse tab after it has been sorted based on the primary key. Location, an OGC POINT entity, is in binary form and cannot be displayed.

The variables defined in the table can be examined with the Structure tab (see Figure 18.14). The actions associated with each variable include browsing distinct values (columns), changing or deleting the data values, and changing the primary key.

The phpMyAdmin SQL tab allows for data query. Figure 18.15 shows the results

FIGURE 18.13. The phpMyAdmin Browse window for the us_capitals table. The table has been sorted based on "name," the primary key. The location variable, a combination of latitude and longitude, is stored in binary and cannot be displayed. Copyright 2013 First-Class Web Hosting.

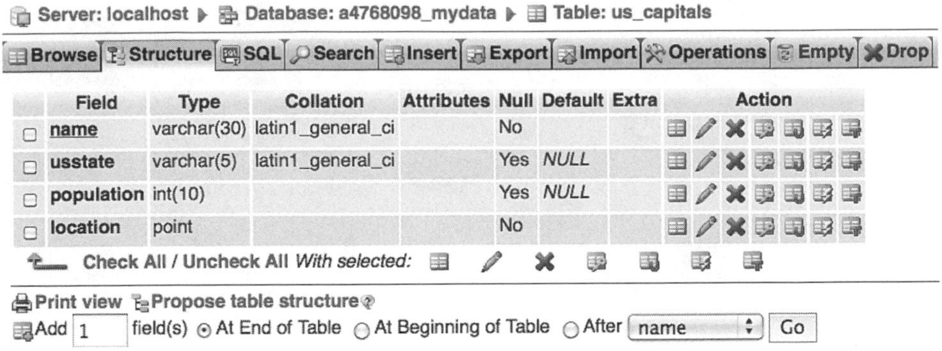

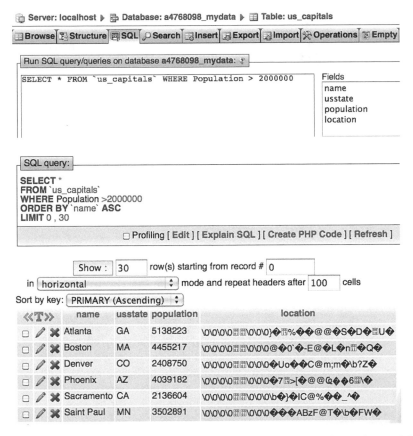

FIGURE 18.14. The phpMyAdmin Structure window for the us_capitals table. Copyright 2013 First-Class Web Hosting.

FIGURE 18.15. A query of the database is possible with the SQL tab in phpMyAdmin. The SQL defined query is selecting all state capitals that have a population in excess of 2 million. The six cities that meet the criteria are listed in the bottom illustration, sorted by name in ascending order. Copyright 2013 First-Class Web Hosting.

of a query that selects all population values greater than 2 million. The resulting list has been sorted in ascending order by city name. The same result can be obtained through the Search tab that offers a more structured query approach (see Figure 18.16).

The Insert tab is also very useful, allowing for the addition of new rows to the table. In contrast, the Export tab is used to extract data and this is possible in a wide variety of formats including CSV, Excel, Word, PDF, SQL, and XML. The Operations tab allows the table to be renamed or otherwise altered.

18.4   Connecting to the Database

To this point, a MySQL database has been defined containing a single table. Operations from this point forward require that a connection to the table be established. This connection is accomplished by referencing a file that includes the information on the MySQL table. Figure 18.17 depicts the `mysql_connect.php` file that includes the names of the MySQL host, database, user, and password. MySQL provided this information at the point when the database was being created (see Figure 18.9). The last line establishes $mysql_database as the name of the database. Any PHP reference to the MySQL database uses the information defined in this file.

18.5   Querying the Database

Figure 18.18 shows the PHP code for reading the data and mapping the points. A marker and infowindow have been defined for each data point. Note how the Google Maps API code is being "echoed" to create the correct JavaScript code for the map. A while statement is then used to process all rows in the table.

Different types of queries can be performed to map a subset of the points. In Figure 18.19, only the capitals north of Omaha with a population less than 500,000 are mapped. This is done with the `$sql = "SELECT name` command that selects only those records from the table that have a latitude greater than 41.25 and the specified population. Many different types of queries could be performed in this way.

In addition to incorporating different geometry type features, the MySQL spatial extension also allows queries on these features. The example in Figure 18.20 illustrates how only state capitals within a specified rectangle are selected. The

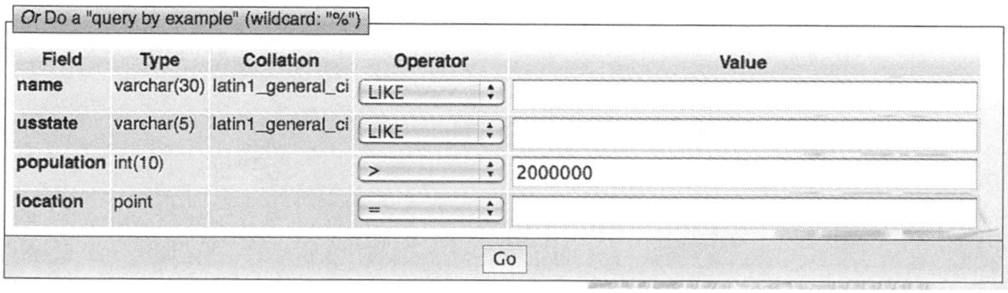

Field	Type	Collation	Operator	Value
name	varchar(30)	latin1_general_ci	LIKE	
usstate	varchar(5)	latin1_general_ci	LIKE	
population	int(10)		>	2000000
location	point		=	

Or Do a "query by example" (wildcard: "%")

Go

FIGURE 18.16. The Search tab in phpMyAdmin is an alternative method for querying the table. Copyright 2013 First-Class Web Hosting.

```php
<?php
    $mysql_host = "mysql1.000webhost.com";
    $mysql_database = "a4768098_mydata";
    $mysql_user = "a4768098_mpeters";
    $mysql_password = "mydata54";
    $conn = mysql_connect($mysql_host, $mysql_user, $mysql_password) or die ('Error connecting to
mysql');
    $mysql_select_db($mysql_database);
?>
```

FIGURE 18.17. The `mysql_connect.php` file that defines the location of the MySQL database, the user, and the associated password. This defines the connection within PHP to the database.

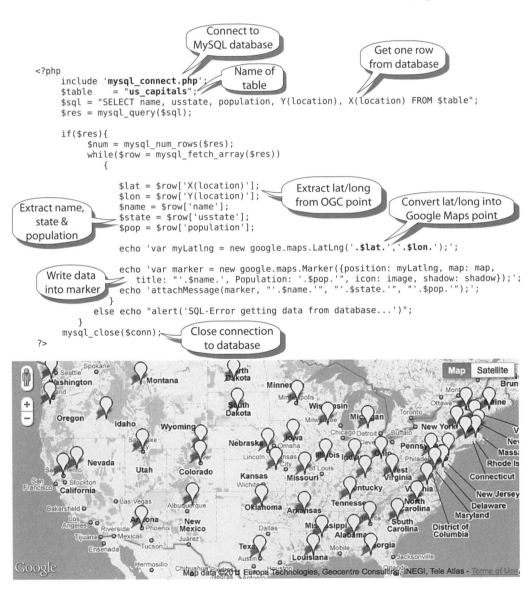

FIGURE 18.18. PHP and Google Maps API code to read and map the U.S. capitals. Copyright 2013 Google.

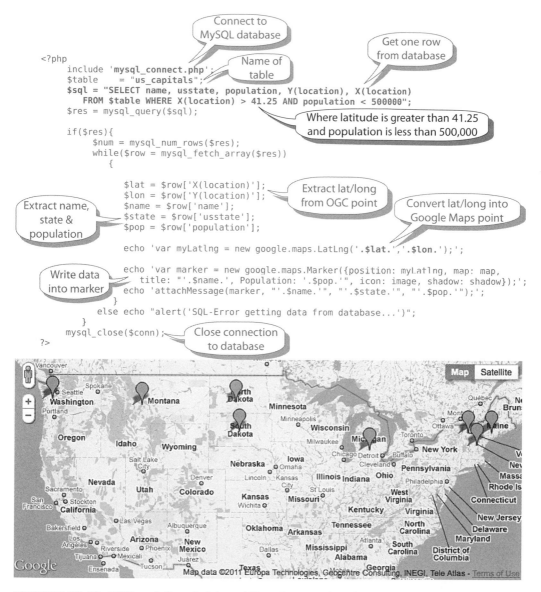

FIGURE 18.19. PHP and Google Maps API code to read and map the U.S. capitals that are north of Omaha with a population less than 500,000. Copyright 2013 Google.

WHERE Intersects query shows how a query can be extended to the spatial domain.

18.6   Querying a Database of Lines

Lines are another spatial entity defined with two points in a MySQL database. In this example, a database of lines representing flight routes is defined using multiple

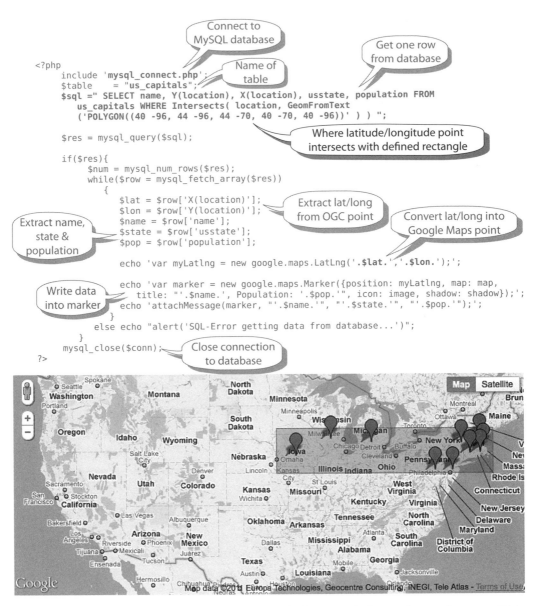

FIGURE 18.20. Example of a spatial query. Only those capitals are shown that are within the specified rectangle. Copyright 2013 Google.

tables. Figure 18.21 shows the portions of the three tables—cities, dl_airports, and dl_routes—that are entered using SQL statements. The data represent selected flight routes for Delta Airlines.

Data for these tables are added to the existing MySQL database using the php-MyAdmin Import tab (see Figure 18.22). Three separate tables are imported. The first defines the locations of the cities, and the second their associated three-letter airport codes. The third table defines the Delta Airlines connections between these cities.

```
DROP TABLE IF EXISTS cities;

create table cities (
        city VARCHAR(30),
        location GEOMETRY NOT NULL,
        SPATIAL INDEX(location),
        PRIMARY KEY (city)
);

INSERT INTO cities (city, location) VALUES ("Omaha", GeomFromText('POINT(41.25 -96)'));
INSERT INTO cities (city, location) VALUES ("Atlanta", GeomFromText('POINT(33.755 -
84.39)'));
INSERT INTO cities (city, location) VALUES ("Lincoln", GeomFromText('POINT(40.809722 -
96.675278)'));


DROP TABLE IF EXISTS dl_airports;

create table dl_airports (
        city VARCHAR(30),
        airport VARCHAR(30),
        code VARCHAR(3),
        FOREIGN KEY (city) REFERENCES cities(city),
        PRIMARY KEY (code)
);

INSERT INTO dl_airports (city, airport, code) VALUES ("Omaha","Omaha Eppley Airfield",
"OMA");
INSERT INTO dl_airports (city, airport, code) VALUES ("Atlanta","Hartsfield-Jackson
International Airport", "ATL");
INSERT INTO dl_airports (city, airport, code) VALUES ("Lincoln","Municipal Airport",
"LNK");


DROP TABLE IF EXISTS dl_routes;

create table dl_routes (
        airportCode VARCHAR(3),
        destinationCode VARCHAR(3),
        FOREIGN KEY (airportCode) references dl_airports(code),
        FOREIGN KEY (destinationCode) references dl_airports(code)
);

INSERT INTO dl_routes (airportCode, destinationCode) VALUES ("OMA","ATL");
INSERT INTO dl_routes (airportCode, destinationCode) VALUES ("OMA","DET");
INSERT INTO dl_routes (airportCode, destinationCode) VALUES ("OMA","MEM");
```

FIGURE 18.21. Part of the SQL code for entering three tables that provide city location, airport information, and airline connections.

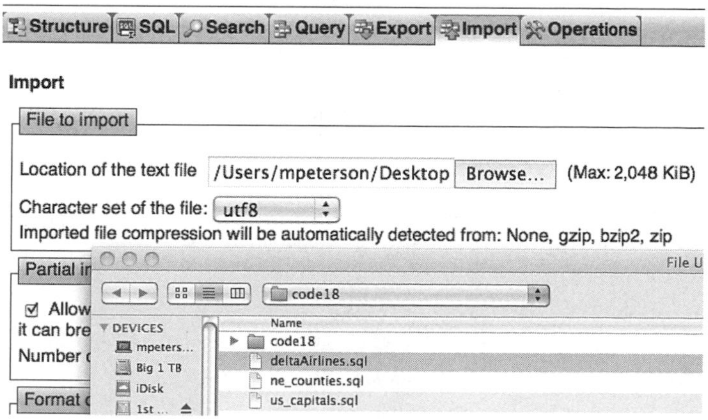

FIGURE 18.22. Importing the deltaAirlines.sql file into the MySQL database. The resulting MySQL database would consist of three separate tables that define the cities, the airports, and the Delta Airlines connections between them. Copyright 2013 First-Class Web Hosting.

A variety of query options exist for this MySQL database. For example, a query could ask for all the flight routes defined in the database, as shown in Figure 18.23. For this query, the AS option is applied in the SELECT command to rename variables to be used in the search. At the end of this command, we have defined the variables `departure`, `latDep`, `lngDep`, `destination`, `latDest`, and `lngDest`. Those field names are then used in a series of INNER JOIN commands to select the desired elements for mapping.

SQL INNER JOIN is used to relate fields between the three tables. It is the most common JOIN operation done with SQL. It compares rows between two tables to find all pairs that satisfy the JOIN condition. The command returns rows where there is at least one match between tables. A new "result table" is created by combining the column values of the two tables.

Figure 18.24 shows how point data is extracted from the route table using four INNER JOIN commands. The example selects only those connections from Omaha to other cities. The first INNER JOIN converts OMA to Omaha. The second matches ATL to Atlanta. The third finds the latitude and longitude for Omaha, and the fourth does the same for Atlanta. Figure 18.25 shows the results for all connections from Omaha.

FIGURE 18.23a. Part of the SQL code for entering three tables that provide city location, airport information, and airline connections. Copyright 2013 Google.

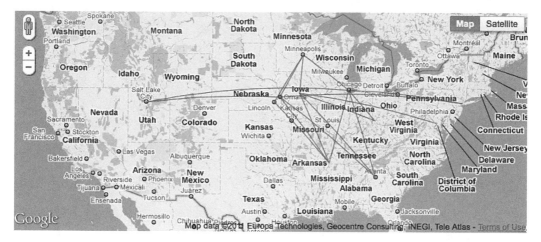

FIGURE 18.23b.

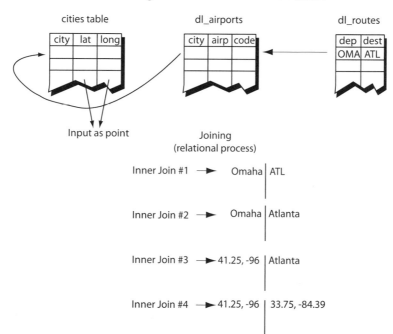

FIGURE 18.24. Three tables are defined for the flight routes. The dl_routes table shows flight connection pairs between the cities. A series of INNER JOIN commands relates these codes to the city in the dl_airports table, which then relates to the latitude and longitude of each city in the cities table.

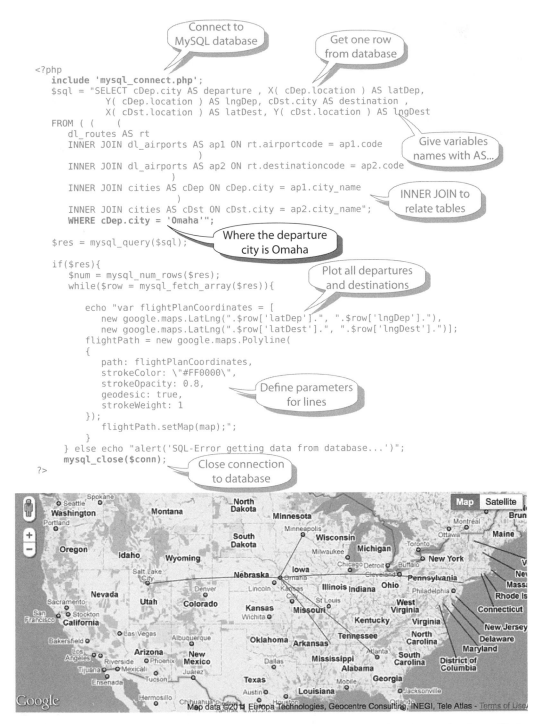

```php
<?php
  include 'mysql_connect.php';           // Connect to MySQL database
  $sql = "SELECT cDep.city AS departure , X( cDep.location ) AS latDep,
            Y( cDep.location ) AS lngDep, cDst.city AS destination ,
            X( cDst.location ) AS latDest, Y( cDst.location ) AS lngDest   // Get one row from database
  FROM ( (
      dl_routes AS rt
      INNER JOIN dl_airports AS ap1 ON rt.airportcode = ap1.code
            )
      INNER JOIN dl_airports AS ap2 ON rt.destinationcode = ap2.code       // Give variables names with AS...
            )
      INNER JOIN cities AS cDep ON cDep.city = ap1.city_name
            )
      INNER JOIN cities AS cDst ON cDst.city = ap2.city_name";             // INNER JOIN to relate tables
      WHERE cDep.city = 'Omaha'";                                          // Where the departure city is Omaha

  $res = mysql_query($sql);

  if($res){                                                                // Plot all departures and destinations
    $num = mysql_num_rows($res);
    while($row = mysql_fetch_array($res)){

      echo "var flightPlanCoordinates = [
        new google.maps.LatLng(".$row['latDep'].", ".$row['lngDep']."),
        new google.maps.LatLng(".$row['latDest'].", ".$row['lngDest'].")];
      flightPath = new google.maps.Polyline(
      {
        path: flightPlanCoordinates,
        strokeColor: \"#FF0000\",
        strokeOpacity: 0.8,
        geodesic: true,                                                    // Define parameters for lines
        strokeWeight: 1
      });
        flightPath.setMap(map);";
      }
    } else echo "alert('SQL-Error getting data from database...')";
    mysql_close($conn);                                                    // Close connection to database
  ?>
```

FIGURE 18.25. PHP and MySQL code to query a database by selecting only those flight routes that depart from Omaha. Copyright 2013 Google.

```
create table ne_counties (
        strokecolor VARCHAR(7),
        strokewidth INT(5),
        strokeopacity FLOAT(5),
        fillcolor VARCHAR(7),
        fillopacity FLOAT(5),
        popdata INT(15),
        name VARCHAR(30),
        geom GEOMETRY NOT NULL,
        SPATIAL INDEX(geom)
);

INSERT INTO ne_counties (strokecolor, strokewidth , strokeopacity , fillcolor, fillopacity ,
popdata, name,  geom) VALUES ("#008800",1,1.0,"#FFCC00",0.06674,33185,"county",
GeomFromText('POLYGON((40.698311157 -98.2829258865,40.698311157 -98.2781218448,40.3505519215 -
98.2781218448,40.3500181391 -98.3309663027,40.3504184759 -98.3344358884,40.3504184759 -
98.7238301514,40.6413298855 -98.7242304882,40.6897706386 -98.7244973794,40.6989783851 -
98.7243639338,40.6991118307 -98.7214281306,40.6985780482 -98.686198492,40.698311157 -98.2829258865,
40.698311157 -98.2829258865))'));
INSERT INTO ne_counties (strokecolor, strokewidth , strokeopacity , fillcolor, fillopacity ,
popdata, name,  geom) VALUES ("#008800",1,1.0,"#FFCC00",0.01334,6931,"county",
GeomFromText('POLYGON((41.9149346991 -98.2956032185,42.0888143169 -98.2954697729,42.0888143169 -
98.3004072602,42.3035282886 -98.3005407058,42.4369738893 -98.300140369,42.4377745629 -
97.8344152223,42.2326686746 -97.8352158959,42.0897484361 -97.8346821135,42.0347688486 -
97.8341483311,41.9164026008 -97.8332142119,41.9152015904 -98.0647423292,41.9149346991 -
98.2956032185, 41.9149346991 -98.2956032185))'));
```

FIGURE 18.26. MySQL commands, attributes, and coordinates for placing two county polygons for Nebraska into a MySQL database.

18.7   Querying a Polygon Database

Polygons can also be defined within a MySQL database as the following example demonstrates using the ne_counties.sql table. Figure 18.26 shows how the coordinates for the first two county polygons for the state of Nebraska are entered, including variables that define the color of the stroke and fill, and the population of each county. The population values could have been placed in a separate attribute table.

Figure 18.27 shows a map of Nebraska that simply displays the county polygons.

Figure 18.28 shows the mapping of population data. A range-graded opacity has been calculated to match the population of each county. This is accomplished by finding the minimum and maximum data values and calculating the corresponding opacity values between 0 and 1.

Figure 18.29 shows the selection of only those counties that have a population of less than 50,000 people—all but four counties. Two of these counties neighbor each other.

Figure 18.30 depicts the use of a spatial operator. Here, the location of three cities are defined, and the INTERSECTS operator finds the polygons to which those points belong.

18.8   Point Input from a Map

In previous sections, we have been placing points, lines, and polygons on the map from a MySQL database. We now examine how to input points from a map and store them in a database. The general process involves capturing the latitude and longitude from a click event, requesting an attribute for the point from the user

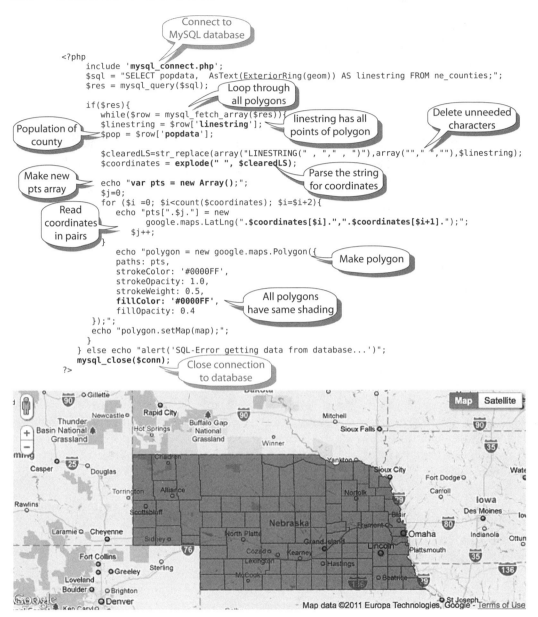

FIGURE 18.27. Nebraska county polygons mapped from a MySQL database. The PHP code shows how the polygon data is extracted and how Google Maps API codes are integrated to display the county polygons. Copyright 2013 Google.

(a text or numeric ID), and writing these to a MySQL database. Subsequently, the points are read from the MySQL file and plotted on the map.

A new point is added through an HTML form:

```
<form action="markers.php" method="GET">
   <div id="map_canvas" style="font-size: 9px;"></div>
   <div id="LongLat">
```

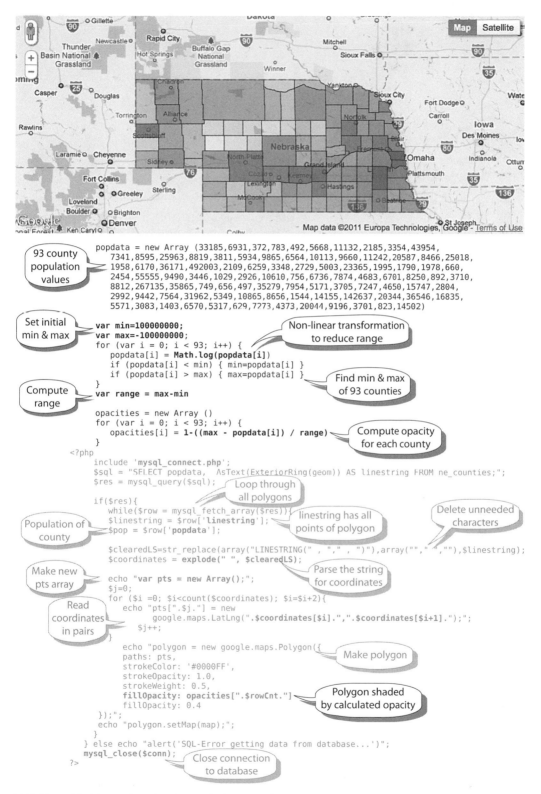

FIGURE 18.28. Nebraska choropleth map of population based on polygons from a MySQL database. Copyright 2013 Google.

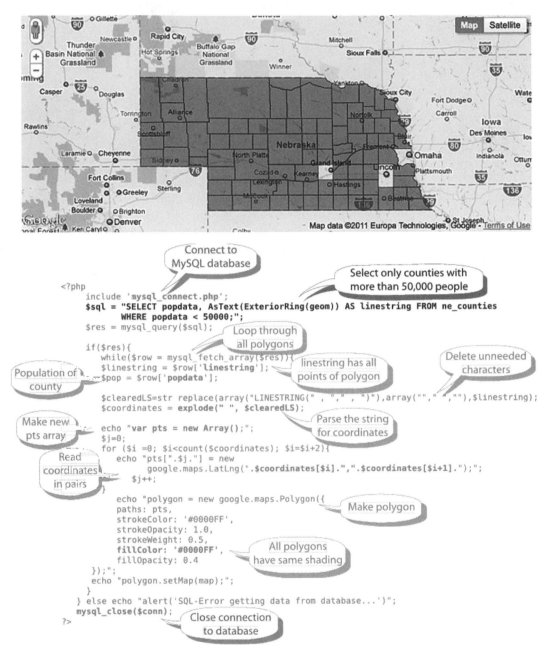

FIGURE 18.29. Results of a query of a MySQL database shading only those counties with less than 50,000 people. Copyright 2013 Google.

```html
<strong>Latitude:</strong>
  <input name="latitude" id="lat" maxlength="30">
<strong>Longitude:</strong>
  <input name="longitude" id="lng" maxlength="30">
<strong>Attribute:</strong>
  <div>  <input type="text" name="name" size="19" maxlength="30">
</div>
```

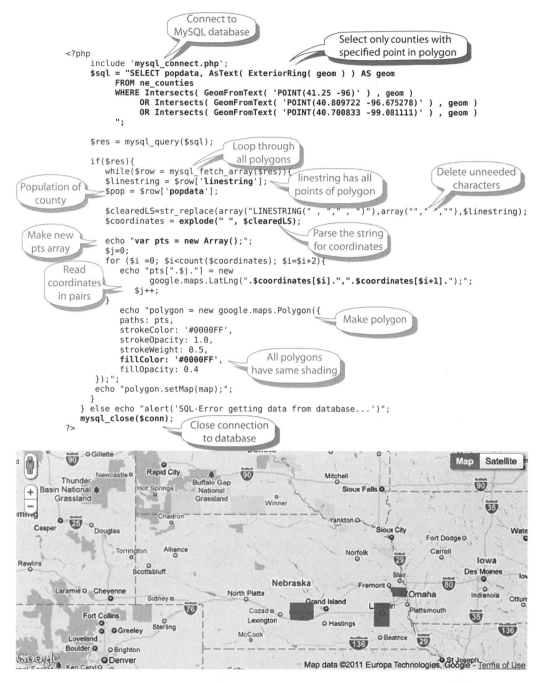

```php
<?php
    include 'mysql_connect.php';
    $sql = "SELECT popdata, AsText( ExteriorRing( geom ) ) AS geom
        FROM ne_counties
        WHERE Intersects( GeomFromText( 'POINT(41.25 -96)' ) , geom )
            OR Intersects( GeomFromText( 'POINT(40.809722 -96.675278)' ) , geom )
            OR Intersects( GeomFromText( 'POINT(40.700833 -99.081111)' ) , geom )
        ";

    $res = mysql_query($sql);

    if($res){
        while($row = mysql_fetch_array($res)){
            $linestring = $row['linestring'];
            $pop = $row['popdata'];

            $clearedLS=str_replace(array("LINESTRING(" , ",", " )"),array(""," ",""),$linestring);
            $coordinates = explode(" ", $clearedLS);

            echo "var pts = new Array();";
            $j=0;
            for ($i =0; $i<count($coordinates); $i=$i+2){
                echo "pts[".$j."] = new
                    google.maps.LatLng(".$coordinates[$i].",".$coordinates[$i+1].");";
                $j++;
            }
            echo "polygon = new google.maps.Polygon({
            paths: pts,
            strokeColor: '#0000FF',
            strokeOpacity: 1.0,
            strokeWeight: 0.5,
            fillColor: '#0000FF',
            fillOpacity: 0.4
        });";
            echo "polygon.setMap(map);";
        }
    } else echo "alert('SQL-Error getting data from database...')";
    mysql_close($conn);
?>
```

Callouts:
- Connect to MySQL database
- Select only counties with specified point in polygon
- Loop through all polygons
- linestring has all points of polygon
- Delete unneeded characters
- Population of county
- Make new pts array
- Parse the string for coordinates
- Read coordinates in pairs
- Make polygon
- All polygons have same shading
- Close connection to database

FIGURE 18.30. Results of a spatial query that finds those polygons that contain a specified point, in this case, the three cities of Omaha, Lincoln, and Grand Island. Copyright 2013 Google.

An action in the form results in a call to markers.php that executes the following PHP code to enter the point into the database. The $_GET function transfers the point and attribute text from the HTML form input to PHP. The INSERT INTO places the attribute (name) and points into the database. The OGC GeomFromText function converts the latitude and longitude values into a single spatial object in the MySQL database.

```php
<?php
  $lat = $_GET["latitude"];
  $lng = $_GET["longitude"];
  $name = $_GET["name"];
  include 'mysql_connect.php';
  $table    = "locations";
  $sql = "INSERT INTO locations (name, location) VALUES ('{$name}',
          GeomFromText ('POINT({$lat} {$lng})'))";
  $res = mysql_query($sql);
  mysql_close($conn);
?>
```

Figure 18.31 shows the combination of JavaScript and PHP code that displays all of the points that have been entered into the database. Additional functions are used to delete either individual points or all of the points.

18.9 Summary

Querying databases is the foundation of our interaction with computers and the Internet. When we use the Internet, we are most often conducting a query of an online database. In almost all cases, this is a relational database and there is a high probability that it was built with SQL and stored in a MySQL file.

The language used to extract data from the MySQL database and create the resultant web page is almost always PHP. As a server-side scripting language, PHP provides the mechanism to both extract data and create web pages on the fly. Most web pages that we examine have been created by PHP.

The open-source world of Linux has provided an inexpensive option for creating and manipulating databases. A variety of associated open-source tools, including phpMyAdmin, facilitate the process. The INNER JOIN operation is a basic SQL command for relating fields from different tables in a relational database. In using a web resource and querying a database, we are requesting that a series of INNER JOIN commands be executed to derive a specific piece of information.

18.10 Exercise

Upload the code18.zip file to your server and make the suggested changes. Use MySQL and phpMyAdmin to create and administer the database.

```
map = new google.maps.Map(document.getElementById("map_canvas"),mapOptions);
map.setTilt(1);
map.setOptions({draggableCursor:'crosshair'});          Change cursor to
map.setOptions({draggingCursor: 'move'});               moveable cross-hair

google.maps.event.addListener(map,'click',function(event) {
  var lat = document.getElementById('lat').value = event.latLng.lat()
  var lng = document.getElementById('lng').value = event.latLng.lng()
})

google.maps.event.addListener(map, 'click', function(event) {    Add listener
  addMarker(event.latLng);                                       function
);

var image = 'http://maps.google.com/mapfiles/ms/micons/blue.png'  Get blue
                                                                  marker
<?php
  include 'mysql_connect.php';          Connect to
  $table    = "locations";              database
  $sql = "SELECT name, X(location), Y(location) FROM $table";
  $res = mysql_query($sql);

  if($res){
    $num = mysql_num_rows($res);
    while($row = mysql_fetch_array($res)){     Get existing        Plot out
      $lat = $row['X(location)'];              points          existing points
      $lng = $row['Y(location)'];
      $name = $row['name'];

      echo 'var myLatlng = new google.maps.LatLng('.$lat.','.$lng.');';
      echo 'var marker = new google.maps.Marker({position: myLatlng, map: map,
        title: "'.$name.'", Lat: '.$lat.',Long: '.$lng.'", icon: image});';
        echo 'attachMessage(marker, "Name:  '.$name.'<br> Latitude: '.$lat.
                          '<br> Longitude: '.$lng.'");';
  }
    } else echo "alert('SQL-Error getting data from database...')";
          mysql_close($conn);
?>
```

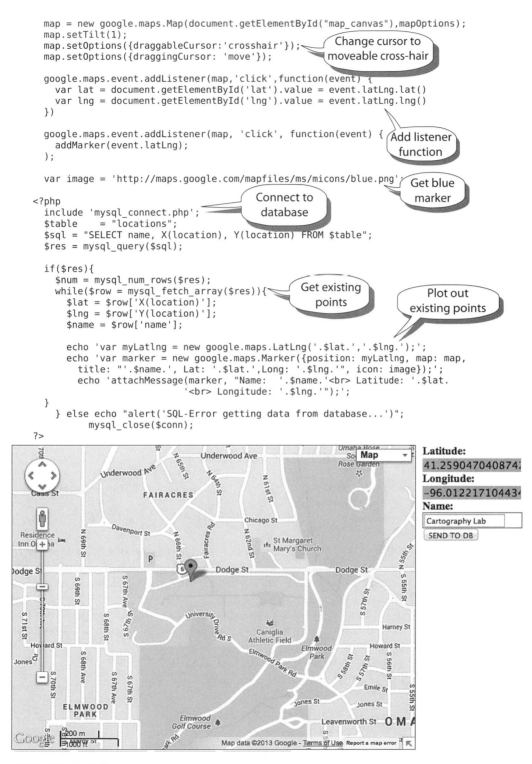

FIGURE 18.31. The input of points from a map through Google Maps. The points are stored in a MySQL database along with a corresponding attribute. Copyright 2013 Google.

18.11   Questions

1. Explain the advantages of Linux in all of its various versions.

2. What does creating web pages "on the fly" mean?

3. What is the importance of concatenating variables in PHP, and how is it done?

4. How can the MySQL user name and password be used to distribute databases?

5. List some functions of phpMyAdmin.

6. What is the function of the "INSERT INTO" SQL statement?

7. Explain the query performed by these lines of PHP code:

```
//connection to database and querying
<?php
        include 'mysql_connect.php';
        $table    = "us_capitals";
        $sql = "SELECT name, usstate, population, Y(location),
            X(location) FROM $table WHERE X(location) > 41.25 AND
            population < 500000";
        $res = mysql_query($sql);
```

8. What happens when a triangle instead of a rectangle is defined with the WHERE Intersects query?

9. Explain the INNER JOIN command.

10. Explain the input of lines and polygons in SQL.

18.12   References

Deek, Fadi P., and James A. M. McHugh (2007) *Open Source: Technology and Policy.* Cambridge, UK: Cambridge University Press.

Hong, Bryan J. (2008) *Building a Server with FreeBSD 7.* San Francisco, CA: No Starch Press.

Krafft, Martin F. (2005) *The Debian System: Concepts and Techniques.* San Francisco, CA: No Starch Press.

Thomas, Keri, and Andy Channelle (2009) *Beginning Ubuntu Linux.* New York: Springer-Verlag.

Tyler, Chris (2007) *Fedora Linux.* Sebastopol, CA: O'Reilly Media.

Whittaker, Roger, and Justin Davies (2008) *OpenSuse 11 and SUSE Linux Enterprise Server.* Indianapolis, IN: Wiley Publishing.

W3Techs. (2013) Usage of Operating Systems for Websites. August 2011. [http://w3techs.com/technologies/overview/operating_system/all]

CHAPTER 19
Mobile Mapping

Watching something on your mobile phone seems like crazy talk to me.

—MATT THOMPSON

19.1 Introduction

The growth in map delivery on mobile devices was particularly strong in the first decade of the 21st century. Cellphone screens became larger, positioning technologies improved, and all types of location-aware applications developed. Location Based Systems (LBS) emerged with the overall goal of providing information specific to the current location of the mobile user.

Cellular telephone technology developed parallel to but separately from the Internet. Much of the early development can be traced to the Nordic countries, particularly Finland and Sweden, where the first international mobile phone system was introduced in 1982 (Klemens 2010, p. 66). By 1987, 2% of the population of Norway, Denmark, Sweden, and Finland were mobile phone users. In the two decades that followed, the mobile phone spread to every part of the world. The United Nations Telecommunications Agency reported in 2010 that about 4.6 billion or 67% of the world's population were mobile phone users. This figure was only 11% in 2002. The number of users expanded rapidly in the highly populated areas of China and India as cellphone ownership became affordable (Virki 2008). In an example of leapfrog technology, many poorer countries that never had a widespread wired telephone system developed a wireless telephone network that was more advanced than those of even the most developed countries.

Initially limited to voice communication, text messaging or short message service (SMS) was added as a commercial service on mobile phones in 1993. Mobile email service soon followed as well as other services that were typically associated with the Internet. Although mobile browsers were available, the common use of mobile devices for the web did not emerge until 2007. This was largely related to the introduction of 3G wireless data communication and the Apple iPhone. Mobile Internet access grew by a factor of 50 from 2006 to 2008 (see Figure 19.1) after the introduction of the iPhone in July 2007.

This chapter provides an introduction to mobile mapping and its associated technology. We begin by examining the differences between the development of the Internet and the mobile network.

19.2   Comparison of the Internet and Mobile Technology

The Internet and mobile phones developed during the same time period, but there were major differences in how they developed. The Internet originated through a government program, while the mobile phone network was largely constructed by private interests. Government had little role in building the mobile phone network. In Europe, governments mandated that mobile phone companies share cellphone towers. The more laissez-faire approach in the United States resulted in every company building its own towers. For a country as large as the United States, the significantly greater costs for infrastructure development resulted in slower adoption, as can be seen in Figure 19.2.

In contrast to the mobile network, the infrastructure was already in place for the Internet. Personal computers began to be used in the 1980s. When the graphical World Wide Web was introduced through Mosaic in March 1993, over 1.3 million computers were already connected to the Internet (Kikta et al. 2002, p. 10)—including 225 million personal computers (*Computer Industry Almanac* 2005). The

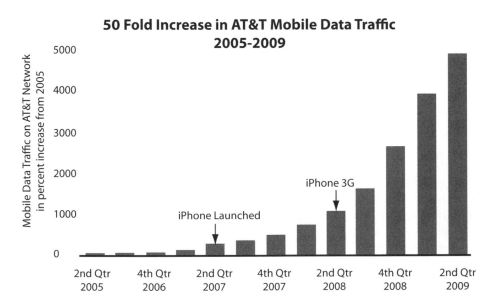

FIGURE 19.1. Increase in mobile data traffic after the introduction of Apple's iPhone in 2007 and 3G data communication in 2008. Source. Paczkowski 2009.

Cellular Subscribers Per 1,000 People				
	1990	1995	2000	2005	2010
USA	21.1	127	388	683	946
W. Europe	9.1	60	634	930	1,008
Asia-Pacific	0.4	7.1	71	230	379
Worldwide	2.1	15.6	123	319	478

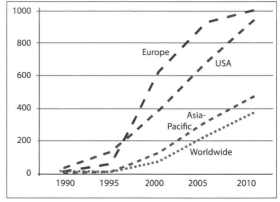

FIGURE 19.2. Cellular subscribers per 1000 people by region. Cellphone usage in the United States lagged behind usage in Europe. Source: The World Bank 2013.

browser software was free, and no additional hardware was required. Initially, most data communication took place over existing telephone lines. A faster, broadband data communication was quickly implemented in urban areas through existing cable television service. Although service costs averaged US $50 a month, broadband access to the Internet quickly surpassed telephone access (see Figure 19.3).

The mobile network developed much differently. The entire wireless infrastructure first needed to be built, requiring the construction of an enormous number of cell phone towers. Figure 19.4 shows the tower locations for just one city. AntennaSearch.com (2011) maintains a database on nearly 1/2 million cell phone towers and 1 1/2 million antennas for the United States and Canada. Although towers can be as far as 20 miles apart, each tower can only handle a limited number of calls requiring a much closer spacing in urban areas. In the suburbs, the spacing is between 1 to 2 miles but more densely populated urban areas require distances of 1/4–1/2 mile to handle the volume of calls.

Mobile phone handsets were another aspect of the infrastructure. Initially fairly large, mobile phones went through a rapid process of miniaturization. Smaller and smaller phones were engineered, partly driven by social pressure to own the latest and smallest mobile phone. The small sizes contributed to small-screen displays that limited their usefulness for anything beyond voice and short text messages.

Despite the expense of owning and using mobile phones, the number of mobile phone subscribers quickly surpassed the number of Internet users. The *Computer Industry Almanac* reported in 2005 that the worldwide number of cellular subscribers surpassed 2 billion—exactly twice as many Internet users at that time and up from only 11 million in 1990 and 750 million in 2000. Following 2005, Internet use increased as a result of smartphones.

19.3 The Mobile Medium

Transforming mobile phones into a true medium of communication required that mobile phones become larger again. In 2004, Funk (p. 44) estimated that doubling

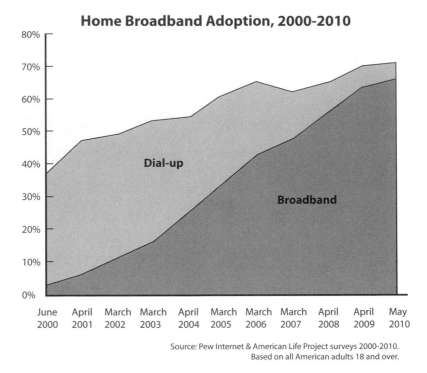

FIGURE 19.3. Broadband adoption in the United States, 2000–2010, compared to dial-up service. Source: Pew Internet 2012.

the display size would double the display weight and increase the weight of the plastic housing by 50%. Similarly, a 20% increase in display area would lead to a 30% increase in price. The issue, as Funk pointed out in 2004, was whether customers would choose a heavier and more expensive phone in order to have a larger display to facilitate better access to Internet content. The answer to that question would become clear by the end of the decade with the proliferation of larger and more expensive smartphones. This had a major influence on the delivery of maps to mobile users.

In order for a new medium to take its initial content from a medium that precedes it, it must incorporate a delivery mechanism that can display that information. The personal computer made this possible for all kinds of maps except large maps—the kind of maps that only a few had access to anyway. In contrast, the screens on initial mobile devices were too small to display any maps—either scanned paper maps or maps designed for the computer.

As Rhoton (2002, p. 117) states, it would be convenient if all the applications that had been developed over the past years on computers "could simply be dropped into a mobile environment and continue to work without any additional effort." To be backward compatible, the mobile device would have to be able to perform the same functions of the previous device in the process of adding new functionality. In addition to the small-screen size, Rhoton (2002, p. 117) pointed out that mobile systems could not be backward compatible because (1) many mobile platforms had

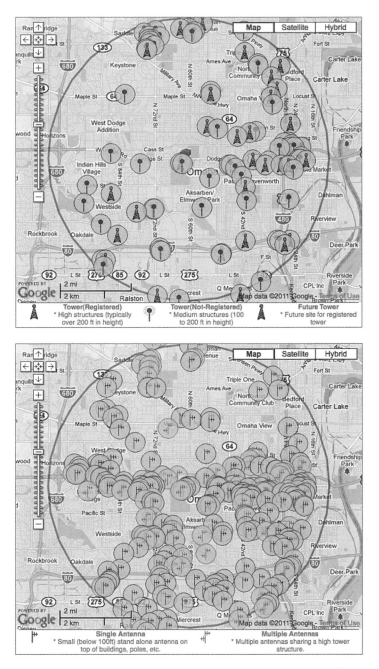

FIGURE 19.4. Cellphone towers (top) and antennas (bottom) for part of Omaha, Nebraska, mapped with Google Maps. There are 92 registered towers and 566 antennas in this area. The Federal Communications Commission does not require that each antenna structure be registered, especially if an existing tower or building is being used. Source: AntennaSearch. com 2012.

closed operating systems, such as the Blackberry™; and (2) there was a great deal of diversity in the machine interfaces of mobile devices. For these reasons and others, the mobile devices that were initially available were not backward compatible with either the paper medium or the web.

With a larger screen, many mobile phones integrated access to the World Wide Web. A study by Nielsen/Netratings showed that there were 40 million mobile subscribers to the Internet in the United States in 2008, roughly 15% of the active mobile user population (Knight 2008). The UK at 14% and Italy at 11% were close competitors. It was already apparent that the mobile web would reach critical mass and develop into a new form of information delivery.

The Nielsen report also stated that about 14% of mobile web users accessed the Internet through an unlimited access plan (Covey 2008). These access plans seemed to be associated with areas that had faster data service. This would mean that 86% of users accessed the mobile web on a pay-by-usage system that tends to limit use (Knight 2008). Most providers have since dropped their unlimited data plans, adopting metered usage plans. Mobile access to the Internet is limited by these restrictive data plans, especially for the larger file sizes for graphic content like maps. Users soon discover that maps quickly "eat up" their data plan allocation.

19.4   Positioning Technology

While the U.S. Federal Communications Agency maintained a laissez-faire relationship with the cellphone industry, it did mandate automatic location identification (ALI) on mobile phones. The mandate was a response to several publicized incidents involving emergencies in which the caller could not provide a location (see Text Box 19.1). Wireless carriers were required to have 95% ALI-capable handsets among their subscriber bases by December 31, 2005. ALI stipulated positioning within 100 m or less to ensure that emergency workers could find cellphone callers. The ALI mandate was the main impetus for the growth of location-aware cellphones, at least in the United States (GPS World 2007).

TEXT BOX 19.1   *The story of an emergency call in 2005 for which the location of the callers could not be determined.*

The wind chill was 10° F below zero when the first call came to the 911 operator in Omaha, Nebraska, in early January of 2005. The couple, Janelle Hornickel and Michael Wamsley, both 20, were caught in a snowstorm, and their truck had slid off the road around 12:30 AM. Over the next several hours they made a series of frantic phone calls for help but the 911 operators had no way of determining their location. At that time, Nebraska was one of only nine US states that had not implemented the latest automatic location identification system for cell phones. Hornickel and Wamsley were in no condition to provide any help to emergency personnel. In one call, Hornickel described her surroundings by saying that there were people taking cars apart and putting them in trees. The couple described elephants skating on ice. In another call, Wamsley said they had encountered people, but he didn't think they spoke English. "We've tried, we've asked for help, we've begged." Wamsley's frozen body was found the next day. It took another six days to find the snow-covered body of Hornickel. It was determined that they had taken methamphetamine. A police investigator speculated that the people they mentioned were actually cattle. (ABC News 2005)

In order to comply with ALI, most carriers initially decided to integrate GPS technology into cellphone handsets rather than overhaul the tower network used to triangulate the position of mobile phones. GPS does not work inside of buildings and is power hungry, quickly draining mobile phone batteries. For these reasons, cell towers were eventually upgraded to support position finding. Of the 3.3 billion cellphones in use in 2008, only 175 million had GPS (Bray 2008).

The cell tower triangulation (CTT) system uses signal analysis data to compute the time it takes signals to travel from the telephone to at least three cell towers, analogous to the operation of GPS (see Figure 19.5). The method does not require an active call. The location of the device can be determined based on the *roaming* signal that is emitted by all mobile phones when they are capable of making or accepting calls. This signal is used to contact the nearest antenna. The method is fairly accurate in urban areas but can be off by more than a mile in rural areas where cell towers are further apart. Most position finding with mobile devices is based on this technology, even if the unit contains a GPS device.

Initially, it was not possible to get direct access to location data. A location-aware mobile phone would only determine its location if an emergency call was made. Later, location data was provided continuously as an aid to navigation, transforming mobile phones into personal navigators.

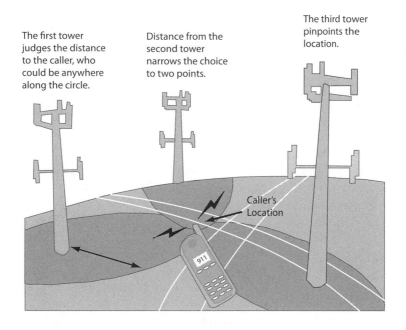

FIGURE 19.5. Triangulation from cell phone towers to find the position of a cell phone user.

19.5 Mobile Phone Navigators

Maps on cell phones were initially limited by extremely small displays. A 160×120 pixel sized screen was typical. Even this small display did not deter early mobile map users. By 2006, it was reported that MapQuest was among the top mobile Web destinations for U.S. subscribers drawing more than 3 million visitors in June of that year. At that time, MapQuest had more visitors than CNN, AOL, and weather and search offerings from Yahoo (Gibbs 2006).

Most mobile phone companies began offering some type of navigation service based on location-enabled devices. Verizon's VZ NAVIGATOR, begun in November of 2005, is an example of a basic $10 a month service. Offered by startup Networks in Motion, the system displays a small map and provides spoken directions through a natural-sounding human voice (Bray 2008). The timing of the directions is sufficiently accurate that drivers can use it as a turn-by-turn navigator. The service would quickly recalculate a route if the user went off-track (Wilstrom 2007). For select cities, the service shows the location of the nearest bank, hotel, or movie theater and gives gas prices at nearby gas stations—along with providing the appropriate phone number. The traffic-tracking center monitors conditions continually along the user's routes, providing proactive alerts on conditions ahead and offering detour options, while maps are displayed to enhance use. These devices also have a pedestrian mode that provides the best route for walking.

Boston-based Where, formerly uLocate Communications, developed over 70 location-based applications for location-aware mobile phones. The applications include GasBuddy which directs you to the location of the nearest gas station with the lowest price. Other so-called widgets list local entertainment events, local news, movies, weather, traffic, restaurants, concerts, sports events, coffee shops, local airports, and a flight-time updater (Yu 2008). The ShopLocal widget shows currently available products and where to buy them in the local area.

Google Maps for Mobile introduced a "my location" feature that utilizes the location of the mobile device. The application plots a blue circle around the estimated position based on the nearest cell tower (see Figure 19.6). This initial position is then updated by triangulating from the signals of surrounding cellphone towers. The map can be easily zoomed in or out by touching the screen with the now ubiquitous two-finger control. Two- and three-finger controls have since migrated to the touchpad's on laptop computers.

Mobile phone navigators effectively competed with car-based GPS navigation devices. The market research firm, Metrics, reported that 30 million dedicated GPS navigation units were sold in 2009, outpacing navigation-enabled mobile phone sales by 50% (RCR Wireless 2008). About 2.5 million GPS devices were sold in North America in 2006. This figure jumped to 10 million in 2007 and 20 million in 2008 (Yu 2008). Sales of these devices have since declined.

Navigation-enabled mobile phones are now used for auto navigation, pedestrian navigation, and many other types of location-based services (RCR Wireless 2008). This helps explain why Navteq, a provider of digital mapping databases and on the verge of being acquired by TomTom, a maker of car-based navigation devices, for $4.25 billion, was instead acquired by telecommunications giant Nokia for $8.1 billion in mid-2008 (RCR Wireless).

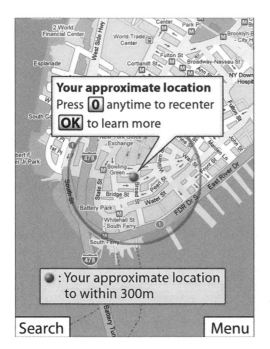

FIGURE 19.6. Google Maps for Mobile showing current estimated position based on the nearest cell phone tower. The location is then computed more precisely by triangulating based on signals from other cell towers. Copyright 2013 Google.

19.6   Location-Based Services and Social Mapping

A variety of spatially aware applications have been developed for Apple's iPhone. In one application, the current location of friends can be placed on a map. An early example was Loopt that would map where friends were located if their cellphones were turned on. A message could even be broadcast by a user that was displayed on the map as a virtual sign. The technology could also be used to locate a child with a cellphone.

Another early example, Pelago's Whrrl, shows users "cool places" that friends have visited and recommended to others. The application can filter recommendations by user. A user can look for a bar and browse to see if any friend has recommended it (Yu 2008). Spatially aware advertisements are also a possibility. A coupon could be delivered to entice a potential customer to a nearby store, restaurant, or movie theater.

Other spatially enabled applications include Wikitude, which provides a Wikipedia description of nearby points of interest. SynchroSpot can automatically bring up a shopping list as the user walks into a supermarket. Traffic alerts and updated weather conditions are the most popular real-time tools for travelers (Yu 2008). Google Maps later introduced public transportation options to get to the user's destination.

Google's Latitude lets users share their position with others and view the locations of all friends on a map. The user can control the accuracy and details of what any user can see—an exact location can be allowed, or it can be limited to identifying the city only. For privacy, Latitude can also be turned off by the user, or a location can be manually entered.

Foursquare is a combined web and mobile application that allows registered users to update their location and connect with friends. Points are awarded for "checking in" at certain venues. Users can choose to have their check-ins posted on their accounts on Twitter or Facebook. Foursquare implements a notification of updates from friends, called pings. Users can also earn badges by checking in at specific locations, and checking in at a certain time or a certain number of times.

19.7   Mobile Data Communication

The single major limitation of mobile mapping is the speed of data communication. In most of the United States, data communication is either not possible or only a slower data communication standard called EDGE/GPRS is available. EDGE/GPRS is designed for email, not web pages with graphics. Faster data communication is limited to urban areas and surrounding suburbs (see Figure 19.7). While networks are capable of speeds up to 100 megabits per second (Mbps), these speeds are normally not achieved in real-world testing. A slower data communication network is used in large parts of the world that is rated at speeds of between 75 and 384 kilobits per second (Kbps). Download speeds vary based on distance from a cell site, general load on the network, intermediate links to the core network, and a host of other factors.

Mobile mapping is not possible with the slower data communication speeds, especially when the user is moving through the environment. When the user is traveling, the map needs to be updated faster than the rate of data communication will allow. If a map cannot be downloaded, all that is left are a blank screen and a dot that indicates the current position. But a dot that indicates the current position is useless without the map. Any network connection slower than 3G renders the device of no value for the display of online maps. This would be the case for most of the United States.

With a tile-based mapping system, the map is divided into 256 × 256 pixel

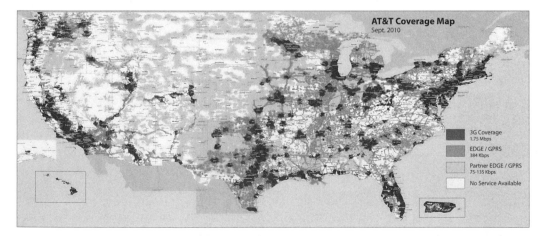

FIGURE 19.7. The map shows AT&T 3G coverage in 2010. The use of online applications for maps effectively requires 3G service. Data from AT&T. Map copyright 2010 by Microsoft.

pieces. Usually, these tiles are sent so quickly that the user rarely notices that the map is composed of pieces. With slower data communication, it is painfully obvious that the map is composed of tiles. Data transfer is so slow that one can see the tiles appear, usually one by one. Once downloaded, the tiles are stored locally for some time, making the panning and zooming process almost instantaneous. The Maps application on Apple's iOS caches a large number of map tiles. Android devices have a similar caching feature. It is thus possible to preload the tiles at multiple scales before traveling and to rely on the cached files for map display. One may not even need an Internet connection.

The total number of bytes that need to be downloaded for one map display can easily be estimated for a tablet computer like the Apple iPad. Initial versions of the iPad have a display size of 1024×768 pixels. If the point of interest (POI) falls perfectly in the middle, the device would need only four 256×256 pixel tiles in one dimension by three tiles in the other, for a total of 12 (see Figure 19.8). This perfect centering of the map would never happen, and the device would need to download one more row and column of tiles along each dimension for a total of 20 tiles. Each tile is stored in the PNG format and would average around 20 kb each for land areas. That means that one iPad map display would be about 300 kb.

At 300 kb per map, it is possible to calculate the cost of each screen-sized map

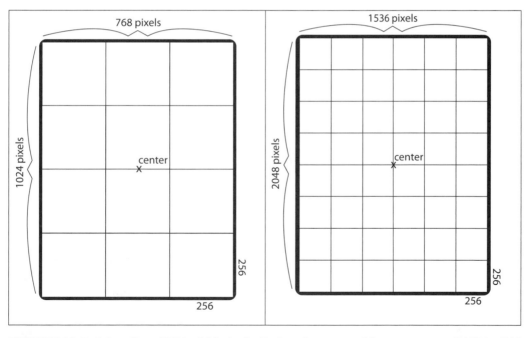

FIGURE 19.8. Map tiles of 256×256 pixels displayed on two tablet computers of 1024×768 pixels and 2048×1536. In the tablet on the left, a total of 12 tiles would be needed if the center were perfectly located between the two middle tiles. Additional tiles would need to be downloaded along both dimensions, resulting in a total of 20 tiles (4×5). At an average of 20 kb per tile, this would be 300 kb for one map view. The newer version on the right would require 48 tiles. The additional rows and columns would bring the total to 63 tiles, for a total of 1.23 mb.

relative to the available data plans. For a $15/250 mb plan, approximately 833 screen displays can be downloaded—about 1.8 cents for a screen-sized map. For the $25/2 gb data plan, one would get 667 maps at about 0.37 cents a map. Newer versions of the iPad have a 2048-by-1536-pixel resolution. This would require a total of 8 tiles in one dimension by 6 tiles in the other for 48 total tiles, if the center point is perfectly in the middle. An additional row and column would be needed, resulting in a total of 63 tiles (9 rows by 7 columns). At 20 kb per tile, the total data communication requirement is 1.23 mb per map display. The 250 mb data plan would result in 203 maps for a cost of 7.4 cents per map display. The 2 gb plan would result in 1665 map displays, or about 1.5 cents per display.

One might wonder if all of the map tiles could simply be stored locally on the mobile device. At the 20th zoom level there are over one trillion tiles for the entire world. At an average of 15 kb per tile (tiles over the ocean are about 10 kb), a map of the world would require 15 petabytes. At disk drive prices of US $100 per terabyte, the cost of storage would be over US $2 million. A semitruck would be required to carry this many hard drives. It is clear that a map with this many levels of detail can only be stored at data centers.

Mobile phone maker, Nokia, uses an encoded vector method to store an entire world map. At approximately 10 gb, the map can easily be stored within the memory of a smartphone. No connection is needed to display a map. This method of map storage and presentation has a number of advantages, not the least being the ability to display the map without incurring data connection charges. The disadvantage is that any updates to the map on the mobile device are difficult to make. Replacing the 10 gb file would be almost impossible at current mobile data communications speeds. In addition, the display of frequently updated information such as current traffic conditions is more complicated.

19.8   Locational Privacy

The sensitive issue of tracking via GPS devices—and for what purposes—is still a topic of debate (Yu 2008). The International Association for the Wireless Telecommunications Industry (CTIA) has issued guidelines for location-based services that stress consumer notice and consent, and data security. The main problem is that there is no oversight mechanism. Technically, if your cellphone is on, someone can find you (Nakashima 2008).

The term *locational privacy* refers to the concept that a person's location should not be made available without consent. This concern seems to have a generational component. It has been noted that younger people are less concerned about being tracked while the older generation sees this as an invasion of privacy.

Methods of tracking are not limited to cellphones. Using an EZ-Pass system for tolls on highways means that someone can find out where you are driving and when. It has been proposed that toll booths be eliminated by installing a GPS transmitter in every car and assessing toll charges based on the car's recorded path (Blumberg and Chase 2005). This idea raises similar concerns about locational privacy.

An entire industry has developed around people tracking. Several service providers advertise the capability to track people on a map. Most commercial trucks are constantly tracked. It is possible to track the location of children, retirees, parolees, or even pets. Police regularly track cell phones to follow suspects. It has been suggested that the location of every single individual could be tracked using cellphone technology.

Monmonier (2003) examines how mapping is becoming a tool to help invade our privacy. He points out that online mapping is "potentially threatening—because it not only greatly expands the audience of potential watchers but also allows for unprecedented customization of maps that describe local crime patterns, warn of traffic congestion and inclement weather, disclose housing values, or—thanks to the global positioning system (GPS) and the new marketplace for 'location-based services'—track wayward pets, aging parents, errant teenagers, or unreliable employees" (p. 98). He concludes that as "society and government work through the significance of locational privacy and decide what legal limitations, if any, are appropriate and permissible, the debate will turn to possible restrictions on Internet cartography" (p. 111).

19.9 Summary

When it comes to navigation, our movements through the environment are now often directed by the reassuring voice of a GPS device. The user simply follows these directions to find a location and will often need to rely on the device again for the return trip. While a map is often displayed, the user has little time to examine it. The real danger, of course, is that people will continue to use the device and not create a mental map of their surroundings. Unless a mental map develops, we will continue to be dependent on these devices and have the permanent sense of being lost.

The expansion of the mobile web means that people have access to Internet maps via small mobile devices. New types of user input—including voice and gestures—will be introduced and will cross over to laptop and desktop devices. For example, Apple's MultiTouch system developed for its mobile touchscreen devices has already migrated to Apple's laptop touchpad. Other innovative methods of user input for mobile devices will certainly be developed. The way we use computers in the future may well be based on how we use mobile devices now.

The most disappointing aspects of the mobile Internet are the cost and slowness of data communication. While faster methods of data communication are being introduced in larger cities, most areas are not well served, if at all. By area, most of the world either has no coverage or a slower data communication system that cannot be effectively used to display maps. Outside of the major cities, the usefulness of mobile devices for mapping is extremely limited.

The amount of memory in mobile devices is increasing, making it more possible to store maps on the devices. This can either be done by caching a large number of map tiles or storing a compressed vector version that is then drawn by the device. In either case, you can zoom in and out while the device indicates the current location.

19.10   Questions

1. In 2010, approximately 67% of the world's population were cellphone users according to the United Nations Telecommunications Agency. What is the current percentage?

2. What single device led to major increases in wireless data communication beginning with its introduction in 2007?

3. What were some factors that limited the growth of mobile phone usage in the United States?

4. Describe the purpose and implementation of automatic location identification (ALI).

5. What are the advantages and limitations of cell tower triangulation (CTT)?

6. How did mobile phone navigators develop, and what is their current status?

7. Describe the current state of social mapping.

8. What are the data communication requirements for mapping with a tablet computer?

9. Describe the issues related to locational privacy.

10. What are current methods for tracking?

19.11   References

ABC News (2005, Mar. 3) Meth Makes Winter Deadly for Young Couple. [http://abcnews.go.com/Health/Primetime/story?id=549455&page=1#.UcZjkPac5g8]

AntennaSearch.com (2011) Free online source for 1.9 million antennas in the US. [http://antennasearch.com]

Blumberg, Andrew J., and Chase, Robin (2005) Congestion Pricing That Respects Driver Privacy. *Proceedings of the IEEE Intelligent Transportation Systems* 2: 725–732.

Bray, Hiawatha (2008, Apr. 17) GPS Turns Cellphones into Powerful Navigators. *Boston Globe.*

Computer Industry Almanac (2005) China Tops Cellular Subscriber Top 15 Ranking. [http://www.c-i-a.com/pr0905.htm]

Covey, Nicholas (2008) Critical Mass: The Worldwide State of the Mobile Web. *The Nielsen Company.* [http://www.nielsenmobile.com/documents/CriticalMass.pdf]

Funk, Jeffrey L. (2004) *Mobile Disruption.* Hoboken, NJ: John Wiley & Sons.

Gibbs, Colin (2006, Sept. 11) Mobile Maps on the Map. *RCR Wireless News.*

GPS World (2007, Sept. 12) FCC to Require Full E911 Adherence by 2012. *GPS World.*

Kikta, Roman, Al Fisher, and Michael Courtney (2002) *Wireless Internet Crash Course.* New York: McGraw-Hill.

Klemens, Guy (2010) *The Cellphone: The History and Technology of the Gadget That Changed the World.* Jefferson, NC: McFarland & Company.

Knight, Kristina (2008, July 10) Nielsen: US leads in mobile web adoption. *BizReport.*

[http://www.bizreport.com/2008/07/nielsen_us_leads_in_mobile_web_adoption.html]

Monmonier, Mark (2003) The Internet, Cartographic Surveillance, and Locational Privacy, in Michael Peterson (Ed.), *Maps and the Internet*. Oxford, UK: Elsevier.

Nakashima, Ellen (2008, July 12) When the Phone Goes with You; Everyone Else Can Tag Along. *The Washington Post.*

Paczkowski, John (2009) Time to Cut AT&T Some Slack, iPhone Users? AllThingsD.com, Nov. 18. [http://allthingsd.com/20091118/time-to-cut-att-some-slack-iphone-users/]

Pew Internet (2012) Broadband Changed Us as Internet Users. Pew Internet & American Life Project. [http://pewinternet.org/]

RCR Wireless. (2008, Jan. 7) Cell Phones to Dominate for Navigation. *RCR Wireless News.*

Virki, Tarmo (2007, Nov. 29) Global Cellphone Penetration Reaches 50 pct. *Reuters News Service.* [http://investing.reuters.co.uk/news/articleinvesting.aspx?type=media&storyID=nL29172095]

Wilstrom, Stephen H. (2007, Jan. 29). Easy Ways to Find Your Way. *Business Week.*

World Bank. (2013). Mobile Cellular Subscriptions. [http://data.worldbank.org/indicator/IT.CEL.SETS.P2]

Yu, Roger (2008, July 8) GPS Becomes a Vital Tool for Frequent Travelers. *USA Today.*

CHAPTER 20

Local Mapping

By the end of this month [July 2006], more than 40 high-risk sex offenders in the Antelope Valley region will be issued the GPS electronic bracelets that will make it easier for law enforcement officers and parole agents to track their whereabouts and determine their location at the time of a crime.

—JIM TILTON

20.1 Introduction

Location-aware devices are becoming increasingly common. Virtually all mobile phones can be located within a few meters. Smartphones have the additional ability to display their current location on a map. Tablet computers based on Apple's iOS and Google's Android can usually do the same but with the added benefit of displaying a much larger map. Hand-held GPS devices and the "electronic bracelets" worn by parolees and registered sex offenders are specifically designed to provide location, with continual updates sent to authorities.

There are many different types of mobile devices and many different ways of determining location. To provide a standardized approach, the World Wide Web Consortium (W3C) created a freely available, geolocation API (application programming interface). Supported by nearly all browsers, the API uses multiple methods to find the location of the computer or mobile device (Svennerberg 2010, p. 235). This chapter describes the mapping of location using the W3C Geolocation API. The examples are demonstrated within the browser of a mobile device and work from any desktop or laptop computer.

20.2   W3C Geolocation API

The Geolocation API attempts to determine location in various ways. Common sources of location information include the global positioning system (GPS) and location information inferred from the IP address, Wi-Fi and Bluetooth addresses, cell telephone IDs, and cell tower triangulation. The geolocation API will not specifically identify how the location was determined, and there is no guarantee that the API will return the device's actual location.

Without access to GPS, cellphone positioning, or a wireless signal, a desktop or laptop computer is left to determine position based on the computer's IP address. Finding location through the computer's IP address is accomplished through the WHOIS service, which retrieves a registered physical address that corresponds to an IP number such as 137.48.16.54. Every computer connected to the Internet is assigned an IP address with an address like this. The associated geolocation data can include the country, region, city, postal/zip code, time zone, and, in some cases, the actual latitude and longitude of the computer. The WHOIS service was designed, in part, to assist law enforcement in finding the location of criminals but IP geolocation usually provides only a rough estimate of the actual location.

Identifying location through a wireless signal, such as Wi-Fi or Bluetooth, is done through its media access control (MAC) address. Every device on the Internet has a unique MAC identifier. To determine location, the MAC address is identified in reference to a nearby hotspot and compared to a database of Wi-Fi locations. Location software developed by the Boston-based company, Skyhook, uses a massive reference network made up of the known locations of over 700 million Wi-Fi access points and cellular towers. To develop the database, Skyhook deployed drivers to survey every single street, highway, and alley in tens of thousands of cities and towns worldwide, scanning for Wi-Fi access points and cell towers and plotting their precise geographic locations (see Figure 20.1). These locations are updated on a regular basis. Skyhook's coverage area includes most major metro areas in North America, Europe, Asia, and Australia. The company offers a software development kit (SDK) available for Android, Linux, Windows, and Mac OS X.

Privacy is a major concern with geolocation. Although the API only retrieves the geographic location of the device, this information usually discloses the location of the user, thus invading the user's privacy. The geolocation API must provide a mechanism that protects the user's privacy. This mechanism should ensure that no location information is made available through the API without the user's express permission. A statement by Skyhook states that "your exact physical location is inherently private and that it should only be used in very limited ways over which you have control" (Skyhookwireless.com 2013). User control is usually implemented through a dialog in which the user must agree to "Share Location" before the device is located, as shown in Figure 20.2. In most cases, law enforcement can determine your location even without your consent.

The call to the geolocation position object and the values that are returned are shown in Figure 20.3. The items returned depend on the way the position is determined. If GPS is used, the altitude, altitude accuracy, and current time are also included; otherwise they are listed as "null." If altitude has a value, the GPS inside

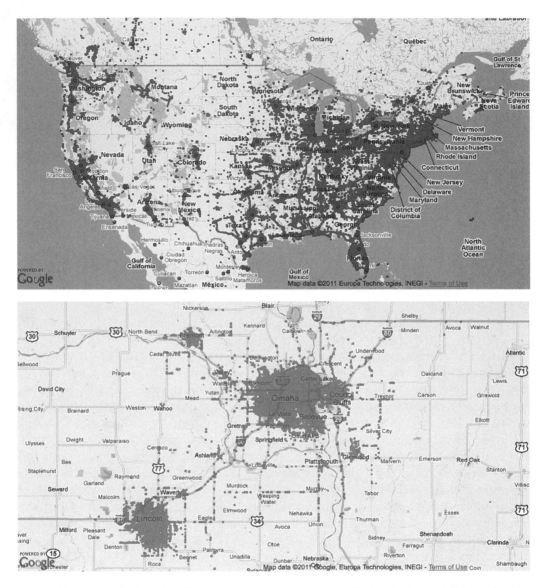

FIGURE 20.1. Wi-Fi and cell-tower access points in the Skyhook, Inc., database for the United States and the area around Omaha, Nebraska. The database of 250 million Wi-Fi locations was developed through survey methods by driving the streets in thousands of cities around the world. Copyright 2011 Skyhook.

🌐 maps.unomaha.edu wants to know your location. Learn More...   [Share Location]   [Don't Share]   ☐ Remember for this site   ⊠

FIGURE 20.2. W3C Geolocation API locational privacy dialog presented by the Safari browser on the iPad. Pressing "Don't Share" aborts any attempt to determine location. Copyright 2013 Google.

```
navigator.geolocation.getCurrentPosition(function(position) {
    initialLocation = new
    google.maps.LatLng(position.coords.latitude,position.coords.longitude);
}
```

Property	Type	Notes
coords.latitude	double	Decimal degrees
coords.longitude	double	decimal degrees
coords.altitude	double or null	meters above the <u>reference ellipsoid</u>
coords.accuracy	double	meters
coords.altitudeAccuracy	double or null	meters
coords.heading	double or null	degrees clockwise from <u>true north</u>
coords.speed	double or null	meters/second
timestamp	DOMTimeStamp	like a Date() object

FIGURE 20.3. Call to the W3C geolocation position object and the variables that are returned.

the device has been consulted because this information cannot be determined through other means. Whether location is determined by GPS or other means, accuracy estimates are returned for the location. Values for heading and speed will be returned if the unit is moving and the numbers can be calculated. Many mobile devices incorporate an electronic compass that provides the current direction of movement.

20.3   Me on My Map

Mapping the current location of a device through a browser using the W3C API is shown in Figure 20.4. The navigator.geolocation.getCurrentPosition (function(position) statement resolves the current position of the device. If the position cannot be determined by GPS or other means of directly determining latitude and longitude, the API pursues the location of the wireless or Bluetooth network, or the IP address. This example presents an infowindow at the current location.

The example in Figure 20.5 replaces the infowindow with a clickable marker. The contentString for the bubble text displays the latitude and longitude.

More variables from the positioning process are displayed in Figure 20.6 and shown with a screenshot from the iPad. Here, accuracy of latitude and longitude, elevation, accuracy of the elevation, and current time are displayed, the last three requiring measurement by GPS.

20.4   Updated Positioning

When determining a position either through cellphone triangulation or GPS, a rough estimate is initially returned and then refined as more measurements are taken. If the position changes, either through movement or more accurate

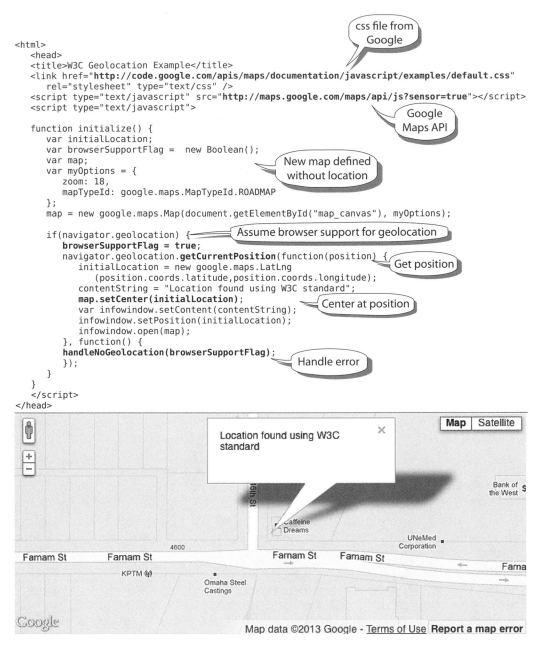

FIGURE 20.4. Implementation of the W3C geolocation API. Copyright 2013 Google.

positioning, the `watchPosition()` function will update the position. This function uses the same input parameters as `getCurrentPosition`, but it continuously updates the position, allowing the browser to either change the location with movement or provide a more accurate location as different ways are used to determine location. The `watchPosition()` function does not require an initial call to `getCurrentPosition()`.

```
if(navigator.geolocation) {
    browserSupportFlag = true;
    navigator.geolocation.getCurrentPosition(function(position) {
        initialLocation = new google.maps.LatLng
           (position.coords.latitude,position.coords.longitude);
        contentString = "Pt:   "+position.coords.latitude+",
           "+position.coords.longitude ;
        map.setCenter(initialLocation);
        infowindow.setContent(contentString);
        infowindow.setPosition(initialLocation);
        var marker = new google.maps.Marker({
            position: initialLocation,
            map: map,
            title:"Hello World!"
        });

        google.maps.event.addListener(marker, 'click', function() {
        infowindow.open(map,marker);
    });
        }, function() {
        handleNoGeolocation(browserSupportFlag);
        });
    }
}
```

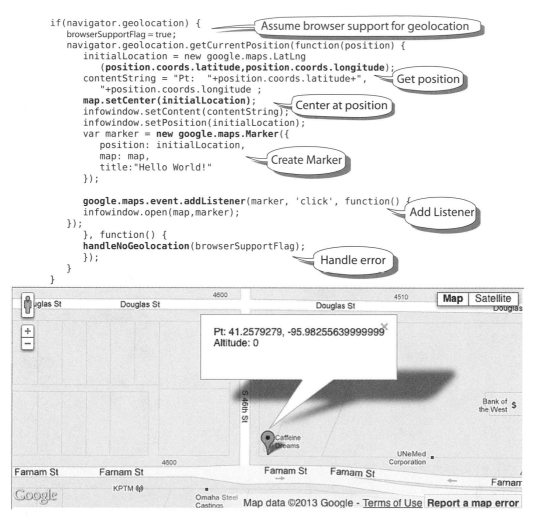

FIGURE 20.5. Displaying a marker instead of an infowindow. The text displayed in the clickable bubble includes the latitude and longitude of the current point. The "+" in the contentString statement is used to concatenate the numbers into a text string. Copyright 2013 Google.

The `navigator.geolocation.watchPosition()` is able to give more accurate readings because of the way mobile devices determine their position. They initially determine their location based on the nearest cellphone tower. The location is subsequently updated by triangulating with other towers. Depending on the proximity of the towers, the accuracy is 10–15 meters. If the device has a clear view of the sky for five minutes or more, the device may begin to use the location as determined by the internal GPS. The example in Figure 20.7 is built on Mapstraction and uses `watchPosition` to update the position, but only if the position changes by more than 0.01 km (10 m).

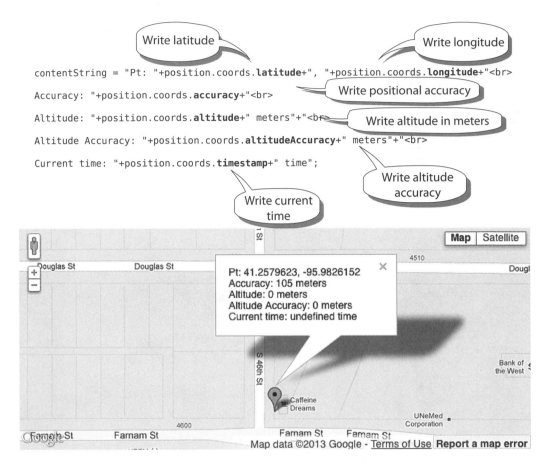

FIGURE 20.6. The current location depicted on the screen of the Apple iPad through Safari. The infowindow displays variables that have been placed into the contentString. The latter three values—altitude, altitude accuracy, and time—are null because they are dependent on a GPS signal. A GPS signal was not available because the device was indoors. Copyright 2013 Google.

20.5   Reverse Geocoding to Find the Street Address

Once the latitude and longitude of the current position have been determined, it is also possible to use a reverse geocoding service to provide a nearby street address. The series of functions depicted in Figure 20.8 resolve the location down to the country, city, and street address.

20.6   Customizing the Geolocation Request

The positioning request to the Geolocation API can be customized by setting properties using the `PositionsObject`. These properties are passed to the getCurrentPosition function. The `PositionsObject` has three properties: (1) `enableHighAccuracy`; (2) `timeout`; and (3) `maximumAge`. The first of these properties, `enableHighAccuracy`, forces the geolocation API to use GPS and not use

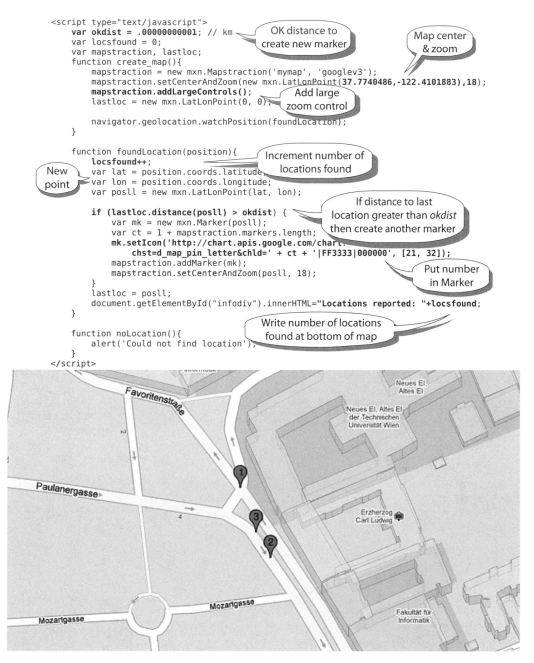

```
<script type="text/javascript">
    var okdist = .00000000001; // km          OK distance to
    var locsfound = 0;                        create new marker
    var mapstraction, lastloc;
    function create_map(){                                     Map center
        mapstraction = new mxn.Mapstraction('mymap', 'googlev3');  & zoom
        mapstraction.setCenterAndZoom(new mxn.LatLonPoint(37.7740486,-122.4101883),18);
        mapstraction.addLargeControls();          Add large
        lastloc = new mxn.LatLonPoint(0, 0);      zoom control

        navigator.geolocation.watchPosition(foundLocation);
    }

    function foundLocation(position){         Increment number of
        locsfound++;                          locations found
        var lat = position.coords.latitude;
 New     var lon = position.coords.longitude;
 point   var posll = new mxn.LatLonPoint(lat, lon);
                                                         If distance to last
        if (lastloc.distance(posll) > okdist) {       location greater than okdist
            var mk = new mxn.Marker(posll);           then create another marker
            var ct = 1 + mapstraction.markers.length;
            mk.setIcon('http://chart.apis.google.com/chart.
                chst=d_map_pin_letter&chld=' + ct + '|FF3333|000000', [21, 32]);
            mapstraction.addMarker(mk);                   Put number
            mapstraction.setCenterAndZoom(posll, 18);     in Marker
        }
        lastloc = posll;
        document.getElementById("infodiv").innerHTML="Locations reported: "+locsfound;
    }
                                        Write number of locations
    function noLocation(){              found at bottom of map
        alert('Could not find location');
    }
</script>
```

FIGURE 20.7. Continuous updating of the user's position using the W3C geolocation function *watchPosition* as implemented through Mapstraction. Copyright 2013 Google.

positioning based on Wi-Fi MAC addresses or cell tower triangulation. The second, `timeout`, allows the user to specify the number of milliseconds before the API ceases trying to determine a location. Finally, `maximumAge` will cache the last known location in memory for a set number of milliseconds and will then use that data the next time the API is called if it is less than `maximumAge` old.

```
<script>
   var mapDiv;
   window.onload = function(){
      mapDiv = document.getElementById('map');
      mapDiv.innerHTML = 'Finding your location...';
      if(navigator.geolocation)
         navigator.geolocation.getCurrentPosition(handleGetCurrentPosition,
            handleGetCurrentPositionError);
   }
   function handleGeocoderGetLocations( addresses, status ){
      if (status != google.maps.GeocoderStatus.OK)
         return maybe_log( 'Could not connect to Google' );

      var city = getCityFromPlacemarks(addresses);
      var country = getCountryFromPlacemarks(addresses);

      var mapOverlay = document.getElementById('found-at');
      mapOverlay.innerHTML = 'You are near <strong>' +
         addresses[0].formatted_address + '</strong>';
      mapOverlay.style.visibility = 'visible';
   }

   function getCityFromPlacemarks( placemarks ){
      return extractNameFromGoogleGeocoderResults('locality', placemarks)
   }

   function getCountryFromPlacemarks(placemarks){
      return extractNameFromGoogleGeocoderResults('country', placemarks)
   }

   function extractNameFromGoogleGeocoderResults(type, results){
      for( var i = 0, l = results.length; i < l; i ++)
         for(var j = 0, l2 = results[i].types.length; j < l2; j++ )
            if( results[i].types[j] == type )
               return results[i].address_components[0].long_name;
         return ''
   }
   function handleGetCurrentPositionError(){
      mapDiv.innerHTML = 'The browser is not compatible with the Geolocation API!';}
</script>
```

Get current position

If status not OK, then exit

Get city and country

Format text string with address

Call function below to extract city

Call function below to extract country

Extracts requested location element

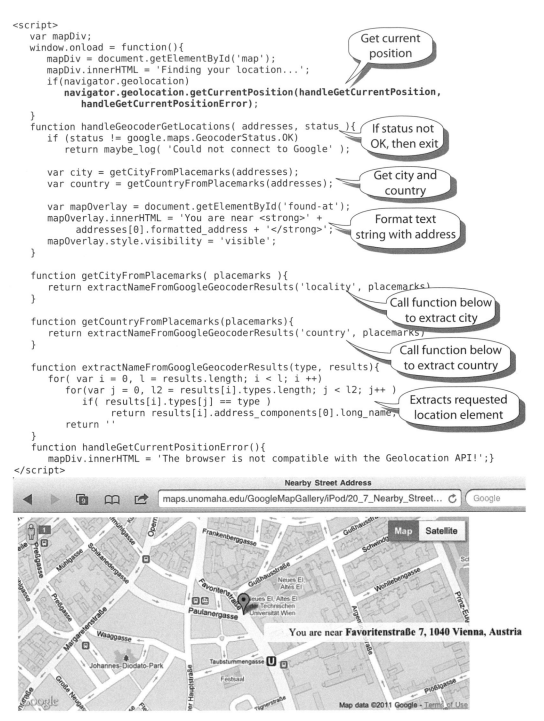

FIGURE 20.8. Reverse geocoding is used to find the street address, city, and country based on the latitude and longitude. Copyright 2013 Google.

Mobile devices often implement separate permission dialogs for low- and high-accuracy positioning. With enableHighAccuracy, it is possible that calling get-CurrentPosition() with enableHighAccuracy:true will fail, but calling with enableHighAccuracy:false would succeed. This is because enableHighAccuracy forces the use of GPS. The mobile device may not have a GPS, or it may be located inside a building where no GPS signal is available.

The timeout property is the number of milliseconds one is willing to wait for a position to be determined. The timer starts after the user gives permission to share location. The maximumAge property allows the device to answer immediately with a cached position if multiple calls are made within a short time to getCurrentPosition. The following call would accept a position if that position had been determined previously within 75 seconds:

```
navigator.geolocation.getCurrentPosition(success_callback, error_callback, {maximumAge: 75000});
```

The code in Figure 20.9 provides values for these three options. It would only work on the browser of a device with a GPS.

20.7 Summary

The W3C Geolocation API, released in 2010, standardizes position finding among different browsers and mobile devices. Not all browsers support the geolocation

FIGURE 20.9. Example application of a call to the geolocation API with values defined for maximumAge, timeout, and enableHighAccuracy.

standard, and the W3C Geolocation API support varies between mobile devices. In addition, while browsers are available on mobile devices, many users avoid using the browser in favor of stand-alone apps. Whether using a browser or an Internet-enabled app, location is determined using identical resources.

Maps that show the user's current location will become increasingly common. While a symbol showing this location makes perfect sense on any type of general reference map, it may also be incorporated on thematic maps. For example, a map showing the rate of unemployment in the United States could be augmented with a symbol that locates the user who is examining the map.

20.8   Exercise

Upload the code20.zip file to your server and capture images of the examples on your mobile device.

20.9   Questions

1. What devices do you use that are location-aware?

2. Describe the location database developed by Skyhook.

3. How many Wi-Fi signals are available where you live? What are the relative strengths?

4. Compare positioning between an urban area where Wi-Fi signals are available and a rural location.

5. What is the WHOIS service, and how does it help determine location?

6. In addition to latitude and longitude, what other values are returned by the W3C geolocation API?

7. How does updated positioning improve locational accuracy?

8. What are some applications of reverse geocoding?

9. How is locational privacy implemented with the geolocation API?

10. Should locational privacy be a major concern?

20.10   References

Svennerberg, Gabriel (2010) *Beginning Google Maps API 3*. New York: aPress.
W3C (2010) Geolocation API Specification. [http://dev.w3.org/geo/api/spec-source. html]

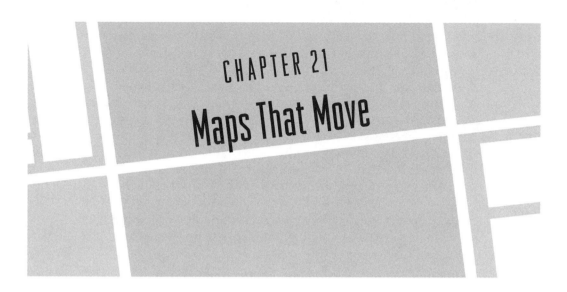

CHAPTER 21

Maps That Move

Animation can explain whatever the mind of man can conceive.
This facility makes it the most versatile and explicit means of
communication yet devised for quick mass appreciation.

—WALT DISNEY

21.1 Introduction

In addition to making more maps available, computers have also made it possible
to create new and better maps—maps that were impossible to make by hand. Interaction with maps has been one of the major benefits of the computer. Another
benefit is the incorporation of various forms of animation, including the simple
sequenced display of maps. Animated maps are typically used to show change over
time but there are a variety of other applications. For example, animating a series of
3D maps creates a terrain flyover where the map changes to give the impression of
flying over an area. Virtual reality takes this a step further by quickly generating a
series of scenes that simulate a real landscape to the point that users begin to orient
themselves within this artificial environment. The Internet has made it possible to
distribute these new map products to a large audience. The purpose of this chapter
is to introduce some applications for animated maps and describe how they are
made and distributed through the Internet.

In many ways, maps have been defined by the paper medium. The single-
map solution to display the world is a result of this medium. The presentation of
maps by computer has mostly adopted the single-map method of display. It may be
argued that all maps presented by computer should incorporate animation of some
kind because the medium easily supports the display of multiple maps. Showing
change over time could be integrated into any map display because an element of

time underlies every map. In addition, aspects other than time can be shown in an animation. Therefore, all maps should now support some form of multiple map display in the form of an animation.

Animated maps are beginning to be a product of a database, but most are still "one-off" productions, made individually—much as a manual cartographer would make individual maps. Some animated maps implement limited forms of interaction in the sense of start-stop, change of speed, and the direction of view, but few examples exist for truly dynamic, database-driven forms of animated map display. Online tools to easily animate maps are also not available, at least in the sense of the application programming interfaces (APIs) that have been used in this book.

This chapter reviews the development of cartographic animation. Actual animations are available in the associated zip file for this chapter and are labeled Animation 21.1 through Animation 21.11.

21.2   The Changing Medium of Animation

Although map animations are often traced to the development of film in the early part of the 1900s, animated maps are probably as old as maps themselves. Because movement is part of the environment, its depiction was likely a part of illustrations by prehistoric humans. The early drawings on cave walls have received much attention. Drawn with charcoal, these pictures of animals and other objects have survived because of the protected environment in which they were made. Lost, of course, are those illustrations done on a less permanent and protected medium. Drawn with a stick in the sand, an early cartographic animation may have depicted the movement of animals or a fast-moving river.

An example of a cartographic animation on film was produced in the late 1930s by the Disney Studio. It depicted the invasion of Poland by Germany in 1939 (see Animation 21.1). In the animation, moving arrows, representing the German Army, are shown advancing toward the capital city of Warsaw and quickly encircling it. Shown as a part of a newsreel before the feature film in movie theaters, the animation was an effective means of both education and propaganda.

Making these early animations on film was time consuming. In addition, animations on film or video could not be easily duplicated, transported, or displayed, thus severely limiting their distribution and use. Computer technology, particularly storage devices such as the CD-ROM, DVD, and the Internet, spurred a resurgence of this method of mapping.

21.2.1   Digital Video

The advent of digital movie file formats in the early 1990s brought cartographic animations to the Internet. Three major digital video formats emerged: MPEG, AVI, and QuickTime. All of these formats essentially encode a series of images in compressed form. Compression is needed because video usually consists of about 30 frames per second. A very large file size would result if each frame were stored separately. To save space, only the changes between frames are encoded. A number of different digital video formats are available, and each can implement a variety of different compression schemes. These compression methods are referred to as *codecs*, short for compression-decompression, and act separately on the graphic and audio portion of the movie.

The most common digital video format is MPEG, short for Moving Pictures Experts Group, an international working group of experts formed by the International Standards Organization (ISO). The movie DVD uses MPEG-2 video encoding. The most common format for online applications is MPEG-4. Microsoft first introduced Audio Video Interleave (AVI) in 1992. AVI with Divx codec is widely used for the distribution of full-length movies through the web because compression is much greater than with MPEG-2. QuickTime is a multimedia format from Apple introduced in 1991 that contains one or more tracks, each of which store a particular type of data, such as audio, video, effects, or text (for subtitles, for example). Each track in turn contains track media, the digitally encoded media stream using a specific codec such as Cinepak, Sorenson, MP4, and DivX.

21.2.2 *Frame-Based Animation*

Frame-based animation is the simplest animation technique, and the resulting files can be distributed using any of the digital video file formats described above. Frame-based animation is analogous to the flipbook method used by grade school children. Here, a series of images are drawn in a small booklet and then "flipped through" to show the animation. The illusion of movement results from displaying the frames in rapid succession. The distinguishing characteristic of this form of animation is that foreground objects are not separable from the background. Rather, everything is reproduced on each frame.

Frame-based map animations can be made with any series of maps and a program for displaying frames in quick succession. This includes anything from a presentation program like Microsoft PowerPoint™ to high-end programs for movie editing (see Table 21.1). In capturing the individual frames, care must be taken to maintain the same location of the map on the screen. Even a slight change in position will be noticeable in the animation. Some programs also provide the possibility for incorporating sound. Another alternative is to use a movie-editing program, such as Avid or iMovie, to assemble a series of frames into a digital movie. A final alternative is to implement the animation in JavaScript or Java (see Figures 21.1 and 21.2).

Frame-based animations generally produce large files. For example, the individual animations constructed for the *Flight Atlas of North America* (Peterson and Wendel, 2007) that depict flights over a 24-hour period use 1440 frames, the number of minutes in a day (see Animation 21.2). Each individual file, stored as a JPEG, is about 500 kb in size. Assembling 1440 of these frames without compression would produce a file of 720 MB. Such a file could not be viewed as an animation because a standard computer cannot transfer this large amount of data to the screen at the speed needed for animation. A moderate amount of compression produces a file of 170 MB—still too large to be made viewable as an animation over a 30-second time span. A particular compression method called DivX was used with Adobe Premiere™, producing file sizes of between 6 and 12 MB,

TABLE 21.1. Methods for Creating a Frame-Based Animation

Presentation Programs
—fast slide show
—possible incorporation of: sound and movies
—example program: Microsoft PowerPoint

Java or JavaScript
—free
—involves calling Java or JavaScript code from HTML
—may include interactive elements such as the possibility of changing the speed of playback

Multimedia Program
—designed for cast-based (sprite) animation but can also be used for frames
—simple interface
—interaction added with scripting language
—example program: Adobe Flash

Movie Editing Programs
—combine video, audio, still images, graphic files, and special effects
—special compression techniques
—example programs: Adobe Premiere, Avid, Final Cut Pro

depending on the amount of variation between the frames. Although such a file is viewable as an animation, it may still be too large for distribution through the Internet.

21.2.3 Cast-Based Animation

An alternative approach to frame-based animation is to direct the movement of objects against a background. In manual animation, this is called cel-based animation because of the use of celluloid plastic sheets. Individual cels are placed on top of each other and then photographed. In computer-based animation, a cel is called a cast member, and computer instructions are used to specify how it is moved against a background. An object is an element that logically can be thought of as a separate entity from the background of an image. For example, in an animation of a deer running through a forest, the trees would be the background and a running deer would be a separate object. Also called sprite animation, this form of animation is the most common form of computer animation.

Almost every video game uses cast-based sprites to some degree. In the classic PacMan game, sprites move along defined paths at a given velocity. The background image remains the same throughout the entire game. Cast-based animations result in a smaller file size than a comparable frame-based animation and are therefore the preferred method for distributing animations through the Internet. However, their construction requires additional expertise and some knowledge of programming, and their display is usually based on a plug-in.

Cast-based animation programs, such as Adobe Flash, are used for a variety of applications from advertising to scientific visualization. The Flash plug-in makes the animation playable within a browser. It is the most widely downloaded program ever created and resides on an estimated 98% of all personal computers.

JavaScript Frame-Based Animation

```
<html>
<head>
<title>JavaScript Animation</title>
<script language="JavaScript" type="text/javascript">
<!--
// Create the image arrays
if(document.images) {
mapSeries = new Array("map00.jpg", "map01.jpg", "map02.jpg",
"map03.jpg")
imagesCache = new Array()
// Cache the images here
for (i=0;i<mapSeries.length;i++) {
imagesCache[i] = new Image
imagesCache[i].src = mapSeries[i]
}
// theCount will keep track of the current image
theCount = 0
}
// advance to the next image
function turnPage() {
if (theCount == imagesCache.length-1)
{theCount = 0
} else {
theCount++
}
// (display a frame of the animation)
document.map.src = imagesCache[theCount].src
// setTimeout inserts a pause between each frame of the animation. In this // case,
there will be a 300 millisecond pause, and then the next frame of // the animation
will be displayed.
setTimeout("turnPage()", 300)
}
// -->
</script>
</head>
// Every line of code so far has been part of the JavaScript code in the
// <HEAD> area of the document. To make the animation appear on the Web
// page, we insert the following lines of code into the <BODY> section of
// the page. The body statement sets the background color and calls the
// turnPage function. The img tag displays the first image.
<body bgcolor="#FFFFFF" onload="turnPage()">
<p align="center">
<img name="map" src="map00.jpg" width="150" height="233" border="0"
alt="map animation">
</p>
</body>
</html>
```

FIGURE 21.1. HTML and JavaScript code used to display a series of images as an animation. The speed of the animation is set in the turnPage function by implementing a pause between the frames of 300 milliseconds.

21.3   Types of Map Animation

Many types of cartographic animation are possible. A basic distinction is often made between temporal and nontemporal cartographic animations. Temporal animations are the most common and simply show change over time (Acevedo and Masuoka 1997; Dorling 1992; Dorling and Openshaw 1992), such as the diffusion of a farming method like irrigation (see Animation 21.3). Temporal animation is essentially a time-lapse. The flight animations presented earlier are also time-based.

A common example of a nontemporal animation is a terrain flythrough that has been popularized by Google Earth. Here, a series of oblique views of a landscape are displayed in quick succession to provide the appearance of flying over an area (see Animations 21.4 and 21.5). The flythrough, usually constructed by draping

---

Java Frame-Based Animation Steps

ANButton, a simple Java applet, available from http://javaboutique.internet.com/
displays up to 10 images when the mouse in located over a start button.
In Javaboutique:
1. Choose Applets by Category | Visual Effects | Sequential Animation.
2. Choose on ANButton
3. Choose anbutton.zip to download the compiled java code and images.
4. Drag the Anbutton.class and the Anbutton.java files to a folder.
5. Replace the images with your own.
6. On the ANButton page, highlight and copy the sample html source that starts with
<APPLET CODE and ends with </APPLET>
7. Paste this code into the supplied html file replacing the words "(Java Applet code
goes here)".
8. Change the html code so it reflects the size of the new images (WIDTH=xx,
HEIGHT=xx)
21. Add the image names after "PARAM NAME="

FIGURE 21.2. Steps to implement an animation using a Java Applet called ANButton.

a satellite image or an air photograph over a digital elevation model (DEM), does not compress time like a temporal animation but only changes the orientation and position of view.

Many other types of nontemporal cartographic animations have been proposed that highlight the different applications of animation with maps:

1. A cartographic zoom would depict a series of maps at increasing or decreasing map scales (von Wyss 1996). This form of animation has been the most difficult to automate because it involves all aspects of the cartographic abstraction process, especially the selection and simplification of features. Online map services have implemented a type of cartographic zoom by quickly switching between tiled maps at predefined scales.

2. A classification animation would show different methods of data classification, such as equal interval, quantile, standard deviation, and natural breaks (see Animation 21.6).

3. A generalization animation depicts maps with a single method of data classification but multiple classes (see Animation 21.7). Sound can be added to an animation to accentuate change in the display (see Animation 21.8). These animations provide a less misleading view of the data than simply relying on one form of data classification, and emphasize the effect of data classification on a mapped distribution.

4. A spatial trend animation would depict a trend in space over time. An example would be an animation of the percentage of population in different age groups within a city (e.g., 0–4, 5–9, 10–14 years of age). Older populations tend to live closer to the center of the city, and younger populations are at the periphery. This type of spatial trend animation for the city of Omaha shows older populations on the right side, closer to the older parts of the city along the Missouri River, and younger populations on the left corresponding to the western suburbs (see Animation 21.9).

21.3.1 Variables of Animation

A set of dynamic variables have been proposed, analogous to the visual variables used in graphic design (DiBiase et al. 1992; Oago and Kraak 2002). Reordering, for example, involves presenting a temporal animation in a different order. Frames depicting earthquake activity could be ordered by the number of deaths so that the more severe earthquakes are shown first. Changing the pace of the animation has also been proposed to highlight certain attributes. In the earthquake example, the duration of each scene in the animation could be made proportional to the magnitude of the earthquake or the number of deaths.

21.3.2 Interactive Map Animation

In addition to developing new forms of cartographic animation, new methods have also emerged for adding interaction to cartographic animation. A program that both automated the production of the individual frames of a cartographic animation and brought interaction to its display was MacChoro II (Peterson 1993). Limited to the display of choropleth maps, the program used dialogs to control the selection of variables and data classification methods. The individual maps were then constructed and stored in memory at a speed of approximately one map per second (late 1980s technology). Once stored in memory, a pop-up control palette was used to change the speed and direction of the animation at speeds up to 60 maps a second.

Alternative methods of adding interaction to cartographic animation have been developed using JavaScript. In one implementation, the legend is used to activate the display of alternate maps. Quickly moving the mouse over the legend creates the appearance of an animation (Peterson 1999). The example in Animation 21.10 shows various causes of death for counties in the state of Nebraska. It is argued that directing the animation in this way improves the human connection to information display through animation.

21.4 Summary

Cartographic animation is an important technique to further our understanding of the spatial environment. It demonstrates that individual maps are only snapshots in time. One should ask: What was before? What will come after? What trends would be evident if the time element were viewed as an animation? The individual map is also a snapshot in reference to other data sets. What nontemporal trends would be evident if the map were viewed along with other related spatial data? Finally, the individual map is a snapshot in the choice of representational forms that were used to depict the data. A cartographic animation can provide a more meaningful view of the data through the use of different symbols or different forms of data classification.

Cartographic animation has been limited by both the difficulty of their construction and distribution, and by a continued fixation on the individual, static map. Viewing static maps without interaction or animation has limited value, particularly in the process of searching for spatial and temporal patterns. Computer technology is making it possible to both create different types of cartographic animations and to distribute these animations to a wider audience.

21.5   Questions

1. What is the single-map solution, and to what degree have computer maps adopted this method of map display?

2. What are some arguments to support the notion that the first map was animated?

3. What is meant by a "one-off" animation?

4. Describe the general process of creating the 24-hour flight traffic animations for the *Flight Atlas of North America*.

5. Describe some flight patterns that are visible in the flight animations available through the "theflightatlas" YouTube channel.

6. Describe the different mediums and methods of distribution that have been used for animated maps.

7. What are some methods for displaying a series of maps as an animation?

8. What is the distinction between frame-based and cast-based animation?

9. Why are frame-based animations usually larger?

10. Provide examples of temporal and nontemporal cartographic animations.

11. How has interaction been incorporated in cartographic animation?

21.6   References

Acevedo, W., and P. Masuoka (1997) Time-Series Animation Techniques for Visualizing Urban Growth. *Computers and Geosciences* 23(4): 423–436.

DiBiase, D., A. M. MacEachren, J. B. Krygier, and C. Reeves (1992) Animation and the Role of Map Design in Scientific Visualization. *Cartography and GIS* 19(4): 215–227.

Dorling, D. (1992) Stretching Space and Splicing Time: From Cartographic Animation to Interactive Visualization. *Cartography and Geographic Information Systems* 19(4): 215–227.

Dorling, D., and S. Openshaw (1992) Using Computer Animation to Visualize Space-Time Patterns. *Environment and Planning B: Planning and Design* 19: 639–650.

Peterson, Michael P. (1993) Interactive Cartographic Animation. *Cartography and Geographic Information Systems* 20(1): 40–44.

Peterson, Michael P. (1999) Active Legends for Interactive Cartographic Animation. *International Journal of Geographical Information Science* 13(4): 375–383.

Peterson, Michael P., and Jochen Wendel (2007) *North American Animated Flight Atlas. Journal of Maps, 1–9, v2007, 98–106.* [http://www.journalofmaps.com/]

Ogao, P. J., and M. J. Kraak (2002) Defining Visualization Operations for Temporal Cartographic Animation Design. *International Journal of Applied Earth Observation and Geoinformation, 4*(1), Aug.

von Wyss, Martin (1996) The Production of Smooth Scale Changes in an Animated Map Project. *Cartographic Perspectives*, no. 23: 12–20.

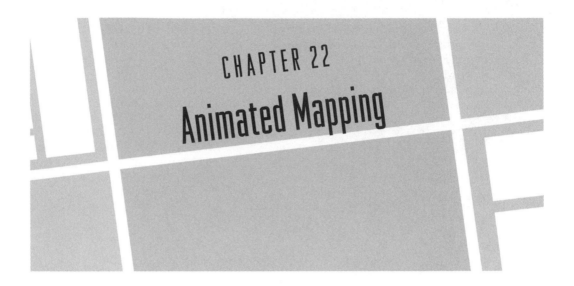

CHAPTER 22
Animated Mapping

Animation is not the art of drawings that move but the art of movements that are drawn.

—Norman McLaren

22.1 Introduction

Animation adds another level of complexity to the making of maps, and their distribution through the Internet is yet a further complication. When using frames, a cartographic animation simply requires making multiple maps. While this process is fairly easy by computer, the size of the resulting animation file can be quite large, making it difficult to distribute through the Internet. File sizes can be reduced considerably using the cast-based method with a program like Adobe Flash™. Unfortunately, these programs do not integrate interactivity through multiscale panable maps (MSPs).

In many ways, animation has been incorporated within the new interface of maps. The now ubiquitous method of zooming into a map could be described as an interactive cartographic zoom. The addition of a scroll-wheel or track-pad with two-finger control further enhances the interactive animation effect. Digital globe programs such as Google Earth incorporate interactive animation by making fly-throughs possible, and this capability is also being incorporated into the browser with Google Maps Earth View.

The application programming interfaces (APIs) associated with online map services have only limited tools for animation. It is clear that animation was simply not a main consideration in designing any of the currently available APIs. In addition, the tile-based approach for map delivery makes the presentation of animations more difficult because the map is composed of pieces and cannot be presented as a series of animation frames. In this chapter, we examine how animation has been incorporated with online mapping APIs.

22.2   Animation with Markers

One limited form of animation that has been incorporated in some online mapping APIs is a drop and bouncing feature for markers. With "drop," the marker drops into place. With "bounce," the marker is made to bounce up and down. Figure 22.1 shows the code for a bouncing icon. Replacing BOUNCE with DROP makes the marker drop-in from the top of the map.

Other animated maps can be made through extensive use of JavaScript. The example in Figure 22.2 uses a number of JavaScript functions to move a marker along a path defined by two addresses. The associated illustration in Figure 22.2 shows one point in the animation as the marker moves between the two end-points (Geocodezip 2012). The moving marker is kept in the middle of the map by continually panning the map.

22.3   Animating List Items

One of the best ways to implement a cartographic animation is to incorporate interaction in the display. Figure 22.3 shows earthquake events for Christchurch, New Zealand, and the surrounding area (Nicholls 2012). The list on the right is dynamically linked to the location of earthquake events on the left, so that moving the mouse over the list causes the corresponding earthquake to be displayed on the map. The hover effect is implemented with onmouseover, which calls a function called quakeHoverOn that displays the circle for the earthquake intensity. The animation is made by quickly moving the mouse over the list of earthquake events.

22.4   Animation with Street View

The panorama offered by Street View may be viewed as a type of animation. To create the panorama, a total of nine pictures are taken at a single instant and then "stitched" together. Spinning the panorama creates the animation. The Google Maps API implements various controls over the panorama. For example, it finds the panorama that is closest to a point provided (see Figure 22.4).

The panorama view may also be combined with a map and symbol to show the current direction of view, as in Figure 22.5. The heading, pitch (up or down), and zoom level is given with the POV settings, where:

- Heading (default 0) defines the rotation angle around the camera in degrees relative to true north. Headings are measured clockwise from north with 90 degrees being east.
- Pitch (default 0) defines the angle variance "up" or "down" from the camera's initial default pitch, which is often (but not always) flat horizontal. An

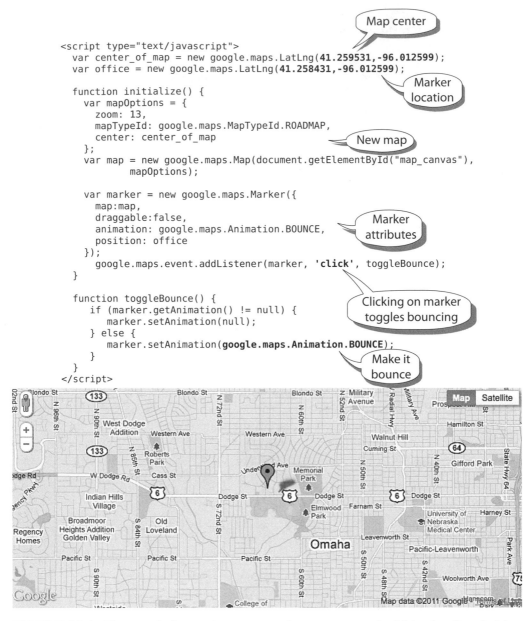

```
<script type="text/javascript">
  var center_of_map = new google.maps.LatLng(41.259531,-96.012599);
  var office = new google.maps.LatLng(41.258431,-96.012599);

  function initialize() {
    var mapOptions = {
      zoom: 13,
      mapTypeId: google.maps.MapTypeId.ROADMAP,
      center: center_of_map
    };
    var map = new google.maps.Map(document.getElementById("map_canvas"),
            mapOptions);

    var marker = new google.maps.Marker({
      map:map,
      draggable:false,
      animation: google.maps.Animation.BOUNCE,
      position: office
    });
      google.maps.event.addListener(marker, 'click', toggleBounce);
  }

  function toggleBounce() {
    if (marker.getAnimation() != null) {
      marker.setAnimation(null);
    } else {
      marker.setAnimation(google.maps.Animation.BOUNCE);
    }
  }
}
</script>
```

FIGURE 22.1. The simple form of animation that is integrated within the Google Maps API either drops or bounces a marker. In this example, the marker bounces until the user clicks on the marker to make it stop.

image taken on a hill will likely exhibit a default pitch that is not horizontal. Pitch angles are measured with negative values looking up (to –90 degrees straight up and orthogonal to the default pitch) and positive values looking down (to +90 degrees straight down and orthogonal to the default pitch).

• Zoom (default 1) defines the zoom level with 0 being fully zoomed out. Most Street View locations support zoom levels from 0 to 3, inclusive.

```
function calcRoute(){
   if (timerHandle) { clearTimeout(timerHandle); }
   if (marker) { marker.setMap(null);}
   polyline.setMap(null);
   poly2.setMap(null);
   directionsDisplay.setMap(null);

   var rendererOptions = {
      map: map
   }
   directionsDisplay = new google.maps.DirectionsRenderer(rendererOptions);

   var start = document.getElementById("start").value;
   var end = document.getElementById("end").value;
   var travelMode = google.maps.DirectionsTravelMode.DRIVING

   var request = {
      origin: start,
      destination: end,
      travelMode: travelMode
   };
   var path = response.routes[0].overview_path;
   var legs = response.routes[0].legs;
   for (i=0;i<legs.length;i++) {
      if (i == 0) {
         startLocation.latlng = legs[i].start_location;
         startLocation.address = legs[i].start_address;
         marker = createMarker(legs[i].start_location,"start",legs[i].start_address,"green");
      }
      endLocation.latlng = legs[i].end_location;
      endLocation.address = legs[i].end_address;
      var steps = legs[i].steps;
      for (j=0;j<steps.length;j++) {
         var nextSegment = steps[j].path;
         for (k=0;k<nextSegment.length;k++) {
            polyline.getPath().push(nextSegment[k]);
            bounds.extend(nextSegment[k]);
         }
      }
   }
}
```

Callouts:
- Reset variables
- Create a renderer for directions and attach to map
- Get beginning and ending addresses
- Method of travel: DRIVING, WALKING, or BICYCLING
- Parameters for travel route request
- Two embedded loops that divide route into separate legs, each having segments
- Marker for start is green
- Moving along segments

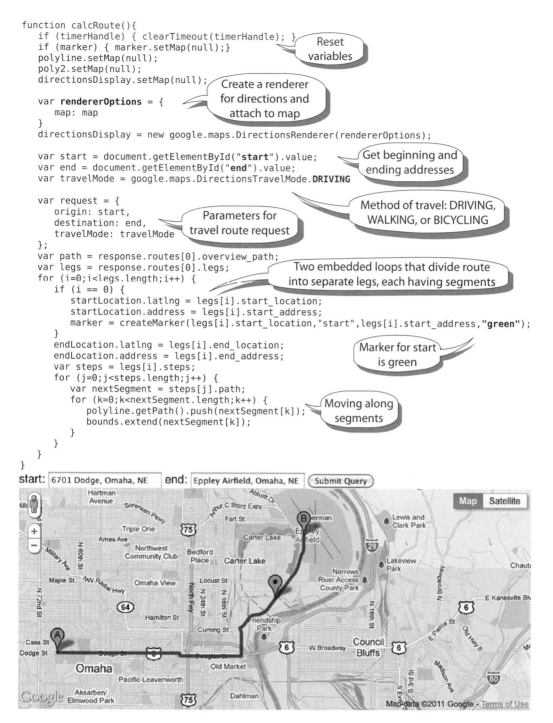

FIGURE 22.2. A red marker is made to move along the line between two locations. This type of animation is not a part of the Google Maps API and is programmed with extensive use of JavaScript. Geocodezip.com 2012.

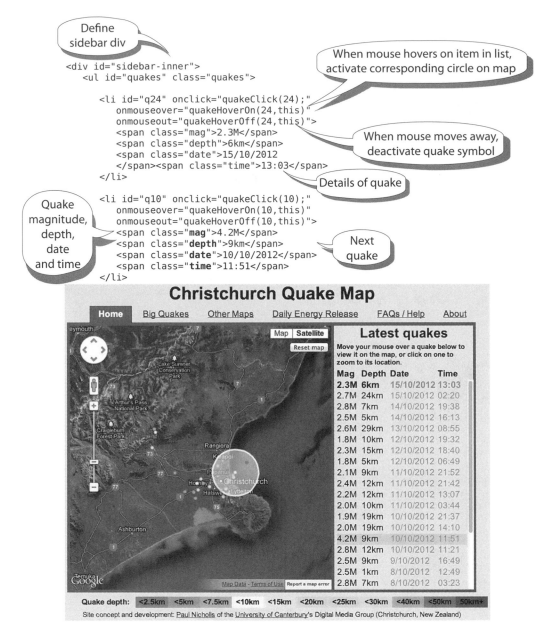

FIGURE 22.3. The list on the right-hand side of the illustration is tied to the map, so hovering the mouse over a list item displays the corresponding earthquake event. Search: Christchurch Earthquake Map, Copyright 2012 Google.

22.5   Animation with a Digital Globe

Digital globes may include a flythrough feature that simulates flying through a terrain. This is implemented through Google Maps Earth View. Figure 22.6 shows the API interface.

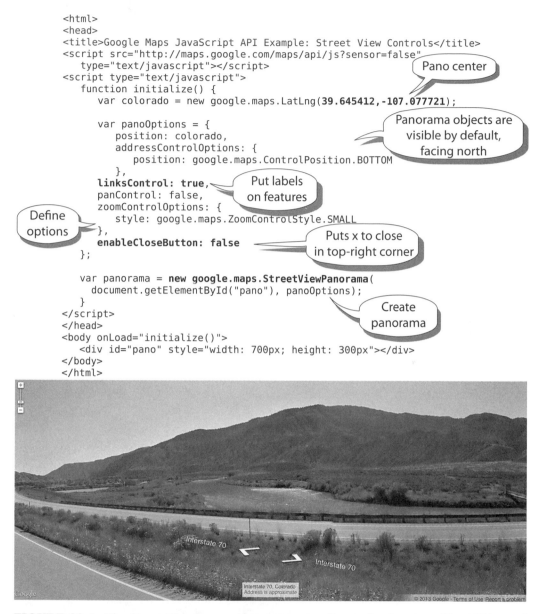

```html
<html>
<head>
<title>Google Maps JavaScript API Example: Street View Controls</title>
<script src="http://maps.google.com/maps/api/js?sensor=false"
    type="text/javascript"></script>
<script type="text/javascript">
    function initialize() {
        var colorado = new google.maps.LatLng(39.645412,-107.077721);

        var panoOptions = {
            position: colorado,
            addressControlOptions: {
                position: google.maps.ControlPosition.BOTTOM
            },
            linksControl: true,
            panControl: false,
            zoomControlOptions: {
                style: google.maps.ZoomControlStyle.SMALL
            },
            enableCloseButton: false
        };

        var panorama = new google.maps.StreetViewPanorama(
            document.getElementById("pano"), panoOptions);
    }
</script>
</head>
<body onLoad="initialize()">
    <div id="pano" style="width: 700px; height: 300px"></div>
</body>
</html>
```

Callouts: Pano center · Panorama objects are visible by default, facing north · Put labels on features · Define options · Puts x to close in top-right corner · Create panorama

FIGURE 22.4. The Street View panorama represents a form of animation. The API is used here to display the Street View scene without the map. The image shown is along Interstate 70 and the Colorado River west of Denver.

Although requiring a considerable amount of JavaScript, movement along a street can be animated in Earth View by moving between successive Street View shots. The example in Figure 22.7 demonstrates the use of the Earth plug-in with a type of drive-through. Here, a car is shown driving down a street. A series of additional JavaScript functions are incorporated to simulate this type of movement because these functions are not implemented within the API or plug-in.

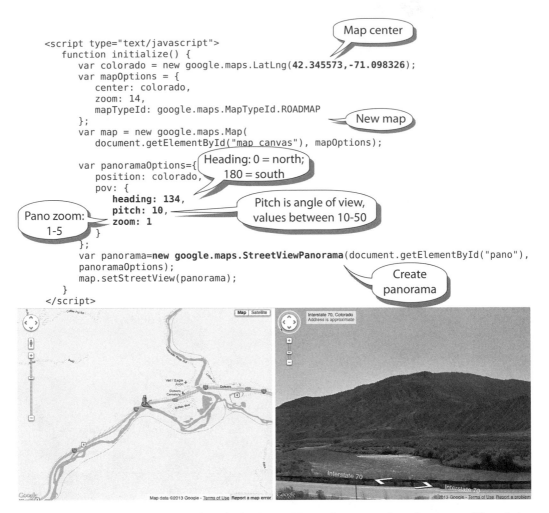

```
<script type="text/javascript">
    function initialize() {
        var colorado = new google.maps.LatLng(42.345573,-71.098326);
        var mapOptions = {
            center: colorado,
            zoom: 14,
            mapTypeId: google.maps.MapTypeId.ROADMAP
        }; var map = new google.maps.Map(
            document.getElementById("map canvas"), mapOptions);

        var panoramaOptions={
            position: colorado,
            pov: {
                heading: 134,
                pitch: 10,
                zoom: 1
            }
        };
        var panorama=new google.maps.StreetViewPanorama(document.getElementById("pano"),
        panoramaOptions);
        map.setStreetView(panorama);
    }
</script>
```

Map center

New map

Heading: 0 = north; 180 = south

Pitch is angle of view, values between 10-50

Pano zoom: 1-5

Create panorama

FIGURE 22.5. A map combined with the Street View of an area along Interstate 70 and the Colorado River west of Denver.

22.6 Summary

While animation is not well integrated within current mapping APIs, a number of possibilities exist for moving objects along a path or panning through Street View images. Street View panning can be combined with a corresponding map showing the direction of view. It is even possible to create an interactive animation by hovering over a series of items in a list. Unfortunately, a considerable amount of programming is still necessary to implement most types of animation. As mapping APIs develop, more functions for animation will be introduced.

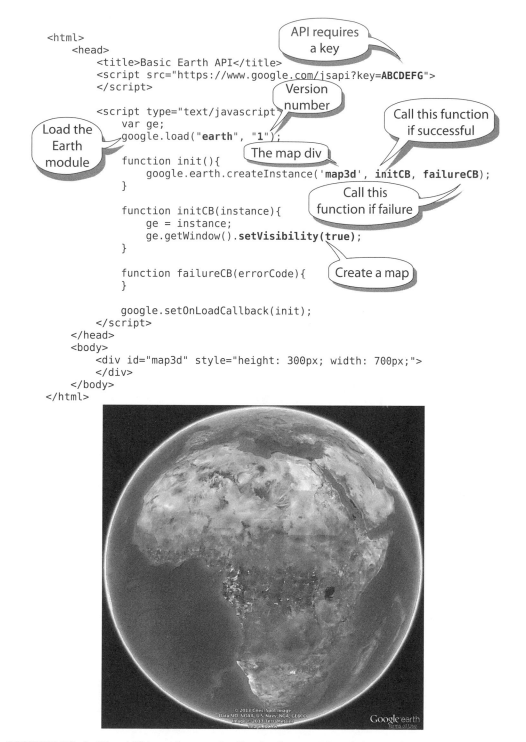

```
<html>
    <head>
        <title>Basic Earth API</title>
        <script src="https://www.google.com/jsapi?key=ABCDEFG">
        </script>

        <script type="text/javascript">
            var ge;
            google.load("earth", "1");

            function init(){
                google.earth.createInstance('map3d', initCB, failureCB);
            }

            function initCB(instance){
                ge = instance;
                ge.getWindow().setVisibility(true);
            }

            function failureCB(errorCode){
            }

            google.setOnLoadCallback(init);
        </script>
    </head>
    <body>
        <div id="map3d" style="height: 300px; width: 700px;">
        </div>
    </body>
</html>
```

Callout: API requires a key

Callout: Version number

Callout: Call this function if successful

Callout: Load the Earth module

Callout: The map div

Callout: Call this function if failure

Callout: Create a map

FIGURE 22.6. The API interface to the Earth plug-in. The globe can be easily rotated and zoomed. The "N" key reorients the map to the north. Copyright 2013 Google.

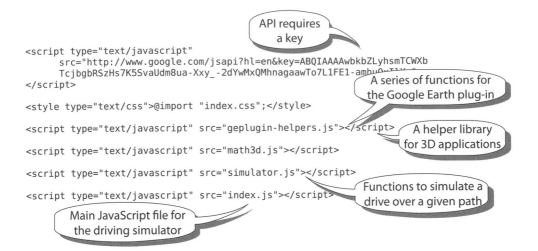

FIGURE 22.7. A drive-through implemented with the Google Earth plug-in. The animation is based on numerous JavaScript functions such as geplugin-helpers, math3d, and simulator. Nurik 2012; copyright 2013 Google.

22.7 Exercise

Upload the code22.zip file and make the suggested changes.

22.8 Questions

1. In what sense is an MSP a form of animation?

2. Why is tile-based mapping not conducive for many forms of cartographic animation?

3. List some reasons why a mapping API would not include separate functions for animation.

4. Describe an organized set of functions that should be part of an API to properly cartographic animation.

5. Describe some applications of list-based, interactive animation.

6. Why would a digital globe always need to be animated?

22.9   References

Geocodezip.com (2013) Animated Directions–Moving Marker. [http://www.geocodezip.com/]

Nicholls, Paul (2012) Christchurch QuakeMap. [http://www.christchurchquakemap.co.nz/]

Nurik, Roman (2012) Driving Simulator. Google Earth API Demo Gallery. [http://code.google.com/p/earth-api-samples/]

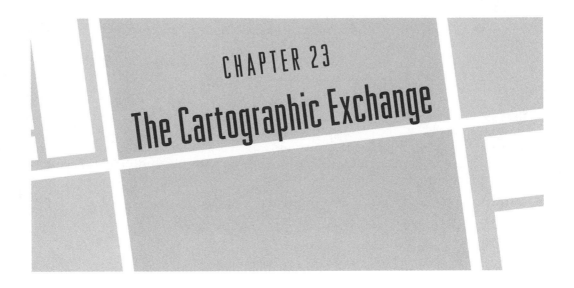

CHAPTER 23

The Cartographic Exchange

Civilization grew in the beginning from the minute that we had communication—particularly communication by sea that enabled people to get inspiration and ideas from each other and to exchange basic raw materials.

—THOR HEYERDAHL

23.1 Introduction

The term *cartographic exchange* may be interpreted in several different ways. First, the exchange of information can be viewed as the basis of cartography. The map serves little purpose if information is not transferred to a user or if it does not contribute to some type of knowledge formation. Second, cartographers have been involved in an exchange with other disciplines, principally psychology, to better understand the communication process, the meaning of interaction, and the influence of the medium in conveying information. Third, the computerization of cartography is also the result of a significant exchange with computer science. The computerization of maps has created a greater understanding of the mapping process and ultimately, through the Internet, a greater availability and a wider distribution of maps. Finally, cartography has served as the foundation for many areas of research and development that emerged during the latter part of the 20th century. In this last chapter, we examine some of the fundamental influences on modern cartography.

23.2 Importance of Interaction

We live in an amazing time for mapping. Within a matter of 20 years, from the 1970s to the 1990s, maps changed from static objects on paper to interactive presentations delivered through an electronic network. In the years since then, maps have become even more interactive in the sense of being able to add information—both thematic information through mashups and the editing of the underlying base map. With the online editing capabilities of OpenStreetMap and Google Map Maker, maps are no longer defined by governments or companies but have become collective representations created by many people.

The incorporation of interaction in all aspects of cartography has been dramatic. But interaction is not new to mapping. It is very likely that the first map ever created, perhaps a drawing in sand, was interactive, with two individuals negotiating a representation of the surrounding world. Questions posed by one individual would have led to immediate changes in the map by the other. This interaction created a collective representation of the environment as it was understood and agreed upon by both individuals. Online collaborative systems like OpenStreetMap implement a similar type of *map in the sand*, with users collectively contributing to a representation of space.

The period of time between the "sand map" and the interactive online map was dominated by a static medium—first stone and then paper. Cartography developed during this period, furthered to a considerable extent by military and other governmental interests. Forms of representation and generalization developed over time that still control how maps are made and presented, even the maps in the cloud that we use today. Methods of cartographic representation that effectively incorporate interaction are still in the early stages of development. In many ways, cartography is still "breaking away" from paper-based, static representations. The process of defining a new interactive cartography that is free of the influence of paper-thinking will take many more years.

Interaction is not unique to this new cartography. A characteristic of any good conversation is that one individual responds to what the other has said. A non-sequitur, a statement that does not follow from the previous statement, effectively stops the conversation. To move forward again, the non-sequitur must be overcome so that the flow of interaction can be reestablished. This can only be accomplished through interaction. Part of the new responsibility of the cloud cartographer is to maintain the user's interaction with the map.

23.3   The Need for Speed

In evaluating all of the factors associated with improving interaction with the Internet map, the single most important is the speed of map display. In the world of cloud mapping, a fast map is a good map. This conclusion is supported by a study on WebGIS systems where it was determined that Internet connection speed is the most significant factor in user interaction and satisfaction (Ingensand and Golay 2011).

A conversation again provides a good analogy to understand the importance of speed in interaction. If too much time goes by between the statements in a conversation, the conversation effectively ceases. Extended gaps result in an odd conversation in which the participants need to continually remind themselves about what is being discussed. Gaps allow the mind to think about other things.

Sustaining any type of interaction is based on timely responses. Interactive mapping is no different.

The speed of map display can be improved in various ways. The use of tiles has significantly improved interaction with maps, especially the process of panning and zooming. Although most tiles are still delivered in a raster format, vector tiles provide many benefits; Google has demonstrated that their display can be as fast, or faster, than raster tiles.

In a mashup environment, attempting to place too many points, lines, or polygons on the map slows its display, especially if the data is being parsed from a text file. Storing the data in the cloud increases the speed of data display, whether the data are points, lines, or polygons—or raster overlays. The server also has a major influence on the speed of interaction. Although a computer server can be used in the "back-office" to serve maps, it is not likely to lead to very fast map display times. Using a server in a data center will generally be both faster and more secure.

Finally, the speed of map display through smartphones and tablets should also be considered. These devices have slower processors and generally rely on a slower mobile communication network. In evaluating any map, it is good to test it on a variety of mobile and desktop devices. This would also serve as a test of map display on a smaller screen.

23.4 The Medium of Mapping

The computer has become much more than a tool to calculate complex equations or store and retrieve data. Computers represent an increasingly dominant medium of communication. The concept of medium is used here in the sense of a carrier of information. Marshall McLuhan argued in the early 1960s that electronic technology is not merely another form of communication but an influence in reshaping and restructuring all aspects of our lives (Kay and Goldberg 1977). The new social media is certainly a further indication of this concept. McLuhan (1967, p. 30) argued that all media alter the sensory mix and force changes on the individual.

McLuhan asks us to recognize the way in which all media, including and perhaps especially electronic media, create psychological environments—environments to which we subordinate ourselves without clearly recognizing the price we pay in doing so (Miller 1971, p. 8). In short, we are effectively shaped by the means we use to communicate. As he famously put it, "the medium is the message." While this motto is certainly an exaggeration, the medium does exert an influence over and above what is carried in the message itself. McLuhan argues that the medium also becomes the massage by kneading, manipulating, and literally reshaping our culture and our relationship to it (Ryan and Conover 2004, p. 15).

McLuhan also argues that we live in a rear-view mirror society (Theall 1971). He states that all new forms of media take their initial content from what preceded them. Not only is the new medium based on the old, but society dictates that the only acceptable way to approach a new medium is by emulating the old—through the rear-view mirror. It is evident from current maps that we have looked in the rear-view mirror while progressing from paper maps to those presented by computer.

23.5 Familiarity Breeds Acceptance

Perhaps one of the more interesting observations in evaluating online mapping relates to the influence of the "familiar map" on map use. Map users seem to gravitate toward maps with which they are the most familiar, and it takes a considerable amount of time for users to fully accept a different-looking map. For example, although Google Maps clearly implemented a more intuitive zooming/panning interface, it took four years, from 2005 to 2009, for it to surpass MapQuest in number of users. A major reason is that users had become accustomed to the look of MapQuest maps, although these maps were clearly not as interactive as Google's.

Maps are made familiar and comfortable to us through both their interface and graphic design. Since most online mapping sites have standardized on the tile-based, zooming-panning interface, the major difference between them is the look of the map, and this look is a function of how the map is designed. Map design includes everything from which color is used to represent the roads to how the features have been generalized. It is the final product of all of the computer-based, decision-making processes that go into the process of rendering the map. The programmed decision-making process reflects principles for how maps have been made for centuries by hand. Although the processes that go into rendering the map are similar among the major providers, the look of the resulting maps varies considerably.

Google and other Internet map providers have adapted the design of their maps to the local market. The large-scale map of San Francisco looks very different from the same scale map of London. By changing the design to match the expectations of each local area, Google has recognized that people are accustomed to a certain look in their maps. This look varies between different parts of the world. As the dominant online map becomes more familiar, people will adapt to this representation and expect all maps to have this look. Any map that does not meet this expectation will not be acceptable. Conformity can be both a strength and a weakness; even Google continually modifies the design of its maps, but only slightly.

23.6 Maintaining the Map

Making and maintaining maps, and the underlying information from which they are based, is extremely expensive. Governments have performed this task in the past to exploit resources, fight wars, and collect taxes. While governments still spend large amounts of money for these purposes, private companies like Google and organizations like OpenStreetMap have assumed this massive responsibility. The most important of these purposes may be the creation and maintenance of the large-scale street map.

Companies have assumed the task of maintaining street maps because they can make money from them, even as they are provided for free. Maps bring users to the Internet, which companies see as a major benefit. While Google and others distribute maps for free, creation, storage, and instantaneous delivery involve significant costs. Of all of the expenditures associated with maps, the primary cost is keeping them up to date. Maintaining maps requires considerable resources, exceeding those of most governments.

Sustainability is a major concern of maps in the cloud. The application programming interface (API) approach presented in this book relies on two fundamental elements: (1) an updated base map and ancillary layers like the satellite and terrain views; and (2) a system to place other information on these maps. In considering which map and API will outlast the others, it is important to keep in mind the revenue streams that will keep these resources available. If a particular mapping resource does not have a system capable of generating sufficient revenue, it is unlikely that it will continue to be updated. In addition, the associated API may also cease to exist.

Competition will be key to the future of maps in the cloud. Further development is based on a competitive environment that encourages and rewards innovation and continued free access to mapping resources. This competition can occur only if multiple mapping applications remain viable.

23.7   Mobile Maps

Smartphones and tablets have already had a major impact on maps. Companies like Google have redesigned their Maps API so that it runs more efficiently on mobile devices. Additional tools will be implemented that make the smaller format map as useful as the larger versions on desktop and laptop computers.

While the display is smaller, the screen resolution is generally greater on mobile devices. For example, the original Apple iPhone had a resolution of 163 dpi (320 × 480). Successive versions increased the resolution to 326 dpi (640 × 960 and 640 × 1136). These resolutions exceed those of most desktop displays, which are generally in the range of 102–128 dpi.

The use of higher resolution displays will require downloading more data to fill the screen. Graphics of any kind—especially map tiles—can be very expensive when using a data plan through a mobile provider. The caching feature on these devices bypasses the need to download the tiles more than once if the user stays in one place, but, when traveling, tiles need to be continually downloaded. Users have learned the hard way that maps can quickly consume an entire month's data plan. Increased availability of caching space, the automatic download of tiles when connected to WiFi, or the permanent storage of maps on mobile devices are all solutions to this problem.

Finally, mobile devices are changing the way we interact with both computers and maps. For example, zooming and panning with touchscreen devices is now done by moving multiple fingers across the display. The pinching and spreading of fingers has become a standard way of changing the scale of a map. New methods of map interaction will certainly develop with mobile devices such as shaking or tilting the device to activate a certain feature. With time, the movements that control display of the map will become standardized.

23.8 Cartographic Spaghetti

In this last section, we return to an illustration presented in Chapter 1 which depicted the development of cartography from 1900 to the present. The right side of the illustration, repeated in Figure 23.1, shows the various areas of research and development that can trace their origins to thematic cartography.

Throughout most of the history of cartography, maps were used to show the location of places. A series of developments, culminating in the 1700s (Friendly and Davis 2009), created a new type of map that moved cartography from simply putting features "in place" toward the depiction of information "about space" (Petchenik 1979; Robinson 1982). A direct link can be made between thematic cartography and the development of geographic information systems (GIS) in the 1970s. Most research in the area of cartographic communication concerned itself with thematic maps. GIS and cartographic communication provided the foundation for the many different areas of research that developed over the following years.

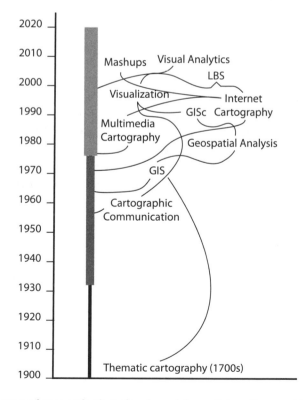

FIGURE 23.1. Areas of research that developed from thematic mapping. The illustration demonstrates the importance of thematic maps in a variety of research areas.

All of these research areas, as well as the growth of mapping through the Internet, represent a renewed interest in the power of maps to convey information. GIS combined with geospatial analysis to form geographic information science (GISc). Cartographic visualization combined cartographic communication with GISc, ultimately leading to visual analytics. Multimedia cartography developed before the introduction of the web through devices such as the analog Laserdisc, the digital Compact Disc (CD), and DVD. It quickly merged with Internet cartography in the latter part of the 1990s. Mashups represent the practical application of bringing information to maps in the cloud. Thematic mapping has inspired a great many ideas and associated areas of research.

23.9 Conclusion

For thousands of years, maps have enhanced our ability to comprehend the world. In addition to assisting in navigation, maps increase our understanding of spatial relationships and help us to comprehend major areas of collective concern, such as global climate change. Recent developments in map delivery by computer have increased our ability to access maps and interact with the underlying information. With new forms of interaction, including the ability to add information to the map, the utility of maps has increased as a tool for both communication and analysis.

Maps have not only had a major influence on the way people think about the world, but have also influenced how scientists in various disciplines conduct research. The many areas of research related to maps provide an indication of their growing importance as scientific instruments in themselves. Used for both presentation and analysis, maps continue to serve as an inspiration to research.

Mapping is a valuable skill. It is a way to both investigate and communicate the spatial world. Although the exact procedure for bringing information to the map will certainly evolve as cloud computing develops, the basic methods for defining location and adding information will remain unchanged. Points, lines, and areas will always be defined with some type of coordinate system, and visual variables like size and color will continue to control the representation. New methods of interaction will be added and new designs will be introduced, yet user expectations will likely dictate that the basic look of maps remain similar to maps of the past.

23.10 Questions

1. Describe the relationship between mapping and interaction, from a hypothetical first map in the sand to modern maps.

2. What is the relationship between interaction and a conversation?

3. What aspects of the paper map are still evident in modern maps presented by computer?

4. Why is speed of map presentation so important for interactive mapping?

5. What is a medium and how has the change in the medium of map delivery influenced cartography?

6. What are some reasons that map users gravitate toward a familiar map?

7. What makes the maintenance of map information so expensive?

8. What is one of the major problems in distributing maps to mobile devices?

9. Describe an entrepreneurial application for mapping in the cloud that would generate an income.

10. Each chapter of the book begins with a quote. How many can you remember and what do they mean?

23.11 References

Friendly, Michael, and Daniel J. Davis (2001) Milestones in the History of Thematic Cartography, Statistical Graphics, and Data Visualization. [http://www.datavis.ca/milestones/]

Ingensand, J., and F. Golay (2011) Remote-Evaluation of User Interaction with WebGIS in Katsumi, Tanaka, Peter Fröhlich, and Kyoung-sook Kim (Eds.), *Web and Wireless Geographical Information Systems*. Berlin: Springer Verlag.

Kay, A., and A. Goldberg (1977) *Personal Dynamic Media*. Computer (USA), pp. 31–41.

McLuhan, M. (1967) *The Medium Is the Massage*. New York: Bantam.

Miller, J. (1971) *Marshall McLuhan*. New York: Viking.

Petchenik, B. B. (1979) From Place to Space: The Psychological Achievement of Thematic Mapping. *The American Cartographer*, pp. 5–12.

Robinson, Arthur H. (1982). *Early Thematic Mapping in the History of Cartography*. Chicago: University of Chicago Press.

Ryan, William, and Theodore Conover (2004) *Communications Today*. Clifton Park, NY: Delmar Learning.

Theall, D. F. (1971) *The Medium Is the Rear View Mirror*. Montreal: McGill-Queen's University Press.

Index

About the Author

Michael P. Peterson, PhD, is Professor in the Department of Geography/Geology at the University of Nebraska at Omaha, where he has been on the faculty since 1982. He has been a visiting scholar or fellow at universities around the world; most recently, he was Fulbright Fellow and Visiting Professor at the Vienna Technical University in Austria (1999, 2011), Visiting Professor at the Munich University of Sciences in Germany (1999, 2000, 2001), Visiting Research Fellow at Carleton University in Ottawa, Canada (2004), and Erskine Fellow at the University of Canterbury in Christchurch, New Zealand (2012). Dr. Peterson is past president of the North American Cartographic Information Society and former editor of *Cartographic Perspectives*. The author of numerous articles and several books, he chaired the Commission on Maps and the Internet of the International Cartographic Association (ICA) from 1999 to 2011 and is currently chair of the ICA's Publication Committee.